Relief Supply Chain Management for Disasters:

Humanitarian Aid and Emergency Logistics

Gyöngyi Kovács
HUMLOG Institute, Hanken School of Economics, Finland

Karen M. Spens
HUMLOG Institute, Hanken School of Economics, Finland

Senior Editorial Director:	Kristin Klinger
Director of Book Publications:	Julia Mosemann
Editorial Director:	Lindsay Johnston
Acquisitions Editor:	Erika Carter
Development Editor:	Joel Gamon
Production Editor:	Sean Woznicki
Typesetters:	Natalie Pronio, Jennifer Romanchak, Milan Vracarich, Jr.
Print Coordinator:	Jamie Snavely
Cover Design:	Nick Newcomer

Published in the United States of America by
    Business Science Reference (an imprint of IGI Global)
    701 E. Chocolate Avenue
    Hershey PA 17033
    Tel: 717-533-8845
    Fax: 717-533-8661
    E-mail: cust@igi-global.com
    Web site: http://www.igi-global.com

Library of Congress Cataloging-in-Publication Data
Relief supply chain management for disasters: humanitarian aid and emergency
logistics / Gyöngyi Kovács and Karen M. Spens, editors.
      p. cm.
  Includes bibliographical references and index.
  Summary: "This book furthers the scholarly understanding of SCM in disaster
relief, particularly establishing the central role of logistics in averting
and limiting unnecessary hardships"--Provided by publisher.
  ISBN 978-1-60960-824-8 (hbk.) -- ISBN 978-1-60960-825-5 (ebook) -- ISBN 978-
1-60960-826-2 (print & perpetual access) 1. Disaster relief. 2.
Humanitarian assistance. 3. Logistics. I. Kovacs, Gyongi, 1977- II. Spens,
Karen M., 1963-
  HV553.R373 2011
  363.34'80687--dc22
                               2011015748

British Cataloguing in Publication Data
A Cataloguing in Publication record for this book is available from the British Library.

All work contributed to this book is new, previously-unpublished material. The views expressed in this book are those of the authors, but not necessarily of the publisher.

Table of Contents

Detailed Table of Contents

Chapter 1

This chapter is about relationship building in relief supply chains. Its primary purpose is to present and discuss the author's actor-based typology of humanitarian relationships. The framework includes relationships among NGOs, as well as between NGOs and UN agencies, military units, and business firms. Examples are used to explore unique issues in the various types of relationships. One particular NGO, Airline Ambassadors International, is offered as an example of an NGO that builds relationships with a wide variety of humanitarian actors. The chapter also examines compatibility and complementarity of organizations across the three phases of humanitarian work: preparation, response and recovery or development. Research opportunities are discussed in the concluding comments.

Chapter 2

Through the use of a case study this chapter discusses the design of a partnership between humanitarian organizations to understand what are the drivers, facilitators, and components of the partnership. This research has been designed using a topical literature review and a case study. The practical implications include a discussion and guidelines for designing partnerships under high uncertainty and limited resources.

Chapter 3

The purpose of this chapter is to provide a framework for the development of relief supply chain systems. An illustrative case study is presented in order to help relief supply chain decision makers in their relief supply chain planning process. Developing simulation models to test proposed relief supply chain response plans is much less risky than actually waiting for another disaster to happen and test the proposed relief supply chain model in a real life situation. The simulated outcome can then be used to refine the developed relief supply chain response model.

Chapter 4

   Anthony Beresford, Cardiff University, UK
   Stephen Pettit, Cardiff University, UK

This chapter contrasts the response to the Wenchuan earthquake (May 2008) which took place in a landlocked region of China with that of the January 2010 earthquake in Haiti, which as an island nation, theoretically easily accessible to external aid provision via air or sea. In the initial period following the Wenchuan earthquake, the response was wholly internal, as a detailed needs assessment was carried out. Once the Chinese authorities had established the scale of response required, international assistance was quickly allowed into the country. Several multimodal solutions were devised to minimize the risk of supply breakdown. Haiti required substantial external aid and logistics support, but severe organizational and infrastructural weaknesses rendered the supply chain extremely vulnerable locally. This translated to a mismatch between the volume of aid supplied and logistics capability, highlighting the importance of 'last-mile' distribution management. The two earthquakes posed extreme challenges to the logistics operations, though both required a mix of military and non-military input into the logistics response. Nonetheless, in each case the non-standard logistics solutions which were devised broadly met the requirements for effective aid distribution in extreme environments.

Chapter 5

   David H. Taylor, Sheffield, UK

The study reported in this chapter was commissioned in 2009 by the charity 'Advance Aid' in order to provide an independent evaluation to compare conventional methods of supplying humanitarian aid products to Africa from outside the continent, with a proposed model of local manufacture and pre-positioned stocks. The evaluation was carried out using 'value chain analysis' techniques based on 'lean' concepts to provide a strategic evaluation of alternative supply models. The findings show that a system of local manufacturing and pre-positioned stockholding would offer significant advantages over conventional humanitarian supply chains in terms of responsiveness, risk of disruption and carbon footprint, and that delivered costs would be similar to or significantly better than current non-African supply options. Local manufacturing would also have important benefits in terms of creating employment and economic growth, which in the long run would help African states to mitigate and/or respond to future disasters and thus become less dependent on external aid.

Chapter 6

Post-disaster housing reconstruction projects face several challenges. Resources and material supplies are often scarce, several and different types of organizations are involved, and projects must be completed as quickly as possible to foster recovery. Within this context, the chapter aims to increase the understanding of relief supply chain design in reconstruction. In addition, the chapter is introducing a community based and beneficiary perspective to relief supply chains by evaluating the implications of local components for supply chain design in reconstruction. This is achieved through the means of secondary data analysis based on the evaluation reports of two major housing reconstruction projects that took place in Europe the last decade. A comparative analysis of the organizational designs of these projects highlights the ways in which users can be involved. The performance of reconstruction supply chains seems to depend to a large extent on the way beneficiaries are integrated in supply chain design impacting positively on the effectiveness of reconstruction supply chains.

Chapter 7

This case study explores the Swedish armed forces' sourcing from local suppliers in the area of the peacekeeping operation in Liberia. The paper discusses why, what, and how the Swedish armed forces develop local sourcing. For the study, a theoretical framework was developed with an industrial network perspective based on three cornerstones: supplier buyer relation development, internationalization, and finally, souring and business development in a war-torn country. The results of the study show that both implicit and explicit reasons to source locally exist. Every operation is unique, and therefore the sourcing needs to be tailored for each operation. Local sourcing was developed in the country based on existing needs and when opportunities arose. Theoretically, new insights of differences between business relations in military operations and normal business to business relations were gained. Practically, this study illustrates the importance to develop and diversify sourcing in international operations.

Chapter 8

The purpose of this chapter is to demonstrate the multitude of activities that military logisticians can provide throughout the various stages in relief supply chains. Most military joint doctrine identifies humanitarian assistance (HA) as one of the "Military Operations Other Than War" (MOOTW) that military personnel are trained to undertake. Part of this HA involves contributing to humanitarian supply chains and logistics management. The supply chain management processes, physical flows, as well as associated information and financial systems form part of the military contributions that play an important role in the relief supply chain. The main roles of the military to relief supply chains include security and protection, distribution, and engineering. Examples of these key contributions will be provided in this chapter.

Chapter 9

Graham Heaslip, National University of Ireland - Maynooth, Ireland

The term civil military coordination (CIMIC) suggests the seamless division of labor between aid workers and international military forces. The media coverage from crises such as New Orleans, Kosovo, the tsunami in Asia, Pakistan, Liberia, Sierra Leone, Chad ,and more recently Haiti, showing humanitarian organizations distributing food and medicines under the protection of military forces, or aid workers and military working together to construct refugee camps, set up field hospitals, provide emergency water and sanitation, et cetera, has heightened the expectation of a smooth interaction. Due to fundamental differences between international military forces, humanitarian and development organizations in terms of the principles and doctrines guiding their work, their agendas, operating styles, and roles, the area of civil military coordination in disaster relief has proven to be more difficult than other interagency relationships. This chapter will identify the many factors that render integration and collaboration problematic between diverse organizations, and especially so between civilian and military agencies. The chapter will conclude with proposals to improve CIMIC within disaster relief.

Chapter 10

Peter Tatham, Griffith University, Australia
Gyöngyi Kovács, HUMLOG Institute, Hanken School of Economics, Finland

Although there is a vast body of academic and practitioner literature championing the importance of trust in long-term business relationships, relatively little has been written which discusses the development and maintenance of trust in networks that are formed at short notice and that often operate for a limited period of time. Some models of trust and trusting behavior in such "hastily formed relief networks" (HFRN) do exist , however, and the aim of this chapter is to consider the theoretical application of one of the most prominent examples –known as "swift trust" – to a post-disaster humanitarian logistics scenario. Presented from the perspective of a HFRN, the chapter presents a discussion of the practical application of the swift trust model.

Chapter 11

Joseph Sarkis, Clark University, USA
Karen M. Spens, HUMLOG Institute, Hanken School of Economics, Finland
Gyöngyi Kovács, HUMLOG Institute, Hanken School of Economics, Finland

Relief supply chain (SC) management is a relatively unexplored field. In this field, practitioners have shown some interest in greening practices, but little practical or academic literature exists to help provide insights into combining the two fields. Adoption of green SC principles in the relief SC requires a systematic study of existing barriers in order to remove these barriers and allow introduction of green practices. The aim of this chapter is to explore barriers to implementation of green practices in the relief

SC. Expert opinions and literature from humanitarian logistics and green supply chain management are used to establish a list of barriers and to propose a categorization of barriers. Further research to evaluate the relationships and importance of these barrier factors is identified.

The study seeks to answer the question whether a country's logistics performance has a correlation with the impacts of a disaster; impact being measured in average amount of affected, the average amount of deaths, the average amount of injured in a disaster, or the average amount of economic damage. This is a quantitative study where the EM-DATs disaster data is analyzed through correlation analysis against the World Bank's logistics performance index (LPI). The findings do not show a significant relationship between countries LPI and the average number of deaths or injured in a disaster. A positive correlation between the variable LPI and the variable economic damage can be found. A negative correlation between the LPI and the average amount of affected can be found for countries with an average ranking LPI. Countries with low LPI and high disaster occurrence are further identified. Findings encourage the identified countries to take into consideration their logistics performance when planning and carrying out humanitarian response operations. Results also encourage humanitarian organizations to pay attention to the receiving countries' logistics performance in planning and carrying out of humanitarian response operations.

Foreword

For some years now, the major insurance and re-insurance companies have been tracking the occurrence of natural disasters. The disturbing findings of all of these analyses point to the fact that these events have been happening with significantly greater frequency and severity in recent years. Graphed over a fifty year time-line, the rate of increase appears to be almost exponential. Whatever the reasons for this increase, the implication is clear: the need to develop a much higher level of capability for the provision of relief and reconstruction will become ever more pressing.

Underpinning the success of any humanitarian aid and relief programme are logistics and supply chain processes that agile and adaptive: agile in the sense that they can respond rapidly to unexpected events, and adaptive in that they can be configured to meet the needs of specific situations and contexts. Surprisingly, it is only recently that the need for higher levels of capability in the practice of humanitarian logistics and supply chain management has been recognised. It could be argued that the shortage of appropriate logistics management skills and supporting infrastructure has meant that many aid and relief programmes in the past have been less effective than they could have been.

For this reason, it is opportune that this book should be compiled and published at this particular time. In the last few years alone, a tremendous amount of knowledge has been gained into how humanitarian logistics and supply chain performance can be made much more effective by the application of new ideas and techniques. The issues addressed by the various contributors to this book are critical to the achievement of the goals of any humanitarian aid and relief programme. The breadth as well as the depth of the analysis contained within these chapters is impressive, and together they provide valuable insights into how current practice can be improved.

The message to be drawn from this is that whilst disasters and existential threats from a multitude of sources will sadly always be with us, at least we can seek to learn how to mitigate their consequences.

Martin Christopher
Cranfield University, UK

Martin Christopher *is an Emeritus Professor of Marketing and Logistics at the Cranfield School of Management in the UK. For many years Martin Christopher has been involved in teaching and researching new ideas in logistics and supply chain management. He has published widely, and his book, "Logistics and Supply Chain Management," has become one of the most widely cited texts in its field. As well as his Emeritus position at Cranfield, Martin Christopher is a Visiting Professor at a number of leading Universities around the world.*

Preface

INTRODUCTION

Relief supply chains are argued to be the most dynamic and agile supply chains, yet research in this area of supply chain management (SCM) is scant. Relief SCM has recently gained attention due to many natural and man-made disasters and the recognition of the central role of logistics in responding to these. Relief supply chains (SC) constitute a substantial industry that responds to over 500 disasters annually resultant in loss of 75 000 lives and affecting over 200 million people. SC costs are also argued to account for over 80% of costs incurred in any disaster relief operation. Due to the fact that relief supply chains so far have received little attention, there seems to be a gap that this book can fill.

The anthology also presents a continuation of a doctoral course in *Supply Chain Management for Disaster Relief* given at Hanken School of Economics in the fall of 2009, as many of the chapters are written by participants, as well as core faculty of this doctoral course. The book is therefore a collection of chapters by researchers, both junior and senior, in the field of humanitarian logistics and relief supply chain management. The chapters were, however, submitted after a broader call for papers and were thereafter peer-reviewed, ending up as a collection of chapters that were accepted. The interest for courses in this field has continued to grow since; therefore, the hope is that this anthology will provide a platform for creating and giving even more courses in the field. More broadly, the anthology is part of a large research project funded by the Academy of Finland, called *Relief Supply Chain Management.*

The overall aim of the anthology *Relief Supply Chain Management for Disasters: Humanitarian Aid and Emergency Logistics* is to further the understanding of SCM in disaster relief. As the first book in this field, the hope is that it will serve scholarly thought as well as provide a textbook for courses introducing this new and exciting area in the field of logistics.

BACKGROUND

Supply chain management (SCM) research has developed rapidly in the past two decades, but is still "a discipline in the early stages of evolution" (Gibson *et al.*, 2005, p.17). The following most commonly used definition of SCM is provided by the Council of Supply Chain Management Professionals (CSCMP):

'Supply Chain Management encompasses the planning and management of all activities involved in sourcing and procurement, conversion, and all Logistics Management activities. Importantly, it also includes coordination and collaboration with channel partners, which can be suppliers, intermediaries, third-party service providers, and customers. In essence, Supply Chain Management integrates supply and demand management within and across companies' (CSCMP, 2006).

Traditional streams of SCM literature encompass different topics, ranging from supply chain modelling and optimisation (Lee *et al.,* 2004; Svensson, 2003) to supply chain performance measurement (Bagchi *et al.,* 2005; Beamon, 1999), supply chain processes (Croxton *et al.,* 2001; Lambert *et al.,* 1998), portfolio models in SCM (Fisher, 1997), and supply chain collaboration and integration (Barratt, 2004; Fawcett & Magnan, 2002; Min *et al.,* 2005). Portfolio models in SCM discuss different types of supply chains, contrasting supply chains for functional products with a focus on cost efficiencies to supply chains for innovative products with a focus on responsiveness to market dynamics (Fisher, 1997). But while this portfolio thinking is at the core of SCM, literature has traditionally focused on efficient (or "lean") supply chains only (Lee, 2004). Therefore, the current trend in SCM literature is towards discussing more innovative and responsive – or "agile" – supply chains that operate in a highly dynamic environment (Christopher et al., 2006; Towill and Christopher, 2002).

Relief supply chain management has recently gained attention due to a number of natural and man-made disasters and the recognition of the central role of logistics in responding to these. Oloruntoba and Gray (2006, p.117) argue that relief supply chains are "clearly unpredictable, turbulent, and requiring flexibility." In essence, relief supply chains can be seen as highly dynamic, innovative, and agile (Oloruntoba & Gray, 2006; van Wassenhove, 2006), and hereby it can be argued that even (traditional) commercial supply chains can learn from the high flexibility of relief supply chains (Sowinski, 2003). Especially in sudden-onset disasters, relief supply chains have to be deployed in situations with a destabilised infrastructure and with very limited knowledge about the situation at hand (Beamon, 2004; Long & Wood, 1995; Tomasini & van Wassenhove, 2004). Relief supply chain management, although arguably much different from business logistics, does also show similarities. Therefore the definitions, techniques, and approaches used within business logistics can often be transferred or altered so they fit the purpose of their context. Notwithstanding the fact that the ultimate goals and purpose of conducting the logistical activities are different, still many of the definitions relating to the field can be extracted from current definitions found in the business context. In the following paragraphs, we are providing an overview of the definitions that the book adheres to. The definitions were provided to the authors of the chapters at the outset and have been used accordingly throughout the book. Admittedly, as in the field of logistics, defining concepts is a difficult task, so authors often tend to use definitions or even define concepts in a way that fits their purpose. The chapters therefore are the sole responsibility of the authors and do reflect their views on particular issues and concepts, however, we argue that the definitions provided in the end of our preface seemingly have gained acceptance among the authors of the chapters of this book.

This anthology is designed to bring together theoretical frameworks and the latest findings from research with their discussion in particular cases. Besides a number of frameworks – of types of relationships in the relief supply chain (ch.1), relief logistics development (ch.3), value chain analysis (ch.5), civil-military co-operation (ch.9), and trust models in disaster relief (ch.10) – cases range from logistical partnerships in the Sudan (ch.2), to a comparison of relief supply chains in different earthquakes (Haiti vs. Wenchuan, ch.4), to local sourcing in Liberia (ch.7), and reconstruction in the Kosovo and the Former Yugoslav Republic of Macedonia (ch.6). This way, insights from theory and practice are combined. The anthology ends with a chapter on one of the most recent areas humanitarian logistics research and practice has embraced: questions of sustainability, and most importantly, the issue of greening the relief supply chain (ch.11).

THE COLLECTION OF CHAPTERS: A SHORT INTRODUCTION

In the foreword, Martin Christopher, Emeritus Professor from Cranfield University, discusses the importance of the topic more broadly. Professor Christopher is undeniably one of the most well-known authors and scholars in the field of logistics who has also recently embraced the field of humanitarian logistics through co-editing a book with Peter Tatham. We hope these two books will complement each other. In the preface, the editors of the book, Gyöngyi Kovács and Karen M. Spens, outline the field, provide some key definitions, and provide an overview of the chapters included.

In the first chapter by Paul D. Larson from University of Manitoba, relationship building in humanitarian supply chains is discussed. The primary purpose of the chapter, named "*Strategic Partners and Strange Bedfellows: Relationship Building in the Relief Supply Chain,*" is to present and discuss the author's actor-based typology of humanitarian relationships. The framework includes relationships among NGOs, as well as between NGOs and UN agencies, military units, and business firms. Examples are used to explore unique issues in the various types of relationships. One particular NGO, Airline Ambassadors International, is offered as an example of an NGO that builds relationships with a wide variety of humanitarian actors. The chapter also examines compatibility and complementarity of organizations across the three phases of humanitarian work: preparation, response, and recovery or development. Research opportunities are discussed in the concluding comments. The chapter serves as a good introduction to following ones that further discuss some of the types of relationships outlined here.

The next chapter takes up the question of partnerships in the relief supply chain. Rolando M. Tomasini, Hanken School of Economics, Finland, in his chapter, "*Humanitarian Partnerships - Drivers, Facilitators, and Components: The Case of Non-Food Item Distribution in Sudan,*" uses a case study to discuss the design of partnerships between humanitarian organizations in order to understand the drivers, facilitators and components, of a partnership. The research was designed using a topical literature review and a case study. The practical implications include discussion and guidelines for designing partnerships under high uncertainty and limited resources.

This is followed by another case study, this time of disaster preparedness and management in Thailand. At the same time, Ruth Banomyong from Thammasat University, Thailand and Apichat Sodapang from Chiangmai University, Thailand present a more general framework for relief supply chain management in the third chapter. Their "*Relief Supply Chain Planning: Insights from Thailand*" builds on and evaluates a general framework for humanitarian logistics. The chapter highlights the need for planning and preparedness prior to a disaster.

Further cases are presented and contrasted in chapter 4, "*Humanitarian Aid Logistics: The Wenchuan and Haiti Earthquakes Compared,*" by Anthony Beresford and Stephen Pettit from Cardiff University, UK. The comparison of a similar disaster in different environments helps to highlight common features in humanitarian logistics and set these apart from contextual factors such as infrastructural weaknesses. Access to a disaster area is contrasted between islands and landlocked countries. Furthermore, as in chapter one, the cases show the importance of co-ordination in the logistics response of humanitarian and military organizations.

Chapter 5, called "*The Application of Value Chain Analysis for the Evaluation of Alternative Supply Chain Strategies for the Provision of Humanitarian Aid to Africa,*" is a prime example of presenting a framework and discussing it on a particular case. David H. Taylor, from Sheffield, UK is an expert in value chain analysis. The study reported in this chapter was commissioned in 2009 by the charity "Advance Aid" in order to provide an independent evaluation to compare conventional methods of supplying humanitarian aid products to Africa from outside the continent, with a proposed model of local manufacturing and pre-positioned stocks. The findings show that a system of locally manufactured and pre-positioned stockholding would offer significant advantages over conventional relief supply chains in terms of responsiveness, risk of disruption, and carbon footprint, and that delivered costs would be similar to or significantly better than current non-African supply options. Local manufacture would also have important benefits in terms of creating employment and economic growth, which in the long run would help African states to mitigate and/or respond to future disasters and thus become less dependent on external aid.

Local sourcing and manufacturing is also at the core of chapter 6, *"Designing Post-Disaster Supply Chains: Learning from Housing Reconstruction Projects."* In this chapter, Gyöngyi Kovács, HUMLOG Institute, Hanken School of Economics, Finland, Aristides Matopoulos, University of Macedonia, Greece and Odran Hayes from the European Agency for Reconstruction, Ireland introduce a community based and beneficiary perspective to relief supply chains by evaluating the implications of local components for supply chain design in reconstruction. The chapter further discusses the challenges of post-disaster housing reconstruction projects on the cases of housing reconstruction programs in the Kosovo and the Former Yugoslav Republic of Macedonia, finding that resources and material supplies are often scarce. Several and different types of organizations are involved while projects must be completed as quickly as possible to foster recovery. The performance of reconstruction supply chains seems to depend to a large extent on the way beneficiaries are integrated in supply chain design impacting positively on the effectiveness of reconstruction supply chains.

Local sourcing is also taken up from a peacekeeping perspective. Per Skoglund and Susanne Hertz from Jönköping International Business School, Sweden, present a case study of the Swedish armed forces in Liberia and compare local sourcing in peacekeeping there with other cases in Afghanistan and the Kosovo. The chapter, *"Local Sourcing in Peacekeeping: A Case Study of Swedish Military Sourcing,"* not only illustrates these three cases but applies the theoretical framework of the Uppsala model of internationalisation to them. Of particular interest is the discussion of psychic distance in local sourcing.

Coming back to different types of actors and relationships in the relief supply chain, Elizabeth Barber, from the University of New South Wales, Australian Defence Force Academy, Australia discusses the role of the military in disaster relief. The chapter, *"Military Involvement in Humanitarian Supply Chains,"* demonstrates the multitude of activities that military logisticians can provide throughout the various stages in the humanitarian supply chains. Most military joint doctrine identifies humanitarian assistance as one of the "Military Operations Other Than War" that military personnel are trained to undertake. The supply chain management processes, physical flows, as well as associated information and financial systems form part of the military contributions to the relief supply chain. The main roles of the military to humanitarian supply chains include security and protection, distribution, and engineering. Examples of these key contributions are provided in this chapter.

Upon outlining the roles and contributions of the military, the next chapter turns to *"Challenges of Civil Military Cooperation / Coordination in Humanitarian Relief."* Graham Heaslip, National University of Ireland-Maynooth, Ireland, goes through the various meanings and definitions of civil military coordination (CIMIC) and the fundamental differences between the principles and doctrines guiding the work of international military forces and humanitarian organizations. This chapter identifies the many factors that render integration and collaboration problematic between diverse assistance agencies, and especially so between civilian and military agencies. It concludes with proposals to improve CIMIC within humanitarian relief.

The challenges to develop relationships, and in particular, trust, between representatives of different humanitarian organizations is also a core theme of chapter 10, *"Developing and Maintaining Trust in Hastily Formed Relief Networks."* In this chapter, Peter Tatham from Griffith University, Australia, and Gyöngyi Kovács, HUMLOG Institute, Hanken School of Economics, Finland, discuss the implications of the practical implication of a "swift trust" model in the ad hoc networks of humanitarian logisticians in the field.

In the following chapter, *"A Study of Barriers to Greening the Relief Supply Chain,"* the authors Joseph Sarkis, Clark University, USA, and Karen M. Spens and Gyöngyi Kovács from the HUMLOG Institute, Hanken School of Economics, Finland reveal barriers to the greening of the relief supply chain. Adoption of green SC principles in the relief SC requires a systematic study of existing barriers in order to remove these barriers and allow introduction of green practices. Expert opinions and literature from humanitarian logistics and green supply chain management are used to establish a list of barriers and to propose a categorization of barriers.

The final chapter, Ira Haavisto from the Hanken School of Economics, Finland, takes a more macro-economic view on disaster occurrence and impact in light of the logistics performance of a country. *"Disaster Impact and Country Logistics Performance"* discusses the links between the states of logistics infrastructure, and hence, country logistics performance, and the various impacts of disasters in terms of loss of life, number of people affected, and economic damage. Not surprisingly, high country logistics performance correlates with the economic damage of disasters, but more interestingly, high country logistics performance shows a negative correlation to the numbers of people affected. At the same time, the analysis points towards an increased need for preparedness in countries with high disaster occurrence and a low logistics performance.

In summary, the topics and chapters provided give a broad overview of the issues relevant and prevailing in the field of relief supply chain management. The actor structure in relief supply chains is, as earlier research has pointed out, complex, due to the fact that there are military, humanitarian, governmental, and for-profit actors involved in delivering relief. Partnerships, coordination, and collaboration are themes found in the chapters that relate more to strategic thinking, whereas value chain analysis and simulation provide a tool for operational types of changes to relief supply chains. The phases of disaster relief are also covered in the text, as some chapters relate more to preparedness, whereas others touch more upon the response phase of the disaster relief cycle. Some topical new issues are also discussed, such as greening the relief supply chain. In business logistics, sustainability and greening has become key, whereas green thinking, at least in academic papers found in the field of humanitarian logistics, are still scarce.

FUTURE RESEARCH DIRECTIONS

The book covers a broad variety of topics relating to relief supply chain management. Many of the chapters identify future directions for research. Relationship building in the relief supply chain is such an area (see ch.1). A significant body of literature has focused on coordination, or the lack thereof, in humanitarian logistics. Turning away from aspects of inter-organizational or inter-agency coordination, the focus is now shifting towards collaboration in the supply chain, i.e. considering partners such as logistics service providers (see ch.2), suppliers, and even beneficiaries (ch.6).

Another important direction is the development of comparative studies (as in ch.4) as to be able to draw on commonalities of the relief supply chain and to learn from previous disasters. The use of logistical concepts and models (ch.5) and the development of generic frameworks for humanitarian logistics (ch.3) aid in unearthing the critical success factors of logistics in disaster relief (cf. Pettit & Beresford, 2009).

The final chapter of the book indicates a further future research direction, that of considering the sustainability of aid. There are multiple meanings of sustainability in the humanitarian context. Ch.11 addresses sustainability from the perspective of greening the relief supply chain, which extends previous considerations of green logistics that were primarily concerned with transportation emissions, beyond organizational boundaries, and to other aspects of environmental impact. Also, ch.5 considers the carbon footprint of humanitarian aid. Greening aspects are of particular importance considering the debate on climate change. Ch.6 considers another aspect of sustainability, involving the community of beneficiaries in supply chain design, while ch.7 highlights the social side of sustainability in local sourcing. Further research is still needed in these areas to address questions of long-term development and sustainable exit strategies of humanitarian aid.

CONCLUSION

The field of humanitarian logistics and relief supply chain management is receiving increasing attention among academics, as well as practitioners. The number of related publications has been increasing steadily (Kovács & Spens, 2008), and a number of journals have dedicated special issues to this field. This book is, however, the first compilation of chapters dedicated to relief supply chain management. As such, it provides an overview of some of the topics covered by academics on the topic of *Relief Supply Chain Management* in a variety of countries around the world. However, the topics certainly do not cover all the research done in this field as we are well aware that there are a multitude of ongoing projects and research being conducted which would have been interesting to include. Our sincere hope is that this book, nevertheless, fills a gap and can be used in courses that aim to introduce academic readers to this new and emerging field. We are very grateful to all the authors who took the time to contribute and we are also indebted to the reviewers who took the time to comment on the chapters. As the editors of this book, and also, the editors of an academic journal (the *Journal of Humanitarian Logistics and Supply Chain Management*) that is to be launched in 2011, our hope is also that this book will inspire even more authors so that the field continues to grow and mature.

Gyöngyi Kovács
HUMLOG Institute, Hanken School of Economics, Finland

Karen M. Spens
HUMLOG Institute, Hanken School of Economics, Finland

REFERENCES

Bagchi, P. K., Ha, B. C., Skjoett-Larsen, T., & Soerensen, L. B. (2005). Supply chain integration: A European survey. *International Journal of Logistics Management, 16*(2), 275–294. doi:10.1108/09574090510634557

Barratt, M. (2004). Understanding the meaning of collaboration in the supply chain. *Supply Chain Management: An International Journal, 9*(1), 30–42. doi:10.1108/13598540410517566

Beamon, B. M. (1999). Measuring supply chain performance . *International Journal of Operations & Production Management, 19*(3), 275–292. doi:10.1108/01443579910249714

Beamon, B. M. (2004). Humanitarian relief chains: Issues and challenges. *Proceedings of the 34th International Conference on Computers and Industrial Engineering,* San Francisco, CA, USA.

Christopher, M., Peck, H., & Towill, D. (2006). A taxonomy for selecting global supply chain strategies. *International Journal of Logistics Management, 17*(2), 277–287. doi:10.1108/09574090610689998

Croxton, K. L., García-Dastugue, S. J., Lambert, D. M., & Rogers, D. S. (2001). The supply chain management processes. *International Journal of Logistics Management, 12*(2), 13–36. doi:10.1108/09574090110806271

CSCMP. (2006). *Supply chain management/logistics management definitions.* Council of Supply Chain Management Professionals. Retrieved on August 1, 2010, from http://cscmp.org/ aboutcscmp/ definitions.asp

Fawcett, S. E., & Magnan, G. E. (2002). The rhetoric and reality of supply chain integration. *International Journal of Physical Distribution and Logistics Management, 32*(5), 339–361. doi:10.1108/09600030210436222

Fisher, M. L. (1997). What is the right supply chain for your product? *Harvard Business Review,* (March-April): 105–116.

Kovács, G., & Spens, K. (2008). Humanitarian logistics revisited . In Arlbjørn, J. S., Halldórsson, Á., Jahre, M., & Spens, K. (Eds.), *Northern lights in logistics and supply chain management* (pp. 217–232). Copenhagen, Denmark: CBS Press.

Lambert, D. M., Cooper, M. C., & Pagh, J. D. (1998). Supply chain management: Implementation issues and research opportunities. *International Journal of Logistics Management, 9*(2), 1–19. doi:10.1108/09574099810805807

Lee, H. L. (2004). The triple-A supply chain. *Harvard Business Review, 82*(10), 102–112.

Lee, H. L., Padmanabhan, V., & Whang, S. (2004). Information distortion in the supply chain: The bullwhip effect. *Management Science, 50*(12), 1875–1886. doi:10.1287/mnsc.1040.0266

Long, D. C., & Wood, D. F. (1995). The logistics of famine relief. *Journal of Business Logistics, 16*(1), 213–229.

Min, S., Roath, A. S., Daugherty, P. J., Genchev, S. E., Chen, H., Arndt, A. D., & Richey, R. G. (2005). Supply chain collaboration: What's happening? *International Journal of Logistics Management, 16*(2), 237–256. doi:10.1108/09574090510634539

Oloruntoba, R., & Gray, R. (2006). Humanitarian aid: An agile supply chain? *Supply Chain Management: An International Journal, 11*(2), 115–120. doi:10.1108/13598540610652492

Pettit, S., & Beresford, A. (2009). Critical success factors in the context of humanitarian aid supply chains. *International Journal of Physical Distribution & Logistics Management, 39*(6), 450–468. doi:10.1108/09600030910985811

SCM. IJ. (2010). Author guidelines. *Supply Chain Management: an International Journal*. Retrieved August 1, 2010, from http://info.emeraldinsight.com/products/journals/author_guidelines.htm?id=scm

Sowinski, L. L. (2003). The lean, mean supply chain and its human counterpart. *World Trade, 16*(6), 18.

Svensson, G. (2003). The bullwhip effect in intra-organisational echelons. *International Journal of Physical Distribution and Logistics Management, 33*(2), 103–131. doi:10.1108/09600030310469135

Thomas, A., & Mizushima, M. (2005). Logistics training: Necessity or luxury? *Forced Migration Review, 22*, 60–61.

Tomasini, R. M., & van Wassenhove, L. N. (2004). Pan-American health organization's humanitarian supply management system: De-politicization of the humanitarian supply chain by creating accountability. *Journal of Public Procurement, 4*(3), 437–449.

Towill, D., & Christopher, M. (2002). The supply chain strategy conundrum: To be lean or agile or to be lean and agile? *International Journal of Logistics: Research and Applications, 5*(3), 299–309. doi:10.1080/1367556021000026736

van Wassenhove, L. N. (2006). Humanitarian aid logistics: Supply chain management in high gear. *The Journal of the Operational Research Society, 57*, 475–589. doi:10.1057/palgrave.jors.2602125

ADDITIONAL READING

Relief supply chain management is a rather new field of research. Nonetheless, there is a steady rise in the number of relevant published articles. Whilst noting some of the most important works we would also like to refer to *Peter Tatham's Bibliography* that is constantly updated and can be obtained from the first author or chapter 9.

To be noted are the following special issues in scientific journals:

Transportation Research Part E: Logistics and Transportation Review, Vol.43 No.6 (2007) on "Challenges of Emergency Logistics Management"

International Journal of Services Technology and Management, Vol.12 No.4 (2009) on "Coordination of Service Providers in Humanitarian Aid"

International Journal of Risk Assessment and Management, Vol.13 No.1 (2009) on "Managing Supply Chains in Disasters"

International Journal of Physical Distribution and Logistics Management, Vol.39 No.5/6 (2009) on "SCM in Times of Humanitarian Crisis"

Supply Chain Forum: an International Journal, Vol.11 No.3 (2010) on "Humanitarian Supply Chains"

International Journal of Physical Distribution and Logistics Management, Vol.40 No.8/9 (2010) on "Developments in Humanitarian Logistics"

Interfaces, Vol.40 No.(in press) on "Doing Good with Good OR"

International Journal of Production Economics, Vol.126 No.1 (2010) on "Improving Disaster Supply Chain Management – Key supply chain factors for humanitarian relief

and the dedicated *Journal of Humanitarian Logistics and Supply Chain Management,* to be launched in 2011.

FURTHER ADDITIONAL READING

Altay, N., Prasad, S., & Sounderpandian, J. (2009). Strategic planning for disaster relief logistics: Lessons from supply chain management. *International Journal of Services Sciences, 2*(2), 142–161. doi:10.1504/IJSSCI.2009.024937

Beamon, B. M., & Balcik, B. (2008). Performance measurement in humanitarian relief chains. *International Journal of Public Sector Management, 21*(1), 4–25. doi:10.1108/09513550810846087

Carter, W. N. (1999). *Disaster Management: A Disaster Management Handbook.* Manila: Asian Development Bank.

Glenn, R. R. Jr. (2009). The supply chain crisis and disaster pyramid: A theoretical framework for understanding preparedness and recovery. *International Journal of Physical Distribution & Logistics Management, 39*(7), 619–628. doi:10.1108/09600030910996288

Haas, J. E., Kates, R. W., & Bowden, M. (1977). *Reconstruction Following Disaster.* Cambridge: MIT Press.

Heaslip, G. (2008). Humanitarian aid supply chains . In Mangan, J., Lalwani, C., & Butcher, T. (Eds.), *Global Logistics and Supply Chain Management.* Chichester: John Wiley & Sons.

Jahre, M., & Heigh, I. (2008). Does the current constraints in funding promote failure in humanitarian supply chains? *Supply Chain Forum, 9*(2), 44–54.

Kovács, G., & Spens, K. (2009). Identifying challenges in humanitarian logistics. *International Journal of Physical Distribution and Logistics Management, 39*(6), 506–528. doi:10.1108/09600030910985848

Kovács, G., & Spens, K. M. (2007). Humanitarian logistics in disaster relief operations. *International Journal of Physical Distribution and Logistics Management, 29*(12), 801–819.

Long, D. (1997). Logistics for disaster relief: engineering on the run . *IIE Solutions, 29*(6), 26–29.

Maon, F., Lindgreen, A., & Vanhamme, J. (2009). Developing supply chains in disaster relief operations through cross-sector socially oriented collaborations. *Supply Chain Management: An International Journal, 14*(2), 149–164. doi:10.1108/13598540910942019

Ozdamar, L., Ekinci, E., & Kucukyazici, B. (2004). Emergency logistics planning in natural disasters . *Annals of Operations Research, 129*, 217–245. doi:10.1023/B:ANOR.0000030690.27939.39

Pettit, S. J., & Beresford, A. K. C. (2005). Emergency relief logistics: an evaluation of military, non military and composite response models . *International Journal of Logistics: Research and Applications, 8*(4), 313–332.

Rietjens, S. J. H., Voordijk, H., & De Boer, S. J. (2007). Co-ordinating humanitarian operations in peace support missions . *Disaster Prevention and Management, 16*(1), 56–69. doi:10.1108/09653560710729811

Thomas, A. (2003). Why logistics? *Forced Migration Review, 18*(Sep), 4.

Thomas, A., & Fritz, L. (2006). Disaster Relief, Inc. *Harvard Business Review*, (Nov): 114–122.

Tomasini, R., & Van Wassenhove, L. (2009). *Humanitarian Logistics*. Palgrave MacMillan. doi:10.1057/9780230233485

Tomasini & van Wassenhove review of cases

Whiting, M. (2009). Chapter 7: Enhanced civil military cooperation in humanitarian supply chains, In: Gattorna (ed), *Dynamic Supply Chain Management,* Gower Publishing, Surrey, England, pp. 107-122.

KEY TERMS AND DEFINITIONS

Disaster: A disaster is "a serious disruption of the functioning of a community or a society involving widespread human, material, economic or environmental losses and impacts, which exceeds the ability of the affected community or society to cope using its own resources" (UN/ISDR 2009). This definition is also used by WHO and EM-DAT. Disasters can be natural or man-made, as well as complex emergencies (combining a man-made and a natural disaster). Synonyms: emergency, calamity, catastrophe, disruption, conflict.

Disaster Relief: Encompasses humanitarian activities in the phases of disaster preparedness, immediate response and reconstruction. But, if not specified otherwise in a chapter, disaster relief can be seen as synonymous with activities in the immediate response phase. Synonyms: emergency relief, humanitarian aid, humanitarian assistance. Synonyms for disaster relief phases: preparation, planning, prevention / recovery, restoration, rehabilitation. The phases do not need to be seen in a sequential manner as activities from different disaster relief phases can run in parallel, and activities can also be linked to each other in a cyclical manner.

Humanitarian Logistics: "The process of planning, implementing and controlling the efficient, cost-effective flow and storage of goods and materials, as well as related information, from point of origin to point of consumption for the purpose of meeting the end beneficiary's requirements" (Thomas and Mizushima, 2005, p.60). Synonyms as used in this book: emergency relief logistics, relief logistics, disaster relief logistics, humanitarian operations, catastrophe logistics.

Humanitarian Organization: An organization that manages the delivery of aid to beneficiaries, following humanitarian principles. "Humanitarian organization" is an umbrella term for non-governmental organizations and aid agencies regardless of their mandate or organizational structure. Aid can be delivered by the humanitarian organization or through (implementing) partners.

Logistics vs. Supply Chain Management: In this book we adhere to CSCMP's definitions of logistics vs. supply chain management (CSCMP, 2006). Note that activities such as warehousing, purchasing etc. are included in the definition of logistics. We also follow the view of a supply chain extending beyond a dyad, as laid out in the author guidelines of Supply Chain Management: an International Journal (SCM:IJ, 2010).

Relief Supply Chain Management: Encompasses the planning and management of all activities related to material, information and financial flows in disaster relief. Importantly, it also includes co-ordination and collaboration with supply chain members, third party service providers, and across humanitarian organizations. Synonyms: humanitarian supply chain, humanitarian supply chain management. However, relief supply chain management does not include the development aid aspect of humanitarian logistics.

Chapter 1
Strategic Partners and Strange Bedfellows:
Relationship Building in the Relief Supply Chain

Paul D. Larson
University of Manitoba, Canada

ABSTRACT

This chapter is about relationship building in relief supply chains. Its primary purpose is to present and discuss the author's actor-based typology of humanitarian relationships. The framework includes relationships among NGOs, as well as between NGOs and UN agencies, military units, and business firms. Examples are used to explore unique issues in the various types of relationships. One particular NGO, Airline Ambassadors International, is offered as an example of an NGO that builds relationships with a wide variety of humanitarian actors. The chapter also examines compatibility and complementarity of organizations across the three phases of humanitarian work: preparation, response, and recovery or development. Research opportunities are discussed in the concluding comments.

INTRODUCTION

At 4:53 p.m. January 12, 2010, an earthquake of over 7.0 on the Richter scale hit Haiti. It struck 17 km. south-west of Port-au-Prince, the capital city, in an area with more than 2 million people. After the earthquake, electricity was unavailable and communications were difficult. According to initial reports, there was wide-spread damage and many casualties. The Port-au-Prince airport could accommodate radio-assisted, line-of-sight landings only. For now, it would be open only for humanitarian assistance flights. All roads to the capital were partially blocked by debris and other obstacles. United Nations agencies and the International Federation of the Red Cross and

DOI: 10.4018/978-1-60960-824-8.ch001

Red Crescent Society (IFRC) were preparing to deploy teams and material aid, from a regional hub in Panama (OCHA 2010a).

Two days later, access became feasible and needs began to be assessed. The initial priorities included search-and-rescue assistance, as well as teams with heavy-lifting equipment, medical assistance and supplies. Access to people in need remained difficult due to debris and obstacles on the roads. At this point, The United Nations Office of the Coordination for Humanitarian Affairs (OCHA) announced: "*Logistics and the lack of transport remain the key constraints to the delivery of aid.*" Displaced persons were scattered across multiple locations. Temporary shelters were urgently needed. Fifteen sites were identified for distribution of relief items. By the fourth day the World Food Programme (WFP) had reached 13,000 people with food, jerry cans and water purification tablets. The UN Disaster and Assessment Coordination (UNDAC) team and OCHA teams were on the ground in Haiti; and the following announcement was made: "*Coordination of assistance is vital*" (OCHA 2010b).

Like many prior disasters, all over the world, the 2010 Haiti earthquake shows once again that logistics and supply chain management are critical to effective delivery of humanitarian relief. Moreover, supply chain coordination across a wide variety of organizations offering assistance is needed to save as many lives and ease as much suffering as possible, in light of fiscal, material and personnel limits. While logistical considerations are critical, Spring (2006), drawing on information from the Fritz Institute, suggests that humanitarian aid agencies are twenty years behind the large corporations in adopting today's fundamental tools of logistics and supply chain management.

The purpose of this chapter is to discuss a range of relationship issues within humanitarian relief supply chains. The remainder is organized as follows. The second and third sections present a brief contrast of humanitarian logistics vs. business logistics and an abbreviated review of

relationship building literature, respectively. This sets the stage for the author's actor-based typology of humanitarian supply chain relationships, in section IV. Section V discusses relationship building across four phases of humanitarian action, and then section VI provides a summary and conclusions.

HUMANITARIAN LOGISTICS VERSUS BUSINESS LOGISTICS

According to Pettit and Beresford (2005, p. 314), "There are clear parallels between business logistics and relief logistics, but the transfer of knowledge between the two has been limited and the latter remains relatively unsophisticated."

Kovács and Spens (2007) discuss several important differences between business logistics and humanitarian logistics. While business logisticians work with predetermined actors or partners and predictable demand, humanitarians deal with unknown or changing actors and unpredictable demand. Aid agencies receive many unsolicited and sometimes even unwanted donations, such as: drugs and foods past their expiry dates; laptops needing electricity where infrastructure has been destroyed; and heavy clothing not suitable for tropical regions. Compared to their business counterparts, humanitarian logisticians have greater challenges in collaboration and coordination of effort. Coordination of many different aid agencies, suppliers, and local and regional actors, all with their own ways of operating and own structures can be very challenging. Descriptions of relief operations frequently criticize aid agencies for their lack of collaboration, redundancies, and duplicated efforts and materials.

McLachlin, Larson and Khan (2009) offer a framework in which differences between business and humanitarian logistics largely follow from two dimensions: motivation (profit versus not-for-profit) and environment (uninterrupted versus interrupted). In business logistics, actors have the profit motive and generally operate in uninterrupted environments. Interrupted operating environments are rare exceptions. To the contrary, in humanitarian logistics, actors are usually not-for-profit organizations and interrupted environments are the norm, especially in the case of disaster relief as opposed to ongoing development aid operations.

Thus, the framework (see Table 1) contains four quadrants representing four types of supply chains: (1) for-profit, uninterrupted; (2) for-profit, interrupted; (3) not-for-profit, uninterrupted; and (4) not-for-profit, interrupted. Uninterrupted environments are reasonably stable in terms of political and economic conditions; infrastructure is in place; and all the critical actors (customers, suppliers, service providers and employees) are on the stage. Interrupted environments, on the other hand, are characterized by a lack of stability, greater complexity, and special challenges in matching multiple sources of supply with shifting customer (or recipient) demand. Unlike for-profit firms, not-for-profit (NFP) agencies emphasize social rather than economic objectives.

Kleindorfer and Saad (2005) identify two primary categories of supply chain risk: (1) problems in the coordination of supply and demand; and (2) disruptions to normal activities. Supply chain risk analysis is important for humanitarian logistics for at least two reasons. First, supply chain interruptions can cause, or at least contribute to humanitarian crises. Second, humanitarian relief efforts often face multiple risk events simultaneously, including operational sources of

risk, "the interruption" that caused the crisis, and various political and infrastructural issues.

Unlike for-profit firms, NFP organizations emphasize social or environmental objectives, rather than economic ones. The NFP sector serves multiple stakeholders, including two fundamentally different types of "customers:" beneficiaries or recipients (those in need of food, material and services); and donors (those who provide funding, material and/or service support). Donors are customers since their wishes and mission statements can mean restrictions on the use of funds or loss of funds altogether. NFPs face stiff competition for donor support, rather than competition for paying customers. NFPs typically deploy volunteers, in addition to paid staff. Rather than money, strong commitment to "the cause" motivates volunteers (Murray 2006).

Table 2 summarizes the contrast between business and humanitarian logistics. While "time is money" to the business logistician, time is life to the humanitarian. Humanitarians seek social impact rather than profit, though they must be mindful of donor desires and budget limits. Such supply chains must be flexible and responsive to unpredictable events, as well as efficient and able to maximize reach of scarce resources. More effective supply chain management can be the difference between life and death; greater efficiency means serving more people in dire need. There are tremendous opportunities to serve more people in need at lower cost through supply chain best practices. The most pressing humanitarian supply chain challenge may be to balance the conflicting objectives of flexibility and efficiency.

Table 1. Four types of supply chains

Motivation	Environment	
	Uninterrupted	Interrupted
For-profit	*Business as usual*	*Risk management*
Not-for-profit	*Development aid*	*Disaster relief*

Adapted from: McLachlin, Larson and Khan (2009)

AN ABBREVIATED REVIEW OF THE RELATIONSHIP BUILDING LITERATURE

There is a great deal of literature on the nature of supply chain relationships. For the most part, this literature is focused on describing relationships in terms of trust, commitment, coordination of effort, loyalty, shared resources, etc. between the parties. However, there is very little written on relationship building in a supply chain.

The literature on supply chain relationships is also almost exclusively focused on business-to-business relationships, to the neglect of the not-for-profit and public sectors. The literature is laced with ideas and techniques that could be adapted for non-business situations.

For instance, Whipple and Russell (2007) detected three types of collaborative relationships during their exploratory interviews of business people: collaborative *transaction* management, collaborative *event* management, and collaborative *process* management. Transaction-oriented relationships focus on operational issues/tasks. Co-ordination of effort is targeted at solving problems and developing immediate solutions (e.g. expediting late deliveries). Event-oriented relationships are about joint planning and decision-making centered on critical events or issues, such as developing joint business plans or sharing information on upcoming product promotions. Collaborative event management includes problem prevention, such as identifying where supply chain disruptions or bottlenecks may occur. Finally, process-oriented relationships imply a more strategic collaboration, covering both demand (downstream) and supply (upstream) processes. This type of collaboration involves long-term joint business planning and more fully integrated supply chain processes, across functions and organizations.

Table 2. Business vs. humanitarian logistics

Aspect	Logistics Context	
	Business	**Humanitarian**
Purpose	Economic profit	Social impact
Context	Uninterrupted	Interrupted
Perspective on Time	"Time is money"	Time is life (or death)
Source of Funds	Paying customers	Donors

The Whipple-Russell typology can readily be adapted for humanitarian logistics. An example of transaction collaboration is close cooperation in the field; adapting, improvising and overcoming obstacles to get the job done, to help people live to see tomorrow. Joint needs assessment and sharing of assessment information is an example of event collaboration. The non-governmental organizations (NGOs) and others who post updates to "ReliefWeb" are engaged in this type of collaboration. Launched in October 1996 and administered by OCHA, ReliefWeb is the world's leading on-line gateway to information (documents and maps) on humanitarian emergencies and disasters[1] Moving to the third type of collaboration, pre-positioning partnerships are an example of collaborative process management. For joint pre-positioning, two or more humanitarian agencies would have to strategically plan and integrate their upstream and downstream supply chains in preparation for the next disaster.

The above is one example of adapting ideas from business logistics to the humanitarian context. However, reviewing that voluminous literature in supply chain relationships, and adapting it for humanitarian logistics, must be left for another day. That is not the purpose of this chapter.

Schary and Skjøtt-Larsen (2004) suggest that: "management's capability to establish trust-based and long-term relationships with customers, suppliers, third-party providers and other strategic partners becomes a crucial competitive parameter." Relationship building is a critical capability in supply chain management.

Lambert and Knemeyer (2004) offer a reasonably comprehensive framework for building supply chain relationships (or partnerships). Since close relationships require extra communication and are costly to implement, it is apparent that an organization can only "partner" with a select few of the other organizations it interacts with. For instance, out of the 90,000 suppliers used by Procter and Gamble (P&G) globally, only 400 qualify as key partners. These chosen few suppliers earn nearly 25% of the $48 billion per year P&G spends on materials, packaging, supplies, services, etc. (Teague 2008).

According to Lambert and Knemeyer (2004), to build supply chain relationships, organizations need a process for identifying high-potential partners, aligning expectations and finding the most effective level of cooperation. Their model starts with each prospective partner considering its *drivers* or compelling reasons to partner. For business supply chains, the relevant drivers—asset and cost efficiencies; customer service enhancements; marketing advantages; and profit growth or stability—are all about increasing sales and profits. If we can make more money by working together, let's seriously consider it.

Disaster relief supply chains share some common drivers with their business counterparts. It is critical to get the most out of scarce resources and limited budgets. It is also important to reach more beneficiaries in need and to serve them more quickly. However, humanitarian relief supply chains have their share of unique drivers, as well. These drivers fall into additional categories such as: increasing awareness; becoming more prepared for the next disaster; gaining more rapid access to accurate information about what is needed; and providing better security in the field. If we can save more lives by working together, let's do it!

The Lambert model next considers *facilitators* or supportive environmental factors that enhance partnership growth. Lambert and Knemeyer (2004) describe the following four fundamental facilitators: compatibility of corporate cultures; compatibility of management philosophy and techniques; strong sense of mutuality; and symmetry between the two parties. If the possible partners have/use similar cultures, philosophies and techniques, the path to a close relationship will be less steep. Relationships can also accommodate differences, as long as the parties are aware of these differences, along with their implications. The symmetry issue has been explored in great

detail in the literature, by supply chain thought leaders such as Cox (2004).

While the four facilitators were derived with profit-making supply chains in mind, they are also applicable to the NFP, humanitarian context. Issues of compatibility pertain to organizational missions, visions, and guiding principles; as well as operating procedures, information systems, and communication technology. Symmetry between the parties may be somewhat less relevant to humanitarians, as they are trying to save lives rather than make money. "Save money. Live better." This has a very different meaning in a world without Wal-Mart. An additional critical facilitator in humanitarian relief logistics (and business logistics) is the "complementarity" of capabilities each prospective partner brings to the table.

Imagine I have a warehouse full of supplies and a team of logisticians and public health workers in Denver; and you have an airplane, complete with crew and fuel, ready to depart San Francisco with 12 hours notice. If we both want to respond to the earthquake in Haiti, to ease the suffering, our capabilities appear highly complementary.

For purposes of relationship building in humanitarian supply chains, two broad dimensions merit careful consideration: compatibility and complementarity. As noted above, there are strategic and tactical elements to compatibility. Compatibility certainly pertains to things strategic, such as missions, guiding principles, "agendas" (whether overt or covert) and organizational culture. But it also includes things more tactical, e.g. operating procedures, information systems, and training of staff.

Complementarity covers a wide range of capabilities; from administration to operations. There is probably a lot of duplication of effort in the various administrative aspects to humanitarian relief. If two or more agencies combined elements of their administrative functions to reduce redundancies, then more resources should be available to serve beneficiaries in the field. Note that the example above the previous paragraph (my supplies/your airplane) is one of logistical (or operational) complementarity. There are also many opportunities for NGOs to form partnerships for purposes of advocacy, increasing awareness, and even fund-raising.

Let's get back to the Lambert and Knemeyer (2004) partnership model. After full consideration of the drivers and facilitators, the decision is made to create or adjust the partnership—or not. If it's a "go," then the parties next discuss various management *components* or joint activities and processes that build and sustain the partnership. The four key components are: (joint) planning; joint operating controls; communications; and risk/reward sharing.

In humanitarian supply chains, relationship building is very interesting, as well as challenging, as diverse entities such as Canadian Forces, CARE Canada, the IFRC, World Vision International, and the UN World Food Programme think about working together. The next section outlines an actor-based typology of humanitarian supply chain relationships.

ACTOR-BASED TYPOLOGY OF RELATIONSHIPS

In this typology, the focal organization is a NGO, such as Médecins sans Frontières (MSF) or World Vision International (WVI). The focus is also inter-agency, i.e. on relationships between a NGO and other organizations, including other NGOs.

However, it is important to recognize that there are also opportunities for closer relationships at the intra-agency or within organization level. Relevant intra-agency relationships occur between relief and development teams, between headquarters and field staff, between various national or regional divisions, or across functional units.

In the business world, internal relationships are noted as an important and challenging aspect of collaboration. Teague (2008) reports that closer relationships with suppliers begin with close

working relationships with internal business units at P&G. Business units share their strategies and goals with purchasing, which then works with unit leaders on how purchasing can support unit strategies. Whipple and Russell (2007) observe how internal issues can become obstacles to collaborative success with external relationships.

During Kirby's (2003) interview with leading supply chain authorities, the importance of intra-organizational relationships clearly emerged. One panelist suggested at many large corporations, different functions do not know what the others are doing. Another panelist cited an Association of Alliance Professionals survey which found the number one concern to be creating alliances internally between the silos, rather than creating strategic alliances with other companies. This panelist went on to state: "We don't deal with our own internal integration. How do we integrate externally if we can't do it internally?" Again, while this chapter focuses on building external relationships, let's not forget about opportunities to build close internal relationships.

The actor-based typology of relationships is shown in Figure 1. It includes the following four types of unique relationships: (1) humanitarian – NGO to NGO; (2) humanitariUN – NGO to UN agency; (3) humoneytarian – NGO to commercial service provider; and (4) humilitarian – NGO to military unit.

Note that the typology in Figure 1 does not explicitly include donors, which can vary from individuals to businesses to other NGOs to government agencies. Figure 1 also excludes governments and government agencies, except for military units. A future iteration of the typology aspires to include donors and governments more generally, but that must be left for another day.

Humanitarian Relationships

Two early responders to the earthquake in Haiti were the Mennonite Central Committee (MCC) and Médecins Sans Frontières (MSF). Informa-

Figure 1. A typology of relationships

tion for this section has been excerpted from their respective web-sites: MSF[2] and MCC[3]. The question is: Could MCC and MSF be compatible and/or complementary in a supply chain relationship?

MCC medical teams and structural engineers provided relief soon after the January 12, 2010 earthquake in Haiti. Distribution of food and other relief supplies to the people is ongoing. MCC has worked in Haiti since 1958; before the earthquake, this work focused on reforestation and environmental education, human rights and advocacy for food security. A multiple year response to the earthquake is planned, melding relief work with ongoing development work.

By mid-March, 2010, MSF had 348 international staff in Haiti, working with over 3,000 Haitian staff. Since the earthquake, MSF teams have done over 3,700 surgeries, provided psychological counseling to more than 22,000 people, and treated nearly 55,000 patients. Its teams have also distributed over 18,000 non-food item (NFI) kits—kitchen and hygiene kits, Jerry cans, blankets and plastic sheeting—and 10,500 tents.

MCC's motto is: "Relief, development and peace in the name of Christ." The following words are excerpted from MCC purpose and vision statements:

Mennonite Central Committee (MCC), a worldwide ministry of Anabaptist churches, shares God's love and compassion for all in the name of Christ by responding to basic human needs and working for peace and justice. MCC envisions communities worldwide in right relationship with God, one another and creation. MCC's priorities in carrying out its purpose are disaster relief, sustainable community development and justice and peace-building. MCC approaches its mission by addressing poverty, oppression and injustice – and their systemic causes; accompanying partners and the church in a process of mutual transformation, accountability and capacity building; building bridges to connect people and ideas across cultural, political and economic divides; and caring for creation. MCC values peace and justice. MCC seeks to live and serve nonviolently in response to the biblical call to peace and justice. MCC values just relationships. MCC seeks to live and serve justly and peacefully in each relationship, incorporating listening and learning, accountability and mutuality, transparency and integrity.

MSF was established in 1971 by a small group of French doctors who had worked in Biafra. Upon their return, they were determined to find a way to respond rapidly and effectively to public health emergencies. The following is excerpted from the MSF Charter:

Médecins Sans Frontières offers assistance to populations in distress, to victims of natural or manmade disasters and to victims of armed conflict, without discrimination and irrespective of race, religion, creed or political affiliation. Médecins Sans Frontières observes neutrality and impartiality in the name of universal medical ethics and the right to humanitarian assistance and demands full and unhindered freedom in the exercise of its functions. Médecins Sans Frontières' volunteers promise to honour their professional code of ethics and to maintain complete independence from all political, economic and religious powers. As volunteers, members are aware of the risks and dangers of the missions they undertake and have no right to compensation for themselves or their beneficiaries other than that which Médecins Sans Frontières is able to afford them.

Is the MSF mandate for independence from all political, economic and religious influences compatible with the MCC inspiration: in the name of Christ? Should MCC and MSF build a relationship to improve the effectiveness and/or efficiency of their relief work in Haiti?

HumanitariUN Relationships

Harr (2009) quotes an aid worker in Chad: "This can be a well-paying business, depending on the organization you work for. If you work for the U.N.—and just about everybody wants to work for the U.N.—your salary is not taxed, you get hardship pay, time off, and a lot of your expenses are covered." He also notes that with a few exceptions (e.g. MSF and IFRC) most aid organizations in Eastern Chad are both funded by and monitored by the United Nations High Commissioner for Refugees (UNHCR). UNHCR, in turn, is funded by "wealthy" governments, led by the United States.

Wealthy may be an inappropriate term, since 49 million Americans now face chronic "food insecurity" or not having enough food for an active healthy lifestyle. Indeed, almost one in four children in the USA lives on the brink of hunger (Jackson 2009). "The primacy of the United States is slipping" (Srinivasan 2009); once the world's largest lending nation, currently it is the biggest debtor.

While the United States generously funds a variety of UN activities, it reserves the right to ignore or at least loosely interpret UN Security Council resolutions. Authority over critical security decisions is not ceded to the United Nations. The United States spends about five percent of its gross domestic product (GDP) on its military (Srinivasan 2009). If the UN is dependent (financially) on the United States, what does this mean for humanitarian NGOs that are dependent (financially) on the UN?

Due to its sheer size, the UN is dominant in terms of distributing funds and sharing information in the humanitarian space. Adopting Cox's (2004) nomenclature, it is one of a very few large "suppliers" in a world of many "buyers," i.e. the much smaller NGOs. This dominance makes the NGOs dependent; there is an absence of interdependence, from which close relationships can be built. Of course, it is possible for the elephant to dance with the mice, but there is no question about who is leading.

The Inter-Agency Standing Committee (IASC) is promoted as the primary mechanism for inter-agency coordination of humanitarian assistance. It is a unique forum involving the key UN and non-UN humanitarian partners[4]. However, the non-UN partners appear limited to a very few, very large NGOs. It is far from clear how the IASC will effectively facilitate meaningful *humanitariUN* relationship building.

Another tool for inter-agency (and possibly UN/NGO) coordination is the *cluster approach*. It was endorsed by the IASC in September 2005. The World Food Programme (WFP) is the global Logistics Cluster lead for the UN. While WFP also usually assumes leadership at the field level, in circumstances where WFP is unable to fulfill this role, another agency or NGO can be given the responsibility[5].

The Logistics Cluster provides information sharing and coordination, such as assessments of infrastructure, port and transport corridor coordination, information about carriers and their rates, customs clearance, and equipment supplier information. Regular coordination meetings are held involving "all" stakeholders (UN and Government agencies, international and local NGOs).

A February 10, 2010 blog exclaims: "One of the cluster system's greatest strengths is bringing divergent groups of people together to find the most efficient response in the circumstances." The Cluster encourages agencies and NGOs to help each other with issues like transportation capacity gaps. If such gaps persist, the Cluster lead is considered the provider of last resort. To support the Haiti earthquake operation, or other disaster relief efforts, the Logistics Cluster posts updates and situation reports on logcluster.org and reliefweb.int.

Humoneytarian Relationships

These are relationships between humanitarian NGOs and business firms, with the firms acting as suppliers of goods and/or providers of for-hire services (e.g. logistics services) to humanitarian buyers. Figure 2 reveals a barrier to relationships between businesses and the NGOs—demand uncertainty. While business-to-business (B2B) buyers often have reasonably stable and reliable demand streams, humanitarian buyers offers erratic, uncertain demand, in terms of when, where and how much. Such uncertainty is not usually the stuff of close buyer/supplier relationships.

Businesses engage the humanitarian community in two ways: (1) as suppliers of goods or for-hire provider of services to NGOs, UN agencies, etc., as noted above and (2) as possible donors of goods, services and expertise. For instance, commercial entities can support humanitarian operations by assisting in the fund-raising effort or by offering expert advice on logistical issues. When an NGO or the UN is a buyer, they should try to get the best deal in terms of cost, quality and delivery, similar to their counterparts in the business world or the public sector. More detail on building such relationships is a task for another day, beyond the scope of this chapter.

Figure 2. Business vs. humanitarian demand streams

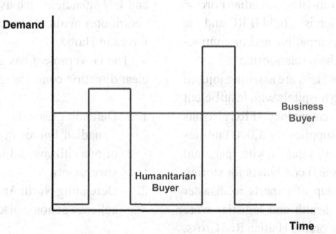

Humilitarian Relationships

Munslow and O'Dempsey (2009) demonstrate how the American-led invasion and occupation of Iraq, commencing in 2003, has been a disaster for the humanitarian community. They suggest: "Humanitarian agencies lost their indispensible shield of neutrality;" they became targets. The military increasingly became involved in humanitarian aid delivery and humanitarian agencies were pressured to integrate their efforts with the "war on terror." Perhaps the ultimate example is the Provincial Reconstruction Teams (PRTs), with military and civilian personnel working together in Afghanistan to provide security, reconstruction and relief. The result has been a loss of *humanitarian space*, "an operating environment conducive to independence, neutrality and impartiality in the relief of human suffering." For humanitarians in the field, it has been deadly; e.g. the murders of International Rescue Committee (IRC) aid workers Mohammad Aimal, Shirley Case, Nicole Dial and Jackie Kirk on August 13, 2008 in Afghanistan. The purpose here is not to suggest the military has no role in the delivery of humanitarian relief. However, the loss of humanitarian space surely compromises relationship building between humanitarian agencies and military units.

Two additional early responders to the earthquake in Haiti were the Canadian Forces and IFRC. Information for this section has been excerpted from their respective web-sites: Canadian Forces[6] and IFRC[7]. The question is: Could IFRC and the Canadian Forces be compatible and/or complementary in a supply chain relationship?

IFRC volunteers in Haiti are assisting injured people and supporting hospitals with insufficient capacity to handle this emergency. IFRC also has pre-positioned relief supplies for 3,000 families in Haiti, consisting of kitchen kits, personal hygiene kits, blankets and containers for storing drinking water. A group of experts in disaster response, emergency health and logistics were also dispatched to support the Haitian Red Cross,

and to coordinate international assistance from members of the International Movement and other organizations. Another IFRC team led a damage assessment together with the Humanitarian Aid Office of the European Union.

Operation HESTIA is the Canadian Forces participation in the humanitarian response to the earthquake that struck Haiti on January 12, 2010. Joint Task Force Haiti (JTFH) has a mandate to deliver a wide range of services in support of the Government of Haiti and the Canadian Embassy in Port-au-Prince, such as emergency medical services; engineering expertise; mobility by sea, land and air; and defense/security support. About 2,000 personnel have been assigned to JTFH, including the Disaster Assistance Response Team (DART) with three Reverse Osmosis Water Purification Units (ROWPUs). Canadian Forces have also deployed: the destroyer HMCS Athabaskan, the frigate HMCS Halifax, six CH-146 Griffon helicopters, an urban rescue and recovery team, a detachment of Military Police, and one Field Hospital.

DART's water supply section can provide up to 50,000 liters of drinking water per day. In the earthquake's aftermath, water bodies are often infested with many different kinds of bacteria and it is extremely difficult to find good drinking water. Canadian troops do not directly distribute water to the people; their mandate is only to produce drinking water. Distribution falls to NGOs and UN agencies, such as UNICEF. DART also contributes medical capacity and security to CF forces in Haiti.

The Government has given Canadian Forces clear direction concerning their three roles:

1. Defending Canada – First and foremost, the Canadian Forces must ensure the security of our citizens and help exercise Canada's sovereignty.

2. Defending North America – Delivering excellence at home also helps us contribute to

the defense of North America in cooperation with the United States, Canada's closest ally.

3. As a trading nation in a highly globalized world, Canada's prosperity and security rely on stability abroad. As the international community grapples with numerous security threats, Canada must do its part to address such challenges as they arise. This will require the Canadian Forces to have the necessary capabilities to make a meaningful contribution across the full spectrum of international operations, from *humanitarian assistance* to stabilization operations to combat.

The DART can be deployed on very short notice anywhere in the world in response to natural disasters and humanitarian emergencies. It focuses on two critical needs: primary medical care and production of safe drinking water.

The IFRC is the world's largest humanitarian organization, providing assistance without discrimination as to nationality, race, religious beliefs, class or political opinions. Founded in 1919, the International Federation comprises 186 Red Cross and Red Crescent societies, a Secretariat in Geneva, and more than 60 delegations strategically located to support activities around the world. Following is the IFRC vision and mission:

We strive, through voluntary action, for a world of empowered communities, better able to address

human suffering and crises with hope, respect for dignity and a concern for equity. Our mission is to improve the lives of vulnerable people by mobilizing the power of humanity.

IFRC work is guided by seven fundamental principles, as shown in Table 3.

Are the IFRC principles of neutrality and independence compatible with the three roles of Canadian Forces? Should Canadian Forces and IFRC build a relationship to improve the effectiveness and/or efficiency of their relief work in Haiti?

Airline Ambassadors International[8]

Airline Ambassadors International (AAI) mobilizes "the world's most valuable resource: men and women of goodwill – traveling to make a difference." AAI partners with a wide-range of other organizations and people, including "voluntourists," in the interest of getting the job done.

AAI's mission and vision are as follows:

Mission: Airline Ambassadors provides humanitarian aid to children and families in need and conducts international relief and development to underprivileged communities.

Vision: By establishing a venue for ordinary people to do extraordinary service, we demonstrate that

Table 3. The seven fundamental principles

Principle	Brief Description
Humanity	Prevent and alleviate human suffering wherever it may be found.
Impartiality	Do not discriminate as to nationality, race, religious beliefs, class or political views.
Neutrality	Do not take sides in hostilities or engage in political, racial, religious or ideological controversies.
Independence	Remain autonomous from governments.
Voluntary service	Serve without desire for gain.
Unity	There is only one (IFRC) society per country.
Universality	It is a world-wide movement.

continued friendships between our volunteers and aid recipients is key to fostering sustainable projects. This phenomenon enhances the benefits that accrue to the aid we provide. As the only humanitarian organization emanating from the airline industry, we leverage industry resources and professional expertise for our members so they can 'Travel to Make a Difference.'

Partnering with non-governmental organizations (NGO's), schools, churches and civil society, Airline Ambassadors also works with local government and business, and leverages its contacts with the travel industry to match world resources to world needs and to help build capacity in local communities.

For instance, in 2009, AAI received donations of aircraft capacity to transport humanitarian assistance from American Airlines, JetBlue, the U.S. military, and the Colombian National Police. AAI also partnered with local universities and technical schools in Ecuador and El Salvador to build technical capacity in the CASA program.

On January 13, 2010, AAI headquarters announced the agency's Haiti Earthquake Disaster Plan. Working with major airlines, AAI was preparing to transport needed supplies to Port au Prince as soon as air traffic control would allow it. The plan was to deliver 160,000 lbs. of high priority humanitarian aid. The first shipments would consist of: water bottle filters, family food kits (meals ready to eat), powdered milk, hygiene kits, tents for temporary shelter, canned food, 80 lb. bags of rice and beans, and emergency medical modules. Initially, these supplies, donated by AAI partner LDS Charities, were located in Denver and Salt Lake City. AAI would re-position these supplies at New York's JFK airport or Miami International Airport, where charity flights to Haiti on American Airlines, United Airlines, JetBlue and Spirit would be easier to coordinate.

AAI's main logistics partner on the ground in Haiti would be U.S. Southern Command who was busy coordinating the relief effort in Port au Prince. Southern Command would provide ground support for distribution and store AAI aid, initially with Food for the Poor and Catholic Relief Service (as they had intact warehouse space.) AAI was also busy collecting funds to pay for the fuel to fly aircraft from JFK/MIA to Port au Prince. These flights would deliver food, water, medical supplies, emergency personnel, blankets, sheets, and other relief supplies.

By February, the Canadian-based Belinda Stronach Foundation entered into a partnership with AAI to provide 70,000 lbs. of food and medicine on weekly cargo flights from Toronto for twelve weeks. The Belinda Stronach Foundation is committed to advancing human potential and achievement through individual empowerment and social change.

By February 12, AAI had received three of these flights. With help from the U.S. 82nd Airborne and the Haitian National Police, AAI distributed food in a district of Port-au-Prince where some 70,000 people were expected to be reached. Airline Ambassadors also partnered with the Henry Schein Foundation. These two foundations, Schein and Stronach, along with the LDS Charities, have provided AAI with over $1.3 million in aid for Haiti.

AAI readily partners with other NGOs, secular and faith-based; businesses (airlines); military units; and others to make a difference. This mutual desire to make a difference, to ease the suffering, makes AAI *compatible* with what may seem like "strange bedfellows" on the surface. AAI also seeks *complementary* partners, matching and assembling the right mix of aid supplies, transport capacity, expert personnel, etc.

RELATIONSHIP BUILDING ACROSS THE PHASES OF HUMANITARIAN ACTION

According to Pettit and Beresford (2005), the disaster management cycle consists of three key elements: preparedness, response, and recovery. Kovács and Spens (2007, p. 101) outline three similar elements, observing that "different operations can be distinguished in the times before a disaster strikes (the preparation phase), instantly after a disaster (the immediate response phase) and in the aftermath of a natural disaster (the reconstruction phase)."

For the purpose of this brief section on relationship building across the phases of humanitarian action, an additional phase, advocacy and awareness, leads the list of elements. These phases are relevant to a discussion on relationship building because opportunities for collaboration can be very different across the various phases. While awareness spans the other three phases, it may be temporarily put on hold during disaster response.

There may be tremendous opportunities for NGOs to build closer relationships in the advocacy/awareness phase. However, the more this phase is linked to fund-raising, the more challenging collaboration will be. The rivalry between NGOs must surely be most fierce in the area of fund-raising or courting donors. Relationships built to increase awareness of the global humanitarian situation should be designed to increase the overall size of the donations pie.

The preparation (and pre-positioning) phase is where there may be the greatest opportunity for relationship building. This is a time when NGOs and other agencies could focus on their relative capabilities—and identify ways they complement one another. Whether on a national, regional or global basis, the focus is on who has access to what facets of the supply chain. These facets include transportation capacity, storage capacity and location, pre-positioned supplies, suppliers of critical material, communications technology, personnel with expertise, etc. Kovács and Spens (2007) suggest the preparation phase is when agencies can create collaborative platforms, e.g. the UN Joint Logistics Centre or the Disaster Relief Network run by the World Economic Forum.

In disaster response, temporary relationships can focus on one common point of compatibility—*to make a difference*. Compelled by urgency, the cooperation should come easily. As urgency fades and the focus turns to recovery and development aid, there are opportunities for greater logistical efficiency through coordinated effort. But the relative lack of life and death urgency may cause subtle compatibility issues to stand in the way of relationship building. Once the dust has settled, there is time to carefully consider the complete mission statement prior to looking for partners. Relationships formed to support preparation and recovery are probably more enduring than relationships formed during response.

SUMMARY AND CONCLUSION

Humanitarian supply chain practitioners are advised to consider opportunities for relationship building, as a path to increase their effectiveness and/or efficiency. The NGOs should seek other organizations with complementary capabilities; and they should re-visit issues of compatibility. At what point are compatibility concerns "dealbreakers" among organizations trying to make a difference?

Notable limitations of this chapter are the missing actors in the typology of relationships. Local community councils are critical to the downstream humanitarian supply chain. These actors, and possible *communitarian* relationships, are beyond the scope of the current piece. Future work is planned to add these relationships into the framework. More generally, government actors (other than military units) are excluded from the framework in Figure 1. Again, this omission will be addressed in the future, since various government agencies are important humanitarian actors, as donors, recipients, regulators and facilitators.

Compatibility is a multi-dimensional, higher-order construct. Indicators of the compatibility between two organizations would include their relative rating of the importance of principles such as humanity, impartiality, neutrality and independence. A comparison of organizational missions can also yield insight into compatibility. All humanitarian organizations should be compatible in the desire to ease suffering, to make a difference. Compatibility concerns are likely to be lowest within a single NGO, higher among NGOs, and highest between NGOs and other types of actors (e.g. military units or UN agencies).

Complementarity is determined by the relative capabilities (e.g. transport and storage capacity, equipment, number and expertise of personnel, supplies, information sharing, communication technology, presence in a certain nation or region, relationship building skill, etc.) of two or more prospective partners. Supply chain mapping can enhance the understanding of possible complementarity between organizations. It is apparent that compatibility and complementarity are probably necessary conditions for relationship building among humanitarian actors.

Strong complementarity may make up for weak compatibility—and vice versa. It is likely that relative importance of various elements of compatibility and complementarity vary across phases of humanitarian action. For instance, compatibility is probably more critical during preparation compared to response. Conversely, complementarity may be more important during response compared to preparation. Response is about taking what we have and getting the job done. In preparation, there is more time to think strategically and to ask: why are we doing this work?

REFERENCES

Cox, A. (2004). The art of the possible: Relationship management in power regimes and supply chains. *Supply Chain Management: An International Journal*, *9*(5), 346–356. doi:10.1108/13598540410560739

Harr, J. (2009, January 5). Lives of the Saints. *New Yorker (New York, N.Y.)*, 47–59.

Jackson, H. C. (2009). *Number of hungry Americans increases*. USDA: Food Manufacturing. http://www.foodmanufacturing.com /scripts/ Products-USDA-Number -Of-Hungry-Americans.asp

Kirby, J. (2003). Supply chain challenges: Building relationships. *Harvard Business Review*, *81*(7), 64–73.

Kleindorfer, P. R., & Saad, G. H. (2005). Managing disruption risks in supply chains. *Production and Operations Management*, *14*(1), 53–68. doi:10.1111/j.1937-5956.2005.tb00009.x

Kovács, G., & Spens, K. M. (2007). Humanitarian logistics in disaster relief operations. *International Journal of Physical Distribution & Logistics Management*, *37*(2), 99–114. doi:10.1108/09600030710734820

Lambert, D. M., & Knemeyer, M. A. (2004). We're in this together. *Harvard Business Review*, *82*(12), 2–9.

McLachlin, R., Larson, P. D., & Khan, S. (2009). Not-for-profit supply chains in interrupted environments: The case of a faith-based humanitarian relief organization. *Management Research News*, *32*(11), 1050–1064. doi:10.1108/01409170910998282

Munslow, B., & O'Dempsey, T. (2009). Loosing soft power in hard places: humanitarianism after the US invasion of Iraq. *Progress in Development Studies*, *9*(1), 3–13. doi:10.1177/146499340800900102

Murray, V. (2006). Introduction: What's so special about managing nonprofit organizations? In Murray, V. (Ed.), *The management of nonprofit and charitable organizations in Canada*. Markham, Ontario: LexisNexis-Butterworths.

OCHA. (2010a). Haiti earthquake situation reports #1 & 2. Office of the Coordination for Humanitarian Affairs, New York, January 12-13, 2010. Retrieved from http://www.reliefweb.int

OCHA. (2010b). Haiti earthquake situation reports #3 & 4. Office of the Coordination for Humanitarian Affairs, New York, January 12-13, 2010. Retrieved from http://www.reliefweb.int

Pettit, S. J., & Beresford, A. K. C. (2005). Emergency relief logistics: An evaluation of military, non-military and composite response models. *International Journal of Logistics: Research and Applications, 8*(4), 313–331.

Schary, P. B., & Skjøtt-Larsen, T. (2004). *Managing the global supply chain* (3rd ed.). Copenhagen, Denmark: Copenhagen Business School Press.

Spring, S. (2006, September 11). Relief when you need it: Can FedEx, DHL and TNT bring the delivery of emergency aid into the 21st century? *Newsweek International Edition.*

Srinivasan, K. (2009). International conflict and cooperation in the 21st century. *The Round Table, 98*(400), 37–47. doi:10.1080/00358530802601660

Teague, P. (2008, September 11). P&G is king of collaboration. *Purchasing*, 46.

Whipple, J. M., & Russell, D. (2007). Building supply chain collaboration: A typology of collaborative approaches. *International Journal of Logistics Management, 18*(2), 174–196. doi:10.1108/09574090710816922

KEY TERMS AND DEFINITIONS

Compatibility: Degree to which two or more organizations or individuals share common missions, principles, values, etc.

Complementarity: Degree to which the capabilities of two or more organizations or individuals are complementary.

Humanitarian Principles: Guiding values on which humanitarianism is built. They include at least the three principles of humanity, impartiality and neutrality. Other principles of e.g. IFRC are independence, voluntary service, unity and universality.

Humanity: Prevention and alleviation of human suffering wherever it may be found.

Impartiality: Absence of discrimination based on gender, nationality, race, religious beliefs, class, political views or anything else.

Independence: Remaining autonomous from governments of any sort.

Neutrality: Absence of taking sides in hostilities or engaging in political, racial, religious or ideological controversies.

Partnership: A close relationship with selected organizations.

Partnership Drivers: Compelling reasons to engage in a partnership.

Partnership Facilitators: Supportive environmental factors that enhance partnership growth.

Relationship Building: A critical capability in supply chain management to develop a close working relationship or partnership among two or more organizations.

Relationship Typology: In this chapter relationships are categorized according to the actors they encompass. Other relationship typologies are built around the intensity, intentions and time period of the relationship.

ENDNOTES

[1] http://www.reliefweb.int/
[2] http://www.msf.org/
[3] http://mcc.org/
[4] http://www.humanitarianinfo.org/iasc/
[5] http://www.logcluster.org/
[6] http://www.forces.gc.ca
[7] http://www.ifrc.org
[8] http://www.airlineamb.org/index.html

Chapter 2
Humanitarian Partnerships— Drivers, Facilitators, and Components:
The Case of Non-Food Item Distribution in Sudan

Rolando M. Tomasini
Hanken School of Economics, Finland

ABSTRACT

Through the use of a case study, this chapter discusses the design of a partnership between humanitarian organizations to understand what the drivers, facilitators, and components of the partnership are. This research has been designed using a topical literature review and a case study. The practical implications include a discussion and guidelines for designing partnerships under high uncertainty and limited resources.

INTRODUCTION

Delivering aid in an emergency situation is a complex process given the high levels of uncertainty, capacity and resources that characterize the needs. Humanitarian agencies work hard to fulfill their specific mandates and ensure that beneficiary needs are met in the quickest and most efficient way. Doing so requires in many circumstances joint efforts (partnerships) between different agencies to achieve the common goal. Partnerships between humanitarian agencies can take many forms ranging from informal agreements, to memorandum of understanding, or formal contracts.

DOI: 10.4018/978-1-60960-824-8.ch002

The literature on partnerships between corporations is rather extensive. However, little has been written about partnerships in the supply chain among humanitarian organizations to respond to emergencies. Studies in commercial supply chains have explained the benefits of partnerships to mitigate uncertainty or foster collaboration. However, these differ from the focus of this research in that the final recipient is not the only customer. In the humanitarian supply chain there are two customers: donors and beneficiaries at opposite ends of the supply chain. They each bring high levels of uncertainty for the humanitarian organization at the onset of the operations. In other words, it can take days, if not weeks, until a good picture of how much will be donated from which donors for an inaccurate number of beneficiaries.

Researchers have begun to look into inter-organizational collaboration in the humanitarian supply chain (Samii, 2009; and Schulz, 2009) though a lot of questions remain unaddressed regarding formalized multiparty partnership agreements among humanitarian agencies. Formal partnership agreements from humanitarian organizations have been more the subject of researchers looking at cross-sectoral partnerships (i.e., humanitarian-private, humanitarian-military, humanitarian-governments), still a greater understanding of why these partnerships emerge between humanitarian agencies in logistics remains to be discussed.

The focus of this chapter is on the initial phase of Darfur crisis in 2004 when the humanitarian organizations made the decision to partner for the delivery of aid and the steps and decisions taken to create this partnership. In the case presented here this phase corresponds to a ten-months (February-November 2004) period during which humanitarian agencies present in Sudan became aware of the rising needs of large numbers of people in the Darfur regions. Unable to access the region, yet aware of the logistical challenges they would encounter in responding to such a massive operation, managers from different humanitarian agencies worked together to plan and decide which tasks would be addressed collectively and under what conditions. The result was a set of inter-agency initiatives meant to create cost savings and efficiency. This chapter focuses specifically on the partnership designed to deliver non-food items through a common pipeline.

The arguments are developed as follows: Section 1 explains the context of humanitarian logistics and partnerships. Section 2, describes the methodology used to design and analyze the case study presented in section 3. Section 4 combines the literature to analyze the case and discuss the drivers, facilitators and components for the partnership between the different humanitarian agencies focusing on Darfur. Section 5 lists limitations and recommendations from this research. In closing, section 6 provides some recommendations for future research, and section 7 provides a set of concluding remarks about the application and relevance of this research to the evolving field of humanitarian logistics.

BACKGROUND AND LITERATURE REVIEW

Humanitarian logistics is a relatively new academic field of study and emerging research (Kovacs and Spens 2007) with evolving definitions and concepts. The main difference with commercial supply logistics is that companies aim their "logistics at increasing profits whereas humanitarian logistics aims to alleviate the suffering of vulnerable people" (Thomas and Kopczak, 2005). Unlike actors in a commercial supply chain, humanitarian actors are required to go beyond the profit logic (Ernst, 2003), as this logic is actually absent and replaced by the concepts of speed and cost (Tomasini and Van Wassenhove, 2009). Moreover, humanitarian actors are driven by efficiency given the limited resources as long as performance does not compromise their legitimacy or license to operate as neutral and impartial organizations.

The second most salient difference would be the multiplicity of actors involved. Actors may come from different ideologies and motivations (i.e., United Nations agencies, non-profit organizations, community organizations, religious and faith based organizations, military, governments, media, volunteers, etc), but eventually need to find a way of contributing towards the common goal of meeting the assessed needs of the population. Mandates, organizational capacity, and local expertise combine together to build what Samii (2008) calls 'virtual organizations', where a set of agencies work together as one supply chain to coordinate the use of limited resources in a disaster relief operation.

The lack of a shared profit incentive, combined with the multiplicity of actors and their diverse mandates, capacity, and expertise leads to difficulties in alignment for the humanitarian supply chain. For example, some agencies may have mandates that extend beyond the emergency phase on to the recovery phase, while others may only have the capacity to stay in the emergency phase given their funding and limited staff. At the same time, another agency may only be able to provide specialized aid in the recovery phase that follows the initial emergency window. Determining who will be present where, and for how long at each phase is a challenge for each relief operation that affects the success and efficiency of the supply chain.

Following a longitudinal study of different partnerships, Lambert and Knemeyer (2004) states that most supply chain partnerships failed due to un-aligned expectations and undetermined levels of cooperation. Together they propose a model "developed under the auspices of Ohio State University's Supply Chain Forum from lessons learned from best partnering experiences of that group's 15 member companies". The partnership model for supply chain proposes that there are a set of 'drivers' (compelling reasons to partner), that combined with set of 'facilitators' (supportive environmental factors that enhance partnership growth) lead to a decision to partner. The partnership is then composed of 'joint activities and processes that build and sustain the partnership' to meet a level of performances that matches the expectations of the partners in the first place. See Figure 1.

The Lambert and Knemeyer model has interesting implications for this research given its focus on supply chain performance and the definition of partnership. While the definition is not explicitly stated by the authors, they explain that rather than a supplier-selection tool, a partnership

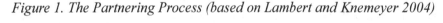

Figure 1. The Partnering Process (based on Lambert and Knemeyer 2004)

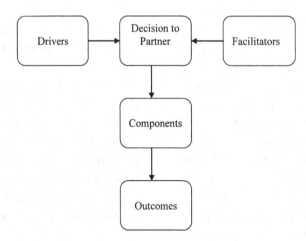

"is a way targeting high potential relationships and align expectations around them" (Lambert and Knemeyer, 2004). Such explanation fits well the intent of the humanitarian agencies in this research who intended this partnership to be yet another interaction point for them to implement their programs in Sudan. Their definition of partnerships presented is not time bound. This is also in line with the humanitarian managers in the non-food items (NFI) pipeline for whom the duration of the partnership was a secondary issue. Instead, the humanitarian managers focused on high-potential relations and alignment as described by the Lambert and Knemeyer model (2004).

METHODOLOGY

Research for this article has been conducted following case study methodology suggested by McCutcheon *et al.* (2002), who adapt Yin (1989) and Eisenhardt's (1989) case study methodology to the field of operations management. McCutcheon *et al.* (2002) also sustain that "case-based research represents the intersection of theory, structures and events" while allowing the researcher to "explore concepts in the real world looking for patterns that are insightful, interesting, and offer the possibility of providing predictive, explanatory power or understanding". Case-based methodology is deemed appropriate for this research as "there is no control over the behavioral events, and the focus is on contemporary events" (Yin 1989). More specific to the context of operations management case-based methodology enables the researcher to observe in complex environments the different factors at play and assess their impact on the outcomes (McCutcheon *et al.* 2002) as it was done through the data collection and analysis. Given the type of research purpose and question, Handfield and Melnyk (1998) suggest that an in-depth case study be used to uncover and explore the topic and uncover areas for research and theory development. The case study here

serves to provide an understanding of a complex subject matter (Meredith 1998) for the analysis and recommendations.

The instrument chosen to gather the data (i.e., study protocol) was a questionnaire designed using the theoretical propositions of the Lambert, Emmelhanz, and Gardner (1996) partnership model. Through open ended questions managers from humanitarian organizations and donor agencies operating in Sudan were interviewed onsite to understand "compelling reasons to partner, the supportive environmental factors that enhanced the partnership growth, the decision to partner, activities that sustained the partnership, and the extent to which performance met expectations" (Lambert and Knemeyer, 2004).

Interviews were carried out in person for approximately one hour each. The list of participants, totaling 40 humanitarians, was initially set to 20 and increased in sets of 10 twice until no new issues were identified in the responses. Participants can be divided into two groups: a focal group was made up of 5 organizations (World Food Program, UNICEF, UN Joint Logistics Center, Office of Coordination for Humanitarian Affairs, and Care International) involved in the design and delivery of the partnership; and an extended group included additional 7 organizations that joined the partnership after it was formed and contributed or benefited from its operations (USAID, DFID, ECHO, World Vision International, Save the Children, Medair, and ADRA). The latter could be considered "clients" or "users" rather than core members as those in the focal group.

Participants were provided with an overview of the research during a weekly meeting that assembled all the partners for updates, and they received a project brief through email prior to the interview meeting. Interviews were transcribed and shared again for corrections and additional comments from the interviewees. To ensure construct and internal validity, interviewees were prompted to provide additional data in the form of support material such as quantitative and qualitative internal reports, bulletins, and opportunities for observation in the field. These investigative observations proved to be useful in validating the focus of the research and the selection of additional participants. Some examples of observations include attendance at coordination meetings, warehouse loading and offloading activities, field visits to the distribution sites, and donor coordination meetings.

Interview transcripts were then analyzed to draw out the drivers to partner from the different organizations and the facilitators to making the partnership work. Comments and quotes from the participants were categorized and coded to develop a list of recommendations for future partnerships of the same type.

Following a period of data analysis the compiled data was shared with the participating organizations in a plenary meeting onsite in Sudan to validate assumptions before drawing final conclusions. Feedback was collected and integrated into the final analysis to describe the benefits of the partnership and critical success factors for its development.

Finally, the results were disseminated in a managerial report to all the agencies which served as an evaluation of the implementation phase of the partnership. Further analysis has been carried out through a more in-depth review of the literature for the purpose of this article.

CASE STUDY: NON-FOOD ITEM COMMON PIPELINE FOR DARFUR (SUDAN)

Early 2004, humanitarian agencies began to focus again on Sudan in anticipation of the signing of the peace treaty that would bring to an end to the long standing battle between South Sudan and the government of Khartoum (Tomasini and Van Wassenhove, 2007). However, the eminent needs in Darfur made everyone switch focus to a different crisis out west where the urgent needs of

approximately a million people, and rising, were suddenly made public. Rumors about violence in Darfur began to circulate in the spring. However, humanitarian presence in the area was too small to confirm the magnitude of the violence. Humanitarian agencies based in Khartoum required a travel permit from the government to travel to the region, which in some cases could take weeks if the purpose was not properly justified and agreed by the government. The bottlenecks and restrictions were eventually lessened as the rumors gained the attention of the international community and the UN Security Council began to pressure the government of Khartoum to collaborate. From one week to the next the number of victims escalated from an initial 200,000 victims to nearly 2 million people by the summer when the international community finally got the details of the fighting and crimes against humanity taking place.

Douglas Osmond, UNJLC Officer in Charge during the initial phases of the Darfur operation, pointed out during an in interview for this research in Sudan, that the mere magnitude of the crisis was beyond the operational capacity of the humanitarian agencies present to respond. He explained that "internally displaced people (IDP) were scattered through a large area with very long distances to travel. These distances were unprecedented. Even in Goma/Bukavu, one of the largest IDP operations in the history of the African region, the people were closer to each other. At the end, resources were exhausted in movement alone."

As most agencies in Sudan were staffed to conduct long term development programs they lacked the capacity to do emergency response as required in Darfur. "Development programs focus on capacity building, using the national staff, cost savings, low budgets and long time frames, where as emergency relief programs are the opposite in all respects" explained Osmond. Staffing highly skilled, qualified, and experienced personnel in short notice was a problem for most agencies resulting in long delays to provide technical and administrative support on the ground.

The sudden emergence of this unanticipated crisis came to test the reactivity of all the humanitarian agencies. On one side, there were lots of bureaucratic bottlenecks by the Sudanese government that made getting information about the nature and magnitude of the crisis difficult. On the other side, humanitarian agencies lacked the readiness and preparedness to roll out such a large program overnight considering the unprecedented scale and intensity.

The combination of these issues created an unanticipated "preparation phase" during which humanitarian agencies were unable to become operational and were forced to discuss a collaborative strategy. This phase raised the expectations of the donors who aware of the magnitude and complexity of the needs were also encouraging innovative solutions to target the needs in Darfur.

At least half of the interview participants explained the added value of the delays. For example, unlike most emergencies, under restricted operational capacity and limited access to resources, the humanitarian actors present had no other choice but to sit down, share information, and collectively plan. They felt the pressure of the donors and the international community who had been lobbying for the Sudanese authorities to give them access. To some extent humanitarians feared they were not ready for the day that access would be granted and felt the obligation to be as ready as possible for that day. In emergency situations, such planning is less frequent than desired due to the urgency of the task, the lack of anticipation, and funds. The preparedness phase ended with Sudanese authorities issuing travel permits to foreigners to access the Darfur region and ramp up their operations and assist violence victims grouped in the camps.

Through the joint planning discussion it became obvious that NFIs such as mattresses, plastic sheets, blankets, soap, jerry cans, etc. would be a common concern for all agencies given the lack of a leader for its sourcing and distribution. None of the humanitarian agencies or NGOs present in Sudan had it in their mandate to provide NFIs to IDPs, let alone funding or the capacity to do it. Even if one agency had the mandate, the magnitude of needs, volume of recipients, and the complexity of the environment revealed that no single agency could undertake the whole responsibility of sourcing, storing, transporting, distributing, and reporting.

In the interviews all donors expressed that by April 2004 they had funded rapid response agencies and had not seen any action or results1. Having exhausted the traditional operating channels innovative approaches seemed more promising. Donors took the lead in recognizing that the best solution was interagency collaboration and promoted it by placing their resources in that direction.

Actors on the Non Food Items Common Pipeline

As the main coordination agency for the humanitarian community, the Office for Coordination of Humanitarian Affairs (OCHA) and the donors requested United Nations Joint Logistics Center2 (UNJLC) support to address the NFI concerns. At the time, this de facto partnership could be defined as a symbiotic relationship between OCHA's mandate, funds, and administrative capacity with the UNJLC's logistics field expertise and experience, access and presence in the field.

Reflecting on their initial conversations with UNJLC, donors and OCHA explained that NFIs demanded significant logistical planning and support, which is UNJLC's core function. UNJLC was staffed with seconded senior logistics staff from other agencies with ample experience in complex emergency and population movement in the region.

The first steps came after a series of negotiations and meetings between OCHA/UNJLC, the donors and the rest of the agencies to sign a few agreements that would set the blueprint for the common pipeline. Under these agreements, UNICEF assumed responsibility for the procurement of the goods for the common pipeline; WFP became the consignee for all in-kind donations; CARE provided warehousing in the Darfur capitals and distribution to the implementing partners at the camps. The agreement meant that agencies in Darfur could request items that had been previously identified as a priority at no cost and receive them from CARE on-site. See Figure 2.

UNJLC and OCHA assumed joint responsibility to chair NFI working groups (NFIWG) in Khartoum and in the three capitals in Darfur (Nyala, Geneina, Obeid). The working groups were open forums for collective decisions attended by donors, UNICEF, and NGOs doing distribution in the camps (also known as implementing partners or IPs). See Figure 3.

In Darfur NFIWGs collected information about beneficiary figures from the field, estimated requests, pending needs, and analyzed the data to suggest priorities per camp to the NFIWG in Khartoum. Then the working group in Khartoum met weekly to identify priorities per state based on the field reports, the participants' input and stock movement reports. Upon reaching consen-

sus on the priorities the working group in Khartoum communicated them to UNJLC/OCHA, who would adjust the quantity of the needs and set the priorities based on the pipeline status.

Having communicated the stated priorities to the rest of the humanitarian community, the IPs issued a request to UNJLC/OCHA to receive the items at a particular camp for distribution. The NFIWG in Khartoum reviewed the requests, sought clarification and assessed adjustments where necessary. Upon satisfactory completion of the review process, the NFIWG would issue an "Action Request" to CARE to dispatch the items to the IPs. The action request was issued within a day. See Figure 2 for a detailed illustration of the material and information flow.

Participant Benefits

For the implementing agencies, having the common pipeline in place allowed them to focus and invest their resources on their core activities. Basically they did not have to worry about procuring their own goods (UNICEF), finding a consignee

Figure 2. Information and material flow

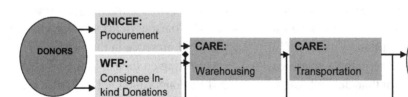

Figure 3. Map of Sudan (from http://www.infoplease.com/atlas/country/sudan.html)

for in-kind goods (WFP), nor warehousing or transportation to the camps (CARE). Their main concern was to be as ready as possible to distribute the items when they were needed and dispatched.

Donors understood the benefits of the model, and did their utmost to support it. Financially, they funded administrative and operating costs of all the agencies, including the secondment of staff to UNJLC and OCHA. They also provided both in-kind donations and funds for procured goods. Politically, they exercised their power on the authorities to remove bottlenecks and reinforce the urgency. In practice, they collectively took a

strong stand on not funding independent initiatives that would compete with the common pipeline. For example, if an appeal came through for an item that is available there, the appeal would be rejected and redirected to the common pipeline. Donors recognized that overall the common pipeline helped to reduce the duplication of efforts thus creating a number of opportunities for cost savings. First, they would fund less administrative and operational activities, by centralizing the activities among just a few agencies. Second, they would minimize redundancies and excesses since they operated with a common assessment of needs, and not each agency's estimates. Third, they reduced the transaction cost of processing the appeals since all NFIs were handled by the pipeline. Lastly, donors increased the impact of their donations by funding more capacity building and reducing lead times for delivery.

In general, the common pipeline instilled some best practices like the standardization of goods whose benefits go beyond the mere logistics, procurement, and inventory management. Having common standards for procurement3 and in-kind donations4 ensured that all implementing partners had access to uniform quality of goods regardless of supplier. It also ensured uniformity of the goods distributed in the region.

Additional benefits from the common pipeline included increased visibility in the supply chain, the possibility to postpone procurement and distribution, and the potential to scale operations according to needs. By consolidating demand from the regions and agencies and matching it with the sourcing process the UNJLC managed to improve visibility in the supply chain for the items in the common pipeline. Under a different set up, each agency would have lacked the incentive to share their information and parallel pipelines would have developed leading to a duplication of efforts and competition for resources like warehousing, transport, and volunteers. Shared information also enabled postponement of distribution from the main warehouse to each of the regions contributing to a lower transportation cost and ensuring that aid was meeting highest ranking needs. Similar arguments hold true regarding postponement for the appeals for donations or procurement since both of these processes were triggered in response to the gaps in the pipeline rather than mere estimates. Having this type of visibility and structure in place provided the delivery agencies with the flexibility to expand and reduce the amount of investment they had allocated for the distribution of the NFIs at different times of the year (i.e., rainy season vs. dry season) and to focus on their core activities.

DISCUSSION: DRIVERS, FACILITATORS, AND COMPONENTS TO PARTNERING

Even though interviews in 2004 were too early to validate the benefits of the common pipeline and to quantify them, most participants indicated positive outlook for the benefits of the collaboration. In relation to the Lambert and Knemeyer model, the collaboration in Darfur can be assessed in terms of the drivers, facilitators and components of the common pipeline to better understand the partnership design.

Drivers

The drivers to partnering were clear and consistent from the beginning of the operation. The magnitude of the needs, the size of the territory to be covered, the distances and cost of delivering aid to Darfur were impossible for one agency to assume alone. Intuitively aware that the length of the crisis would also be significant, all organizations agreed that they needed to find an efficient system to provide aid that could be sustained over time. Some respondents even mentioned that Sudan would be not be able to raise enough funds from the international community over the following few years to cover the expenses of meeting the humanitarian needs in the area. This proved to be true as soon as the Indian Ocean tsunami took place shifting donor attention away from Darfur. Consider as well, as some respondents pointed out, that Darfur had limited infrastructure to operate locally and that population was very dispersed, so beyond the costs of getting to the region there was a significant cost involved in warehousing and distributing aid regionally. Clearly, from the onset the consensus was to find a cost-effective and sustainable way to provide aid to Darfur which would require combining resources among the different agencies.

Facilitators

Convinced that a partnership was necessary to overcome the challenges, the humanitarian agencies worked closely to define and agree on a set of elements (facilitators) that would help the partnership to grow. In retrospect, based the analysis of the interviews, these facilitators were senior level support from each of the participating organizations, donor support to finance all the activities, clear role definitions for each participating organization, and consensus on the items carried by the pipeline.

Senior Level Support: Humanitarian agencies realized the need to express senior level support, preferably in writing, to the agencies (particularly those in a coordinating function) agreeing on their roles. This could be the endorsement of the humanitarian coordinator, agreement by the country directors of the participating agencies (i.e., tri-partite agreements), or special mandates from headquarters. As the most important criterion, support must be given from the initial stage of the crisis to avoid further complications as a result of late deployments. Any lack of support on behalf of the related parties should be formally documented to isolate the effects of the bottlenecks on the whole system. Data from the interviews showed that the initial conversations for the NFI Pipeline were taking place at a rather senior level within each of the organizations, engaging managers who had the capacity to commit their teams for the long run with resources under their responsibility. For the most part these managers were national head of agencies, reporting directly to their regional or global headquarters.

Donor Support: It was also noted the importance of having donors agree and support the model from the beginning. This meant providing funds and staff commensurate to the needs for the pipeline. For example, one of the donors hired and seconded trained staff to the UNJLC directly. In a different case, they funded the cost of staff at one humanitarian agency who were dedicated to

the common pipeline activities. In general the approach by the donors was to minimize the budget impact of the staff for each of the organization participating in the common pipeline. Funding also included the cost of operating the pipeline at its different stages, which one donor claimed would be" cheaper to cover the cost of only one pipeline than funding and monitoring requests from several different agencies which would then compete for staff, trucks and warehouses." Another humanitarian praised the support from the donors "to reinforce funding policies" referring to the fact that if any agency appealed for goods carried by the common pipeline, the donors would automatically dismiss the appeal and request that the agency channel their appeal through the common pipeline system.

Clear Role Definitions: It was also noted the importance of having the roles per agency well defined at all levels (i.e., WFP, UNICEF, CARE, UNJLC/OCHA) with pre-established lines of communication. Each agency should have clear tasks with objectives and indicators per function. These should be communicated to all interested parties to avoid duplication of effort or confusion. For example, in the common pipeline if goods were to move from one point to another, an action request would need to be issued. The latter could only be issued by UNJLC/OCHA to CARE, and CARE is to report on the dispatch to the UNJLC. Having these clear lines of responsibility ensured that decisions were taken by the accountable parties and that other parties would not assume opportunistic behavior to benefit from any confusion in the system. It also ensured that each agency knew what was expected from them (i.e., CARE to do warehousing and transportation), and what level of service was expected.

Consensus on Essential Items: To reduce confusion, some respondents expressed, that it was important from the beginning to define what would be carried by the pipeline (NFI basket) with priority on items that have the highest impact on the livelihood of the beneficiaries. Less recurrent items (office supplies, general equipment) or those requiring special handling (refrigerated drugs, fuel, or hazardous materials) should be processed in a separate model monitored by the agencies5. Among the first discussions held by the participating organization was the composition of the NFI basket which defined the scope of the common pipeline.

Components

Having agreed on the partnership terms and conditions, the respondents agreed that the success of the partnership also depended on the routine activities and management of the processes involved. From the respondents a set of partnerships components were drawn such as having a centralized coordination system, developing transparency and monitoring accountability, and keeping the model open to participation to the different capable agencies.

Centralized Coordination: This enabled agencies to interact with one commonly agreed coordinator as the main interface. To minimize delays and obstacles at the points where goods (for tracking) or information (for reporting) are exchanged, agencies should agree on the reporting criteria among themselves and communicate in a timely fashion to the coordinator. In this case the coordinator was the UNJLC at the different levels, holding regular meeting in Khartoum and the field as well as communicating bulleting and snapshots of the pipeline activity. Having one coordinator also helped to "create a common understanding of the pipeline performance" remarked one respondent, who later explained that as the situation became clearer in Darfur and improvements in the pipeline needed to be made "it was easier to have one focal point to channel the concerns and address them collectively."

Transparency and Accountability: As a public service to the humanitarian community, there needs to be a system that allows interested parties to audit the pipeline's performance. Account-

ability is supported by a reporting system that outlines the agencies' responsibilities, actions taken (goods tracking), report progress (monitoring performance), and defines inter-agency relations (collaboration, flow of goods and data). The system can be used for transparency by having clear communication output (reports) that support periodic meetings and bulletins at the different levels and stages.

Participatory Model: As the common pipeline became the main supplier, and preferably the only, there needed to be an objective criteria for participation. The goal walls that decision making be as inclusive as possible, based on the priority of the needs, and the implementing partners' absorption capacity. For example, the prioritization process was done in an open forum by NFIWG. They identified and ranked needs per state and assigned them to implementing partners based on their requests and absorption capacity in the area. Such system prevented larger agencies imposing their agendas or receiving preferential treatment.

LIMITATIONS AND RECOMMENDATIONS

It is important to note that this research embodies several limitations in relationship to the humanitarian sector, and the academic literature. On the humanitarian side the design and implementation of the relief efforts for Darfur was atypical, as previously noted, in that it included a long window of preparedness during which senior staff were solely dedicated to the task. It is a rare occasion for humanitarian managers to be able to invest such time and resources on-site in preparation for a relief operation. It is also worth noting that most of the key actors in the focal group were UN agencies, and that NGOs involved already had a working relationship with the UN members. Similarly, managers understood the long-term potential of their planning and tacitly acknowledged that their joint efforts could have long-term benefits.

It could be argued that with a shorter term vision, as it is often the case, managers would have not engaged in a partnership negotiation as described in this article. Instead, a version of the partnership model would have evolved over time at a higher cost to the agencies and beneficiaries.

FUTURE RESEARCH DIRECTIONS

Further research is also required to confirm the successful implementation of the NFI common pipeline as designed in 2004. Considering that in 2009 agencies were asked to leave Sudan, it would be interesting to understand how this arrangement operated in current day Sudan. Given the opportunity and data availability, a study could be considered to analyze the costs associated with the pipeline to estimate potential savings and efficiencies.

In relation to the academic management literature the discussion here is limited to organizations operating under high-uncertainty and without a profit incentive. Any conclusions drawn from this case study or analysis into a wider context would require further testing for its applicability.

The literature on supply chain management also indicates that competitive supply chains require alignment, in addition to agility and adaptability (Lee, 2004). An interesting contribution of the partnership model is that it motivates the alignment of the member organizations to form the partnership. Humanitarian supply chains are noted for their agility and adaptability, however the lack of a profit-driven business model seems to present difficulties in creating sustainable alignment between the organizations. It could be argued, and tested through further research, that building high-potential relations could be a substitute for non-profit driven organization to develop alignment. High-potential relations leading to partnerships could be the tool for organizations to discuss their motivations leading to the incentive alignment.

CONCLUSION

Humanitarian agencies are mandate driven organizations whose operational capacities have been designed and adapted to meet the beneficiary needs that fall under their scope e.g., food, medicine, education). As such agencies are specialized in complimentary areas and functions that demand that they interact and work together to respond to the needs of the beneficiaries following a disaster. The need to interact is exacerbated by the limited resources and high levels of uncertainty that characterize humanitarian emergencies.

The mechanisms through which humanitarian agencies interact and set up collaboration have changed numerous times and evolved at different levels. In the 1990s the Department for Humanitarian Action (DHA) was the main body in the UN to foster interagency collaboration. DHA later gained more support and was institutionalized under the UN Secretariat to become the Office of Coordination of Humanitarian Affairs (OCHA). They were responsible for the overall coordination of the humanitarian agencies, including liaisons with non-governmental organizations through the Inter-Agency Standing Committee (IASC). At the turn of the century logistics coordination climbed up in the institutional and political agendas by the creation of the United Nations Joint Logistics Center (UNJLC) (Samii and Van Wassenhove, 2003). The UNJLC was staffed with seconded staff from different agencies and designed to be activated within 24 hours by the IASC to provide logistics coordination support in large scale emergencies. As part of the humanitarian reform started in 2005 a new coordination mechanism was developed better known as the cluster system. Logistics is one of the nine clusters developed, and is led by World Food Program as the lead agency and incorporating UNJLC as a core member. The research presented in this paper dates to back to 2004 prior to the humanitarian reform focusing on the coordination activities of the UNJLC.

Understanding how partnerships between humanitarian organizations are designed and what are the drivers, facilitators and components for them is a still a relevant questions despite any changes in the coordination mechanism of the humanitarian sector6. Most recently we have followed the relief operations in Haiti, where limited resources and efficiency have also dominated the agenda of the participating organizations. The Logistics Cluster, leading the coordination of aid flow, has implemented a similar common pipeline model to the one described here to guarantee the efficient and cost-effective flow of goods into Haiti from the Dominican Republic. They have also worked with a common warehouse system for the humanitarian agencies moving aid into the island. Both examples highlight the importance and relevance of this question and validate that these types of partnerships are an important solution to the management of the humanitarian supply chain.

REFERENCES

Eisenhardt, K. (1989). Building theory from case study research. *Academy of Management Review*, *14*(4), 532–550.

Ernst, R. (2003). The academic side of commercial logistics and the importance of this special issue. *Forced Migration Review*, *18*, 5.

Handfield, R. B., & Melnyk, S. A. (1998). The scientific theory-building process: A primer using the case of TQM. *Journal of Operations Management*, *16*(4), 321–339. doi:10.1016/S0272-6963(98)00017-5

Kovacs, S., & Spens, K. (2007). Humanitarian logistics in disaster relief operations. *International Journal of Physical Distribution and Logistics Management*, *37*(2), 99–114. doi:10.1108/09600030710734820

Lambert, D., Emmelhanz, M., & Gardner, J. (1996). So you think you want a partner? *Marketing Management*, *5*(2), 25–41.

Lambert, D., & Knemeyer, M. (2004). We're in this together. *Harvard Business Review*, *82*(12), 114–122.

Lee, H. (2005). Triple A supply chain. *Harvard Business Review*, *82*(10), 102–112.

McCutcheon, S. D., Handfield, R., McLachlin, R., & Samson, D. (2002). Effective case research in operations management: A process perspective. *Journal of Operations Management*, *20*, 419–433. doi:10.1016/S0272-6963(02)00022-0

Meredith, J. (1998). Building operations management theory through case and field research. *Journal of Operations Management*, *16*, 441–454. doi:10.1016/S0272-6963(98)00023-0

Samii, R. (2008). *Leveraging logistics partnerships: Lessons from humanitarian organizations*. Erasmus Research Institute of Management.

Samii, R., & Van Wassenhove, L. (2003). *The United Nations Joint Logistics Centre (UNJLC): The genesis of a humanitarian relief coordination platform*. INSEAD case study. Retrieved from www.ecch.com/humanitariancases

Schulz, S. (2009). *Disaster relief logistics. Benefits of and impediments to cooperation between humanitarian organizations*. Haupt Berne.

Thomas, A., & Kopzack, L. (2005). *From logistics to supply chain management. The path forward in the humanitarian sector*. Fritz Institute. Retrieved on September 15, 2009, from http://www.fritzinstitute.org/PDFs/WhitePaper/FromLogisticsto.pdf

Tomasini, R., & Van Wassenhove, L. (2007). *UNJLC moving the world: Transport optimization for South Sudan*. INSEAD Case Study. Retrieved from www.ecch.com/ humanitariancases

Tomasini, R., & Van Wassenhove, L. (2009). *Humanitarian logistics*. Palgrave MacMillan. doi:10.1057/9780230233485

Yin, R. (1989). *Case study research design and methods*, 2nd edition. Applied Research Methods Series, vol. 5. Sage Publications.

ADDITIONAL READING

Hoffman, C. A. (2004). *Measuring the Impact of Humanitarian Aid: A Review of Current Practices*. London: Overseas Development Institute.

Minear, L. (2002). *The Humanitarian Enterprise: Dilemmas and Discoveries*. Kumarian Press.

Ogata, S. (2005). *Turbulent Decade*. W.W. Norton, 2005.

Raynard, P. (2002). *Mapping Accountability in Humanitarian Assistance*. ALNAP Online (https://www.alnap.org/ pubspdfs/praccountability.pdf)

Tomasini, R., & Van Wassenhove, L. (2004). Moving the world: TNT- WFP partnership. Looking for a partner. *INSEAD Case Study* (available on www.ecch.com/ humanitariancases).

Tomasini, R., & Van Wassenhove, L. (2005). Managing information in humanitarian crisis – The UNJLC website. *INSEAD Case Study* (available on www.ecch.com/ humanitariancases).

Tomasini, R., & Van Wassenhove, L. (2006). Overcoming the barriers to a successful partnership. *The Conference Board Executive Action Report*.

Van Wassenhove, L. (2005). Humanitarian aid logistics: supply chain management in high gear. *The Journal of the Operational Research Society, 57*,

KEY TERMS AND DEFINITIONS

Accountability: Being able to identify who is responsible for the actions within the processes of the supply chain and how well each of those processes are performed (Tomasini, Van Wassenhove 2009).

Common Pipeline: A joint logistical effort in bringing supplies to a disaster area that is aimed at reducing redundancies and excesses.

Internally Displaced People (IDP): The terms relates to beneficiaries, that affected by crisis, have changed locations but have not crossed the borders of their home country.

Non-Food Items: Essential aid delivered to the beneficiaries during a disaster relief operation, other than food, such as blankets, plastic sheeting, soap, or water buckets.

Partnership: A way to target high potential relationships and align expectations around them (Lambert and Knemeyer, 2004).

Partnership Components: Joint activities and processes that build and sustain the partnership' to meet a level of performances that matches the expectations of the partners in the first place (Lambert and Knemeyer, 2004).

Partnership Drivers: Compelling reasons to engage in a partnership (Lambert and Knemeyer, 2004).

Partnership Facilitators: Supportive environmental factors that enhance partnership growth (Lambert and Knemeyer, 2004).

Refugees: The terms relates to beneficiaries, that affected by crisis, have crossed their national borders seeking assistance.

Transparency: Ability to understand how processes interact within the supply chain to improve performance (Tomasini, Van Wassenhove 2009).

ENDNOTES

[1] Eventually the donors demanded a 90-day implementation plan outlining how each agency intended to roll out their program within that period to address the standing needs.

[2] The activation process for the UNJLC is triggered through the Inter Agency Standing Committee Working Group upon request of one of the agencies or the Humanitarian Coordinator /UN Country Team in Sudan and requires a formal consensus among the participating agencies. UNJLC was activated in Sudan on February 2004 to address the needs of the humanitarian community in the south. Like most they were asked to focus to Darfur a couple of weeks after deployment as the need become evident.

[3] UNICEF enforced their standards in the purchasing process.

[4] Sphere standards were adopted as the norm

[5] Parallel to the common pipeline for NFIs, a common air service operation sponsored by the donors was set up free to the humanitarian community to deliver items that the agencies could not access through the common pipeline and would be inappropriate for land transportation. The service operated by WFP worked under a similar prioritization model chaired by UNJLC taking cargo from Khartoum provided by the interested parties.

[6] The latter is particularly relevant as the humanitarian sector, the UN in particular, reviews their policies and procedures to "deliver as one" following directives from the executive management team.

Chapter 3
Relief Supply Chain Planning:
Insights from Thailand

Ruth Banomyong
Thammasat University, Thailand

Apichat Sodapang
Chiangmai University, Thailand

ABSTRACT

The purpose of this chapter is to provide a framework for the development of relief supply chain systems. An illustrative case study is presented in order to help relief supply chain decision makers in their relief supply chain planning process. Developing simulation models to test proposed relief supply chain response plans is much less risky than actually waiting for another disaster to happen and test the proposed relief supply chain model in a real life situation. The simulated outcome can then be used to refine the developed relief supply chain response model.

INTRODUCTION

The purpose of this chapter is to provide a framework for the development of relief supply chain response model. The proposition of a framework is in itself not sufficient and an illustrative case study is presented in order to help relief supply chain decision makers in their relief supply chain planning process. Emergency or relief supply chain plans and response frameworks have been developed by numerous agencies and governments around the world. However, many of these seem

to be purely theoretical and relatively ineffective in their initial response or subject to unforeseen constraints. It is therefore important to develop and provide a relief supply chain planning framework that key related stakeholders can adhere to.

The proposed planning framework provides a supply chain perspective on relief operations with a focus on responsiveness. Responsiveness is a key issue for relief supply chain as aid has to arrive as quickly as possible, in the right place, in the right condition to help disaster victims. In order to illustrate the propose relief supply chain

DOI: 10.4018/978-1-60960-824-8.ch003

planning framework and its application, insights from Thailand was chosen to illustrate the potential outcome of following such a planning process.

This chapter is separated into four sections. The first section introduces the manuscript, its objectives and its contextual background. The second section discusses key concepts related to the development of a relief supply chain response model and the role of simulation models. The third section describes a proposed Thai relief supply chain response model and its simulated outcome. The summary further discusses lessons learned from the simulation outcome of the proposed Thai supply chain response model and its impact on relief supply chain planning.

BACKGROUND

According to Beresford and Pettit (2009), the aim of relief supply chain is to establish a tailored supply pipeline that fits a particular crisis or natural disaster. The principal leg of the pipeline is usually transport and freight transport is generally a key 'driver' of the relief supply chain in most cases. A variety of transport modes are likely to be used in order for aid to reach a crisis area rapidly.

Over the past few years, the literature related to relief supply chain has greatly expanded. A number of models have been identified which incorporated many of the key stages of the emergency relief cycle and are discussed in detail by Pettit and Beresford (2005). Relevant models include the Disaster Management Cycle (Carter, 1999) and the Recovery Model of Haas et al. (1977). The latter identifies overlaps occurring between each of phases of the full emergency relief cycle. Military involvement in the early stages of an emergency is usually greater due to the capability of military organizations to respond rapidly to severe needs.

A GENERIC RELIEF SUPPLY CHAIN RESPONSE MODEL

The work of Jennings et al. (2000) detailed some of the basic principles surrounding the movement of food and commodities into areas where assistance is required. The authors developed a response model expressed in terms of the selection of transport modes and networks required for effective delivery of assistance to refugees. Pettit and Beresford (2005) expanded the earlier Jennings model with the purpose of developing a better understanding of relief supply chain needs by splitting a specific emergency into different stages or phases. In their model, the focus was on the participation of military and non-governmental organisations (NGOs) in emergency situations. During the initial stages following any disaster, the body playing a pivotal role is the relevant government, often initially activating military resources but as the situation stabilises so the importance of military assistance declines; NGOs then take over, commonly leading specific aspects of the relief operations. Other situational factors that could either facilitate or hinder relief operations were also accounted for in the model such as, for example, the underlying political situation or physical geography/accessibility.

Although each crisis is unique in its characteristics, most crises exhibit similar logistical elements. These elements allow the relief logistician to follow a structured response pattern when dealing with the majority of crisis. This response pattern is illustrated in a generic disaster response model.

Crisis situations share similar logistical elements. Wherever the crisis occurs, the need for first aid and food is immediate and ongoing. According to the World Food Programme it can take on average it approximately 4 months for food aid to reach recipients in crisis area through a fully charged transport pipeline. Therefore the relief agency either has to divert cargo that is already afloat, borrow or buy food from a neighbouring country or geographical area, or even

Figure 1. Generic relief supply chain response model. Source: Adapted from Jennings et al. (2000)

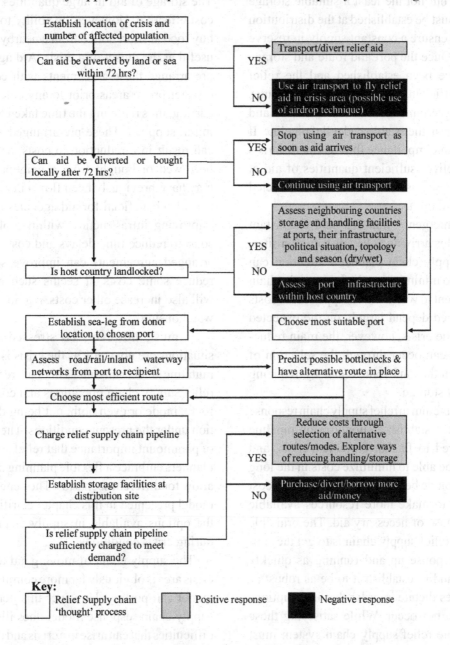

Key:
☐ Relief Supply chain 'thought' process   ▨ Positive response   ■ Negative response

transport first aid and food by air. This method of supplying the first aid and food can continue until aid from various donors arrives.

In order to establish the relief pipeline, the relief supply chain planner must first assess various attributes of the crisis affected location and, in the case of landlocked nations, the neighboring countries as well. As a starting point, a suitable

port will need at least to be chosen. This decision will depend largely on the handling, storage and efficiency of the port in question. The infrastructure from the port to the crisis stricken area needs to be considered and assessed, as well as the political situation, topology and seasonal fluctuations of the weather. All these factors influence the choice of relief route and mode by which aid will be transported. Last but not the least a suitable storage site for aid must be established at the distribution site, so as to ensure a constant supply in reserve at all times. Once the port and route and storage facilities have been established and the relief pipeline is sufficiently charged, alternative routes can then be ascertained in case of bottlenecks and complications in the established relief pipeline. It is of paramount importance throughout the crisis to always deliver sufficient quantities of aid at regular intervals, as the lives of crisis affected people depend on it.

As the emergency matures, and a constant supply of aid is arriving at key distribution sites, the relief supply chain planner's attention can then switch to minimizing relief aid supply chain cost. The extent to which relief supply chain costs can be reduced depend largely on the expected duration of the crisis, however, the main reductions in cost can occur through the selection of alternative modes and routes, and by reducing handling and storage.

The ultimate aim in relief supply chain response planning is to establish a supply chain pipeline that is tailored to fit that particular crisis, and which must be able to minimize costs in the long run. The rationale behind relief supply chain cost reduction is to make more resources available for the purchase of necessary aid. The principle objective of relief supply chain is to get the supply chain response up and running as quickly as possible and to establish it to be as robust as circumstances dictate in order that interruptions in supply do not occur. While satisfying these objectives, the relief supply chain system must also be able to keep costs of operating the supply chain pipeline to a minimum.

A major advantage for any aid agency in a crisis situation is the gift of forethought. If an area is prone to natural disasters and/or civil conflict, then it is beneficial to have equipment needed at the outset of an emergency nearby, for example Strategic Supply chain Stock for Asia's Tsunami areas based in Kuala Lumpur, Malaysia.

The storage of aid in large quantities is not very cost effective. However the ability to borrow or buy food at short notice from a nearby location is useful at the outset of a crisis. Aid agencies can pre-arrange these agreements with countries in disaster prone areas prior to any emergency occurring, thus reducing the time taken to move the initial supplies. These pre-arranged agreements can result in a reduction in costs, as aid that is borrowed or bought from a neighbouring area may have previously been flown in and stocked. It is also beneficial for aid agencies to invest in improving infrastructure within problem areas so as to reduce time delays and cost. While pre-arranged agreements can improve service and reduce some costs, it seems such agreements will also increase other costs, e.g. inventory and warehousing.

It must, however, be stressed that crisis situations are not static as the crisis is constantly mutating and changing. This can result in past relief supply chain operations and cost effective route, mode or even both, not being the best option under the present conditions. Therefore, it is of paramount importance that relief supply chain planners embrace a flexible planning approach to allow for different scenarios. The generic disaster model presented in this chapter clarifies some of the options available in supply chain planning during crisis conditions.

The supply chain of moving aid to an actual crisis area is obviously far more complex than any model can portray. However this generic relief supply chain response model does illustrate key difficulties that can arise in a crisis and the possible thought process that a relief supply chain planner may use. Supply chain pipeline charging, and if necessary re-charging is the key on-going task once supply lines are established and managing the supply chain pipeline in constantly varying conditions is an on-going process. Supply chain pipeline charging describes the activity in which relief aid is being moved and stored within the supply chain supply channel from origin to the disaster area. This pipeline is not continuous and therefore needs to be "re-charged" in order to maintain continuous relief aid to the disaster area.

Meeting customers' requirements is a core supply chain principle but in the case of relief supply chain 'timeliness' and 'responsiveness' are critical performance dimensions. While the price of aid may be a secondary issue in emergency circumstances, the cost should not be as high as via ad-hoc channels which are established by a variety of aid agencies in a post-disaster scenario. Clear time-definite key performance indicators as well as best estimates of the types of goods needed would be clearly defined beforehand (Pettit & Beresford, 2009).

Relief Supply Chain Simulation Model

The use of simulation models is relatively common in management plan formulation to consider the impact of decisions on the management system. Simulation model output can be assessed, interpreted for further refinement or other scenario planning. This avoids the risk of actually implementing a decision without understanding the possible consequences. In developing simulation models for relief supply chain, Monte Carlo simulation can be utilised to simulate the output of any supply chain response model.

Monte Carlo simulation is a computerised mathematical technique that allows people to take into account risks in quantitative analysis and decision making (Mooney, 1997). The technique is used by professionals in such widely disparate fields as finance, project management, energy, manufacturing, engineering, research and development, insurance, oil & gas, transportation, and the environment. To put it simply, Monte Carlo simulation is a method that evaluates iteratively a deterministic model using sets of random numbers as inputs.

Monte Carlo simulation performs risk analysis by building models of possible results by substituting a range of values, a probability distribution, for any factor that has inherent uncertainty. It then calculates results over and over, each time using a different set of random values from the probability functions. Depending upon the number of uncertainties and the ranges specified for them, a Monte Carlo simulation could involve thousands or tens of thousands of recalculations before it is complete. Monte Carlo simulation produces distributions of possible outcome values. By using probability distributions, variables can have different probabilities of different outcomes occurring. Probability distributions are a much more realistic way of describing uncertainty in variables of a risk analysis compared to relying qualitative or perceptual indicators.

During a Monte Carlo simulation, values are sampled at random from the input probability distributions. Each set of samples is referred to as an "iteration" and the resulting outcome from that sample is recorded. Monte Carlo simulation does this hundreds or thousands of times, and the result is a probability distribution of possible outcomes. In this way, Monte Carlo simulation provides a much more comprehensive view of what may happen. It tells the relief supply chain decision-maker not only what could happen, but how likely it is to happen.

Relief supply chain planners need not only to conceptualise and propose relief supply chain response models but also simulate the potential outcomes of their proposed response model. This will enable them to better understand the impact of their planning decisions. Preparedness is a key condition to successful relief operations. The usage of simulation models can refine proposed and developed relief supply chain response model.

RELIEF SUPPLY CHAIN IN THAILAND

The Context

The Southern coastline of Thailand was affected by a tsunami on December 26, 2004 and the relief response was considered to be not appropriate, especially during the first 72 hours of the disaster. After the event, the Thai government revised their emergency plan to respond to the potential threat of another tsunami but shortcomings were identified (Beresford & Pettit, 2009). However, these issues were not addressed as the threat of another tsunami was considered extremely remote by Thai authorities.

In Thailand no detailed relief supply chain response plan has been put in place even though the Thai government has developed a basic protocol for relief operations. As an example a drill was conducted by Thai authorities in the previously tsunami affected provinces during August 2009 to test the readiness of the existing relief protocol. The drill began at 10.20 am when the authorities simulated an earthquake in the Andaman Sea. After being notified of the quake by the Thai Meteorological Department, the National Disaster Warning Centre then sent short messages to department executives, to the governors of the six provinces and to a variety of officials to monitor the earthquake situation. The disaster warning proceeded both in Thai and in English, instructing local residents, as well as Thai and international holidaymakers to evacuate from shoreline areas to safer places on higher ground. However, quite a few areas faced glitches regarding the low-volume sound of sirens. In fact it was the second tsunami drill to occur within a month. A seabed earthquake in the Indian Ocean led to an unplanned tsunami practice in the early morning hours of August 11. Residents of the Phang Nga village of Nam Khem, where more that 800 people perished in 2004, decided to take no chances and evacuated their homes for higher

ground. Elsewhere along the coast, most people continued sleeping unaware that an earthquake had taken place (level one alert) or that those local officials were on standby to begin evacuation (level two alert). The order to move people out (level three) never came because the earthquake did not generate a big wave.

Drills are relatively easy to practice but there is no sign of any genuine preparedness for a tsunami that may comes while the whole of the Andaman sleeps. The key to effective relief response is to be prepared as well as to be able to control the output of the emergency relief supply chain system. The main issue during the relief response itself is coordination to avoid both response gaps and duplication of effort. The relief response plan needs to be disseminated to all stakeholders as it is not uncommon for volunteer organisations to become a hindrance to the management of integrated relief operations.

The Thai Relief Supply Chain Response Model

The focus of this proposed relief supply chain response model is specific for the six southern Thai provinces that were affected by the tsunami in 2004. The focus of the relief response model is related to rapid deployment of resources and aid within the first 72 hours. The phase is often viewed as a period of 'necessary chaos' which rapidly filters-out bad practice based on the 'learning by doing' principle. Figure 2 shows the graphical illustration of this phase of the response model as proposed by Banomyong *et al.* (2009). Water transportation was not taken into consideration in the development of the response model as the geographical conditions specific to the six southern province of Thailand does not provide the possibility for coastal transport supply lines. There is a lack of adequate port infrastructure to cater for such a response model.

The disaster relief response model proposed here is based on the approach of planning, imple-

Figure 2. Proposed Thai relief supply chain response model. Source: Banomyong et al. (2009)

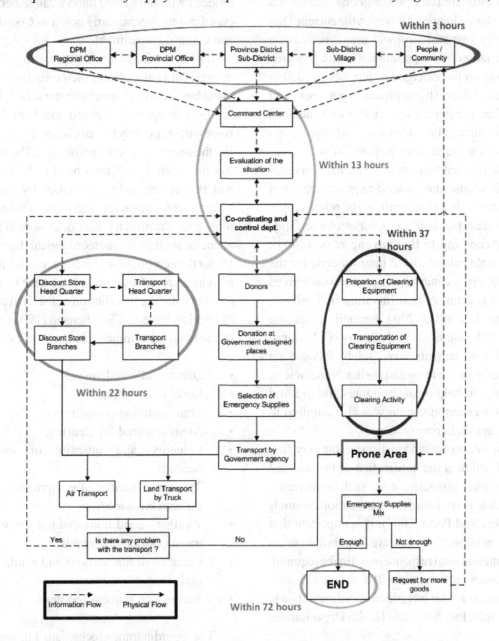

menting and controlling the effective and efficient flows and storage of aid to disaster-struck areas in order to meet the requirement of the affected population within the first 72 hours. When a disaster occurs, waiting for donations to help alleviate the immediate problems faced by the victims does not provide a rapid enough response and causes delays in the delivery of assistance to the victims. Solely relying on donations can also create the dual risk of having insufficient goods to distribute and of delivering aid which is not suitable for those in need of help (Jennings, *et al.*, 2000).

Relief supply chain planners must prepare in advance by making an assessment of the risk area and the population at risk in such area in order to estimate the types of goods needed if a disaster should happen. However, the current Thai emergency response protocol main focus is on the evacuation of the population in the risk area more than on the emergency supplies needed in case of a disaster. This approach is not wrong but even when the population have been evacuated there will always be a need for relief supplies no matter where the affected population is.

What is often forgotten is that each disaster victim is a 'customer' whose requirements need to be met within a time-limit by the relief supply chain system put in place (Oloruntoba & Gray, 2009). According to Banomyong *et al.* (2009), the rationale behind the 72 hour timeline for the proposed supply chain response model was derived from the tsunami-related literature in Thailand. After the December 2004 tsunami it became clear that during the first three days (72 hours), victims' requirements were solely focused on food, medicine, clothing and shelter. Nonetheless, equipment to help repair buildings and to build temporary accommodation was also required to support the sheltering process.

In the event of a sudden disaster, the very first point of action is the notification of the relevant governmental agencies, with such notification being made via existing communication channels (Beresford and Pettit, 2007). It is important that the information is as accurate as possible, as an urgent international response may also be required. This broader requirement will be relayed to other related agencies for integrated coordination (see, for example, Pan American Health Organisation, 2001).

In practice, the cascade of activities can be telescoped into a few hours in recognition of the urgency required in the event of a sudden-onset crisis such as a tsunami strike. Three hours is the agreed upon cycle time for information to flow within the proposed relief supply chain response model at the moment of the disaster. The duty of the national command centre is to establish relief assistance while notifying other related agencies to proceed with the agreed upon arrangements via a coordinating mechanism known as Coordination and Control Department.

It is interesting to note that presently all related agencies in Thailand have their own disaster relief plans but these individual disaster relief plans are not fully integrated or coordinated. Rather they represent the perspective of each agency concerning the delivery of relief assistance. The situation 'on the ground' will then need to be evaluated and this is currently undertaken by numerous governmental agencies such as the Thai ministry of interior, the military, the police as well as other agencies such as the meteorological department. Under the new proposed response model, all relief activities are to be coordinated by the Command Centre under the jurisdiction of the Thai ministry of interior. Based on Stephenson (1993), the role of the national command centre is to evaluate:

- Damage affected areas;
- Level of damage;
- Transportation connectivity;
- Areas required for clearing;
- Category and quantity of assistance needed;
- Types and number of equipments required for area clearance;
- Weather conditions which may affect transportation operations;
- Location of fuel stations and availability of fuel;
- Security on transport routes.

The coordinating mechanism known as the Coordination and Control Department as a subset of the national Command Centre can then take over the duty for prompt transmission of all information related to appropriate implementation of operations. Physical area clearance can then be undertaken by the Royal Thai Army Engineering Regiment as the only large body with the necessary manpower, equipment and tools at their disposal. Their operations can be conducted jointly with the support of the Department of Highways at the Ministry of Transport. Pettit and Beresford (2005) highlighted the main issues which surround parallel working of military and non-military organisations, and this interface is a crucial part of relief supply chain. The Thai Army Engineering Regiment is located in the central parts of Thailand and is thus strategically located to reach any disaster area within 24 hours of receipt of notification from the Co-ordination and Control Department or within 37 hours of the occurrence of the event.

At this point, large retail stores become involved with their remit to prepare aid goods and materials as per their contractual arrangement with the Thai ministry of interior. Their national and regional distribution centres can support local stores in sorting goods destined for crisis areas and local stores can coordinate with local transport providers for aid cargo delivery to affected areas. Road haulage related activities need to be undertaken by retailers' supply chain and distribution systems while governmental aid agencies can handle the coordination of air transport delivery to local stores via the most appropriate transport method in light of conditions (Jennings *et al.*, 2000). Police and allocated security personnel will be responsible for the overall security of the goods while in transit and at destination points. Control of the delivery of items is the responsibility of the Coordination and Control Department which also acts as the data collection and interpretation centre. This department has a duty to inform potential donors of what is needed through various media such as: television, radio or newspapers so that the best match between need and provision can be achieved. Items can thus be delivered at the right time and in the right area, and potential donors will be informed of the location of places where goods donations can be made in order to charge the designed supply chain pipeline.

The final stage of the proposed relief supply chain response model relates to the assessment and evaluation of procedures as the crisis re-sponse develops. If there is a problem during the delivery process, the service providers will need to immediately report to the Co-ordinating and Control Department for remedial action. Upon arrival in the disaster area, the Thai ministry of interior or its representative will need to establish a local coordination centre for the provision of feedback information to the national command centre, to collect data as well as to coordinate the proper distribution of goods to affected residents. An evaluation will then be made to determine whether or not goods delivered to the disaster area are sufficient or not. Finally, a re-evaluation of the situation is carried out. In practice, the response model processes come under rolling re-evaluation as the crisis develops; nonetheless, progress re-evaluation by 'taking a step back' remains important. If goods delivered do not match requirements then it is the duty of the Co-ordinating and Control department to further procure and respond to the need of the affected area. Table 1 is a summary of the proposed supply chain response model parameters.

Based on the proposed relief supply chain response model, there are two critical activities that need to be controlled for optimal implementation of the relief plan. These two activities are the information flows and the physical flows between the affected areas and the Coordination and Control department of the ministry of interior. An indicator of an organisation's capability is the accuracy and timeliness of information

Table 1. Response model cycle times

Activity	Timeline
Information flow to trigger response	Within 3 hours
Coordination mechanism	Within 13 hours
Physical flow	Within 22 hours
Clearance activities	Within 37 hours
Total response time	Within 72* hours

* These activities do not need to be conducted in a sequential manner.

within its system. This is even truer when the supply chain system is responsible for handling relief goods as human lives are at risk. Any error in its information flow can result in a lack of necessary aid to people located in affected areas.

Having accurate data on the nature of the disaster, its extent, the number of affected persons, and the volume and type of goods needed are a prerequisite for successful relief operations. The quantity of goods needed is derived from the relief supply chain planning process and must be confirmed by the Coordination and Control Department before the actual movement of goods from origin to disaster areas.

The period of time required for delivery to the disaster area as set out in the proposed supply chain relief response should be within 22 hours of the disaster occurring or 9 hours after receipt of information needed to prepare the relief goods. During the delivery process, the Coordination and Control Department will have to coordinate with the military area clearing units in order to facilitate access to disaster areas. If the clearing units cannot clear the access to the disaster areas then relief delivery will be delayed.

In this particular proposed Thai relief supply chain response model, a Monte Carlo simulation was utilised to predict the output of the proposed model. The probability distribution for the utilised Monte Carlo simulation was based on the triangular distribution. In the triangular distribution, the user defines the minimum, most likely, and maximum values. Values around the most likely are more likely to occur. Variables that could be described by a triangular distribution include past sales history per unit of time and inventory levels. Figure 3 illustrate an example of triangular distribution utilised in the simulation model for a specific relief activity.

The rationale for selecting triangular distribution for input probability was based on the fact that most of the public and private sector stakeholders that were involved in the Tsunami relief operations had a fuzzy re-collection of the time taken for each specific relief activity as the Tsunami occurred in 2004 and the request for information was done in 2009. No reference data was kept by all key stakeholders previously involved in the 2004 relief operations. It was therefore decided that the best approach was to ask for the cycle time range of each relief activity. Requested time data was then compiled to describe the minimum, the maximum as well as the mode for each relief related activities. This reduced data

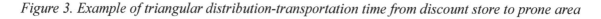

Figure 3. Example of triangular distribution-transportation time from discount store to prone area

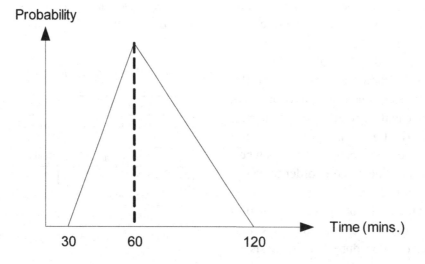

ambiguity. Data ambiguity occurs when there is much discrepancies in the collected data and clarification is needed.

The values are then sampled at random from the input probability distributions. Each set of samples is referred to as an "iteration" and the resulting outcome from that sample is recorded. In this way, the developed Monte Carlo simulation was able to provide a much more comprehensive view of what may happen. It informs the relief supply chain planner not only what could happen, but how likely it is going to happen. The software used for this Monte Carlo simulation was Arena. In Arena, the user builds an experimental "model" by placing "modules" (boxes of different shapes) that represent processes or logic (Altiok & Melamed, 2007). Connector lines are then used to join these modules together and specify the flow of entities. While modules have specific actions relative to entities, flow, and timing, the precise representation of each module and entity relative to real-life objects is subject to the modeller. Statistical data, such as cycle time and WIP (work in process) levels, can be recorded and outputted as reports. Arena also provides the opportunity to test the proposed logistic response model under different types of constraints. Table 2 illustrates the types of constraints used in the simulation model.

The result of the simulated relief supply chain simulation stated that the average time to help victims is 97.4 hours, while the maximum time

is 213.1 hours and the minimum time to help victims is 27.5 hours. These numbers are based from over 30,000 simulation runs. Table 3 illustrates the proposed supply chain response model simulation's output over a number of simulation runs.

From the simulated output, it was observed that the average time to help victims still exceeded the 72 hours limitation relief supply chain operation time. The simulation also indentified that communication activity in the relief supply chain operation system is a critical problem. Other identified constraints in the simulated relief supply chain response model are related to the communication activity between retailers' headquarters and their local branches that are near crisis areas in the South of Thailand, road clearing activity by the Thai army, and the goods distributing activity in the tsunami affected area. After some model refinement by controlling the time variance of mentioned activities within the simulation model, the average time to help victims was being reduced to 17.8 hours. The maximum information flow time is 35.1 hours and the minimum time is 6.7 hours. Table 5 describes the simulated output of the proposed supply chain response model under specific constraints which are related to the transport time from retail stores to prone areas. The reduction in average time was made possible through a reduction in uncertainties

Table 2. Possible types of activity constraints (non exhaustive list) used in the relief supply chain response simulation model

Activity	Low	Medium	High
Information flow from disaster area to trigger response	1 hour <	1 hour	3 hours
Clearance Equipment Preparation	2 hours	2 hours	3 hours
Transport of Clearance Equipment	2 hours	15 hours	18 hours
Clearance activity	1 hour	24 hours	7 days
Transportation Time from Store to Prone Area	1 hour <	1-5 hours	5-10 hours
Vehicle Speed	60 km/h	45 km/h	20 km/h
Aid distribution in prone area	1 hour <	2 hours	2.5 hours

Table 3. Proposed supply chain response model simulated output

Number of simulation runs	Minimum (Hours)	Average (Hours)	Maximum (Hours)
100	29.5	92.0	188.6
1,000	29.6	95.9	198.8
10,000	27.5	97.2	206.9
20,000	27.5	97.4	212.5
30,000	27.5	97.4	213.1

at key bottlenecks in the relief supply chain response model.

The results of Table 4 clearly shows that even if the proposed supply chain response model was able to meet the 72 hours deadline, the relief operations will still be subject to a number of constraints foreseen or unforeseen which will hinder the relief operations. Key bottlenecks observed in the simulated relief supply chain response model are: Discount store communication lead time between headquarters and local branches; Road clearance time; Preparation of supplies for prone area; and Aid distribution in prone area.

SUMMARY

Proposing a new relief or disaster supply chain response model is in itself challenging enough (Kovacs & Spens, 2009) but not sufficient.

What is needed is to be able to predict the model behaviour if it is going to be implemented. This can be done through the usage of a Monte Carlo simulation. This option is much less risky than actually waiting for another Tsunami to happen and test the developed Thai supply chain response model in a real life situation.

It is clear that the proposed supply chain response model by Banomyong *et al.* (2009), for all its good intention, is difficult to implement and control under the 72 hours deadline. Even though the simulated results show that in some instances the output can be achieved in less than 72 hours, the average is still significantly higher.

Other constraints have affected the simulation model outputs and need to be considered in terms of their respective impact on the relief operations. The biggest constraint in the proposed supply chain response model relates to the road clearance activities that enable access to prone areas. The

Table 4. Simulated relief supply chain response output with specific constraint

Type of Simulation	Activities Time (Hours.)		
	Min	Average	Max
Integrated Model			
Basic model*	27.5	97.4	213.1
Additional deliveries required	51.1	127.2	250.9
Low constraint in transport time	28.2	97.9	214.8
Medium constraint in transport time	31.0	100.4	215.8
High constraint in transport time	35.1	104.9	219.8
Separated Model			
Information flow simulation	6.7	17.8	35.1
Physical flow simulation	18.6	86.4	197.9

*The proposed supply chain response model by Banomyong *et al.* (2009)

longer it takes to clear access, the longer will it take for the supplies to get to destination while the alternative would be of course to use a different transport mode to deliver aid such as air drops or even coastal shipping services.

The simulation results enable decision-makers to re-think the proposed supply chain response model and test their performance under a controlled environment. The re-engineering of information, physical and control flows within the supply chain response model can then be conducted and tested to improve the model output.

However, it must not be forgotten that the re-vised and simulated supply chain response model to be designed will never be perfect as there are still other factors and constraints that have not been and can never be included in a simulation model. The simulation model is only a tool that can help emergency supply chain decision makers better understand the dynamics within an emergency supply chain response plan. Simulation results are always subject to limitations but they are a good starting point in any planning process.

Future research should focus on the validity of such an approach for framework development and if this methodology for relief supply chain response model development is applicable within other relief supply chain contexts.

The lessons learned from the research presented in this chapter is that it is always important to not only develop a conceptual model or an action plan of a given relief supply chain response model but also to test and check whether it's application is satisfactory or not. Implementing or executing the relief supply chain response plan on the field is too risky as errors or shortcomings shouldn't occur when trying to serve the need of any affected population. The usage of simulation modeling will help relief supply chain planners to assess the output of developed relief supply chain response model based on different types of assumptions. The capability to develop simulation models becomes then a necessary skill to relief supply chain planning.

REFERENCES

Altiok, T., & Melamed, B. (2007). *Simulation modeling and analysis with arena*. Burlington, MA: Elsevier.

Banomyong, R., Beresford, A. K. C., & Pettit, S. (2009). Supply chain relief response model: The case of Thailand's tsunami affected area. *International Journal of Services Technology and Management, 12*(4), 414–429. doi:10.1504/IJSTM.2009.025816

Beresford, A. K. C., & Pettit, S. (2007, July). *Disaster management and risk mitigation in Thailand following the Asian Tsunami*. Paper presented at the International Conference on Supply Chain Management, Bangkok.

Beresford, A. K. C., & Pettit, S. (2009). Emergency supply chain and risk mitigation in Thailand following the Asian tsunami. *International Journal of Risk Assessment and Management, 13*(1), 7–21. doi:10.1504/IJRAM.2009.026387

Carter, W. N. (1999). *Disaster management: A disaster management handbook*. Manila, Philippines: Asian Development Bank.

Haas, J. E., Kates, R. W., & Bowden, M. (1977). *Reconstruction following disaster*. Cambridge, MA: MIT Press.

Jennings, E., Beresford, A. K. C., & Banomyong, R. (2000). *Emergency relief supply chain: A disaster response model*. Department of Maritime Studies and International Transport. (Cardiff University Occasional Paper No. 64).

Kovacs, G., & Spens, K. (2009). Identifying challenges in humanitarian supply chain. *International Journal of Physical Distribution and Logistics Management, 39*(6), 506–528.

Mooney, C. Z. (1997). *Monte Carlo simulation*. Sage University Paper series on Quantitative Applications in the Social Sciences, 07-116, Thousand Oaks, CA: Sage.

Oloruntoba, R., & Gray, R. (2009). Customer service in emergency relief chains. *International Journal of Physical Distribution and Logistics Management, 39*(6), 486–505. doi:10.1108/09600030910985839

Pan American Health Organization. (2001). *Humanitarian supply management and logistics in the health sector*. Washington, DC: World Health Organization. Emergency Preparedness and Disaster Relief Program, Department of Emergency and Humanitarian Action, Sustainable Development and Healthy Environments.

Pettit, S., & Beresford, A. K. C. (2005). Emergency relief logistics: An evaluation of military, non-military and composite response models. *International Journal of Logistics: Research and Applications, 8*(4), 313–331.

Pettit, S., & Beresford, A. K. C. (2009). Critical success factors in the context of humanitarian aid supply chains. *International Journal of Physical Distribution and Logistics Management, 39*(6), 450–468. doi:10.1108/09600030910985811

Stephenson, R. S. (1993). *Logistics*. Geneva: United Nations Disaster Management Programme, United Nations Development Program (UNDP).

Yin, R. K. (1994). *Case study research: Design and methods*. Newbury Park, CA: Sage Publications.

KEY TERMS AND DEFINITIONS

Monte Carlo Simulation: Performs risk analysis by building models of possible results by substituting a range of values, a probability distribution, for any factor that has inherent uncertainty.

Preparedness: Activities in the beginning of the disaster management cycle that relate to contingency planning in or for a disaster area.

Relief Supply Chain Planning: A process with the aim to establish a logistics pipeline that is tailored to fit that particular crisis, and which must be able to minimize costs in the long run.

Relief Supply Chain Response Model: Logistical elements that allow the relief logistician to follow a structured response pattern when dealing with a disaster.

Responsiveness: A key issue for relief supply chain as aid has to arrive as quickly as possible, in the right place, in the right condition to help beneficiaries.

Supply Chain Pipeline Charging: The activity in which relief aid is being moved and stored within the supply chain from origin to the disaster area.

Chapter 4
Humanitarian Aid Logistics:
The Wenchuan and Haiti Earthquakes Compared

Anthony Beresford
Cardiff University, UK

Stephen Pettit
Cardiff University, UK

ABSTRACT

This chapter contrasts the response to the Wenchuan earthquake (May 2008) which took place in a landlocked region of China with that of the January 2010 earthquake in Haiti, which as an island nation, was theoretically easily accessible to external aid provision via air or sea. In the initial period following the Wenchuan earthquake, the response was wholly internal as a detailed needs assessment was carried out. Once the Chinese authorities had established the scale of response required, international assistance was quickly allowed into the country. Several multimodal solutions were devised to minimize the risk of supply breakdown. Haiti required substantial external aid and logistics support, but severe organizational and infrastructural weaknesses rendered the supply chain extremely vulnerable locally. This translated to a mismatch between the volume of aid supplied and logistics capability, highlighting the importance of "last-mile" distribution management. The two earthquakes posed extreme challenges to the logistics operations, though both required a mix of military and non-military input into the logistics response. Nonetheless, in each case the non-standard logistics solutions which were devised broadly met the requirements for effective aid distribution in extreme environments.

DOI: 10.4018/978-1-60960-824-8.ch004

Copyright © 2012, IGI Global. Copying or distributing in print or electronic forms without written permission of IGI Global is prohibited.

INTRODUCTION

Accessing Disaster Areas

Recent natural disasters have emphasized the importance of emergency relief response logistics. One of the most serious problems affecting the modern world is the vulnerability of nations or regions in relation to natural disasters such as earthquakes, floods, drought or man-made crises: civil unrest, war, political/tribal disturbance (Pettit and Beresford, 2005). Even though modern technology is often used to predict natural disasters, they are still, often, unpredictable. The most unpredictable disasters are natural disasters and they may occur with little or no warning (Wijkman and Timberlake, 1998). For this reason, they cause major damage because of their unexpected impact and the fact that the population is not prepared for them. This results in those in charge of the relief operation primarily focusing on response rather than preparedness, so the system becomes reactive rather than proactive.

There are various difficulties that can occur during a humanitarian aid operation. One of these is to access disasters which occur in landlocked countries, or landlocked regions of maritime countries, making the logistics of the response operation even more complex as, in the first case, it requires a neighboring state to be involved for transit (Pettit and Beresford, 2005). In the case of disasters in landlocked regions, distance, inaccessibility and difficult terrain form the main challenges (Jennings *et al.*, 2002). A different set of problems arise when a country, faced with the consequences of a natural disaster, is unable either through lack of internal capability, or because the disaster has rendered the authorities unable to respond in any meaningful way, unable to provide the necessary response. In such circumstances reliance on third-party countries becomes a necessity. In the recent past there have been several major earthquakes and two are notable because of the problems outlined above.

Disaster Response

In the early stages of an emergency, it is widely acknowledged that the best method in terms of speed and security for distributing food aid is air transport (McClintock, 1997; Jennings *et al.* 2002; Brazier, 2009). The most economic use of air transport for food emergencies is the air drop technique. This technique avoids the need for landing strips, which are often not available or are poorly maintained; or if they are available, they are short, thus restricting the size of aircraft that can carry the aid. The food aid is packed in specially designed parcels that can withstand the shock of being dropped out of the back of a low-flying aircraft (Long and Wood 1995, Jennings *et al.* 2002).

Road transport is flexible, versatile, and relatively inexpensive over short distances and the required infrastructure is usually available in most countries, so roads can normally provide a door-to-door service; roads can also transport almost anything anywhere and at any time (Fawcett *et al.*, 1992). Road transport has the additional advantage that there are often local operators, and it is relatively simple for an aid agency to mobilize and organize a fleet of trucks and to deploy them when and where the need arises (McClintock, 1997). Road transport, however, does have disadvantages as trucks are susceptible to poor weather conditions and the available infrastructure may not be of a suitable quality (Long and Wood 1995, Jennings *et al.* 2002). In floods and earthquakes, however, roads can be vulnerable to surface destruction or bridge/tunnel collapse.

Rail can carry large amounts of cargo cheaply over long distances but it is dependent on a network which very rarely offers a door-to-door service; this means that road transport is needed first and last whenever rail is utilized (Jennings *et al.*, 2002). The major disadvantages of rail transport are its fundamental inflexibility, its lack of gearing to commercial needs and, in the case of many countries, the basic lack of railway infrastructure. It can also be susceptible to flooding and landslides. Waterways are, if deep and wide enough, able to carry large volumes of emergency freight, but their orientation rarely leads directly to the crisis hit region. They are also often segmented by sections of rapids or cataracts; as a consequence, waterways are normally used as part of multimodal solutions rather than as the main method of carriage within either commercial or emergency logistics chains.

Barbarosoglu *et al.* (2002) focused on scheduling in a disaster relief operation whereby tactical decisions are made at the top level, and the operational decisions are made on the ground. Barbarosoglu and Arda (2004) subsequently developed a scenario-based, two-stage model for transport planning in disaster response. They expanded on the deterministic multi-commodity, multimodal network approach of Haghani and Oh (1996) by including network uncertainties related to goods supply, route capacities, and demand requirements. Especially pertinent here is the model of disaster response proposed by Haas *et al.* (1977), further developed by Pettit and Beresford (2005). This highlights the four main phases of

response to disaster as they evolve after an event and includes the build up and decay of military and non-military resources (Figure 1). Typically, emergency response follows a fairly consistent pattern: in the first phase, usually lasting a few days up to 2 or 3 weeks, activity is intense, with high or very high military involvement; the intensity then drops into phases 2 and 3 with military involvement reducing and longer term activities such as rebuilding replacing vital response such as life preservation which normally dominates phase 1.

This chapter therefore contrasts the response to the Wenchuan earthquake (May 2008) which took place in a landlocked region with that of the Haiti Earthquake (January 2010) which took place in a maritime state. Both required significant external input to provide an adequate response. A comparative approach is used to analyze the logistics responses to the two earthquakes. For Wenchuan, a mixture of primary and secondary source data collected in the immediate period following the earthquake is used while for Haiti, secondary source data and information following the earthquake was collated and analyzed. In both

Figure 1. Generic timeline of disaster response phases. Source: Pettit and Beresford, 2005 (adapted from Haas et al., 1977)

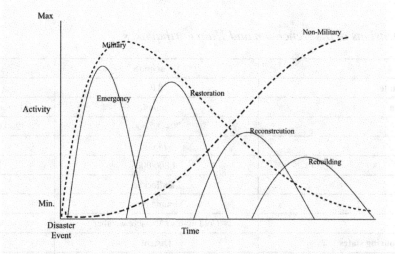

cases the secondary source information was cross-checked by using observations gathered from the aid agencies who were active on the ground. In the case of Haiti, the heavy reliance of the emergency logistics response on 'last mile' performance is highlighted. For Wenchuan, the ways in which transport modes and methods were mixed and matched in order to achieve the single objective of continuous supply was central to the relief effort and forms the main part of the discussion.

WENCHUAN AND HAITI EARTHQUAKES PROFILED

In the immediate aftermath of both the Wenchuan and Haitian earthquakes, very substantial military resources were mobilized, although in the former case this was almost entirely internal capability whereas in the latter virtually the whole rescue effort was external. Key dimensions of the two earthquakes, focusing on their scale and geography, are presented in Table 1.

The response to the earthquake in Haiti should be viewed against a background of the large scale emergencies, notably hurricanes and floods, which have affected the country during the past few years. Haiti is one of the poorest countries in the world with limited internal resources leading to

heavy reliance on external support, especially from the United States government, and other agencies, for economic stability. In contrast, China's economic independence meant that they did not request international assistance until an internal needs assessment had been carried out during the first few days following the earthquake. In this assessment, the Chinese government's mechanisms for response are highlighted, and an appraisal of the armed forces' logistics performance in that period is carried out.

THE WENCHUAN EARTHQUAKE: MAY 2008

Summary of Impact

On May 12, 2008, an earthquake, magnitude 8.0 on the Richter Scale struck Sichuan Province, China. In its immediate aftermath, the State Council Information Office (PRC) reported that 69,197 people had died, 374,176 people were injured and 18,222 people were missing (China.Com, 2008). According to a report from Sohunet (2008), the death toll ultimately reached more than 80,000. Most buildings in the region were damaged or destroyed. Thousands of incidents of damage to roads, bridges, railways and basic infrastructure

Table 1. Key dimensions of the Wenchuan and Haiti earthquakes

	Wenchuan	Haiti
Earthquake Magnitude	8.0	7.0
Earthquake duration	1 minute	1 minute 30 seconds
Fatalities	70 – 80,000	230,000 – 250,000
Injured	374,000	300,000
Homeless	4,800,000	1,000,000
Access	Landlocked	Coastal
Terrain	Mountainous	Flat
Climate	Cold nights / Variable weather	Dry, Hot
Proximity to neighbouring states	Distant	Close
Population	Rural / Urban	Mainly Urban

occurred, making access to the area extremely difficult. Although China itself is not landlocked, the disaster area was located in a mountainous landlocked region, making the relief operation, and especially the logistics and transport, very complex. After the event, two major rivers (Yangtze and Minjiang) in the region (Aldworth, 2009) offered an invaluable transport alternative to the Chinese government.

Emergency Phase: 12th May 2008 to 22nd May 2008

To maintain control of the area, to clear rubble and to assist the aid distribution effort, nearly 20,000 soldiers and armed police units were immediately moved into the disaster area, and another 24,000 soldiers had been air transported to the hardest hit areas. In addition 10,000 more soldiers arrived by rail in the disaster areas (Ku Bing, 2008). In addition, a significant proportion of the navy, the second artillery, the armed police and other units were involved. Extra resources were drawn from the Marines, 15th Air Force and Special Forces from each military region (Gaoyan, 2008).

From May 12th when the earthquake struck, to May 13th, 16,760 soldiers (including 5,000 from the Chinese People's Armed Police Force) became involved in the relief operations. 11,420 soldiers were sent to the vicinity of Chengdu by air on May 13th. On the same day the General Staff Headquarters ordered that the Jinan and Chengdu military regions were to dispatch 34,000 more soldiers to the disaster area by means of air, rail and military mechanized transport (Gaoyan, 2008). By May 13th the Air Force had deployed 32 aircraft in the delivery of troops and equipment to the disaster area in just 17 hours, bringing more than 11,000 soldiers to four airports near to Chengdu. By May 14th, the Air Force had airlifted 6,000 additional personnel and 131.5 tons of relief equipment and medical supplies to the disaster zone.

It was reported by Xinhuanet (2008d) that by May 14th, with 4,000 paratroopers and 2,650

Marines involved, the total military and armed police commitment had reached more than 100,000 personnel. On May 15th, heavy rescue equipment, including excavators and construction vehicles, was airlifted to Yingxiu County. By midday on May 18th, the army and armed police forces had allocated 113,080 personnel to the disaster area (Xinhuanet, 2008e), and by May 28th, a total of 3,688 aircraft sorties had been flown. At peak, during the earthquake relief effort, committed manpower had reached 146,000 troops and 75,000 police and military reservists. (Xinhuanet, 2008f). This level of commitment of human resources and hardware was unprecedented anywhere in the world.

The withdrawal of 60,000 soldiers from the disaster area started on 21st July, 2008, and the second and third phases of withdrawal took place between August 13th and 17th. After this, army, aviation and medical quarantine personnel troops totaling around 3,800 people remained in the earthquake-stricken area to assist both the local government and private individuals in post-disaster reconstruction and rehabilitation (Xinhuanet, 2008g). The pattern of deployment, in terms of headcount, followed a positively-skewed Bell Curve with a rapid increase in military personnel up to the peak two to three weeks after the earthquake, followed by a slower staged reduction as requirements changed. The pattern precisely mirrored that suggested by Pettit and Beresford (2005) as shown in Figure 1.

There were 3 routes to get into Wenchuan County. The first troops to gain access to the central disaster area from the western route were led by the Sichuan Provincial Military District Commander. Starting from Dujiang and moving on foot through Li County, several troop units arrived in Wenchuan County as early as midday on May 13th. Other troops who had marched from Aba, arrived in Wenchuan County at the same time, having taken 21 hours to march 90 kilometers over the hilly terrain. During the night of May 13th / 14th two more military groups arrived in Wenchuan County using the northern and southern routes, arriving in the area on the morning of the 14th. By this time a large-scale, three-dimensional relief operation had been set in motion, and by the evening of May 14th, the military transport system had brought in 12,000 sets of relief rents, stretchers and other equipment, 800 tonnes of food and relief supplies and 6,830 tons of fuel. All of these were transported by a combination of railway, truck, waterway and aviation, with the additional use of shipping for international supplies.

On the northern route, 70% of roads were damaged and aftershocks were frequent, restricting troop movement speeds to just six kilometers per hour. On the southern route, there were two ways to gain access to the disaster area. One was simply by walking into the area carrying relief goods; the other option was to use the Zipinpu reservoir to get to Wenchuan County. This route offered 20 kilometers of waterway and the troops could therefore be transported by assault boats, concluding their journey with a two hour march into Wenchuan County along the hill roads. Assault boats were a valuable means of transport in the relief operation. By May 14th, 19 Marine assault boats had crossed the Sichuan Zipingba reservoir 274 times, transporting 1,710 military relief and rescue workers and road repair technicians. They had rescued 1,770 injured people and shipped a large volume of medicines and daily provisions urgently required in the disaster areas (Zhongguangnet, 2008). Additional troops arrived at midday on May 14th via three military helicopters.

During the first 48 hours, all the relief goods as well as the injured people could only be carried by the soldiers because the damaged roads prevented large vehicles getting into the disaster zone. The added problem of severe weather conditions also meant that the helicopters which were to be used in the rescue operation remained grounded. The military managed to clear one access path on the western route from Li County to

Wenchuan County by the evening of May 15th. Subsequently the heavy rescue equipment was able to be moved into the disaster area. Once the weather conditions had ameliorated relief goods and the injured were able to be transported by both helicopters and trucks, greatly increasing the speed of response.

It was reported by Xinhuanet (2008a) that by 12pm on May 18th, according to incomplete statistics, the army and armed police forces had deployed 113,080 people, 1,069 air sorties had been flown and 92 military trains has been used. More than 110,000 items of equipment including large transport vehicles, cranes, assault boats, portable communications equipment, generators etc. had been deployed; 115 medical, epidemic prevention and emergency psychological intervention teams had been dispatched and around 3 million sets of quilts, drugs, food and tents had been issued.

In the early stages of relief operations, aircraft are often the most important item of transport equipment. Air Force freighters, rescue helicopters and civilian aircraft all played a key role in the relief operations. It was reported by Xinhuanet (2008b) that by 12pm on May 22nd more than 1,000 helicopter sorties had been flown to the remote villages, transporting 786 tons of foods, medicine and tents to Wenchuan and adjacent counties. The helicopters had also shipped 1,537 emergency communications experts and other rescue experts into the area and moved 1,843 critically injured people out.

Second Phase: 22nd May to August 17th 2008

As the relief operation progressed, the volume of goods increased and the transport effort became more intense. To take full advantage of an integrated transport system, on May 27th the Ministry of Transport issued instructions to use four water-road transport routes for relief supplies (China Communication News Net, 2008). Large volumes of emergency cargo could then be shipped via the Yangtze and Minjiang Rivers, and distributed via Zhongqing, Luzhou and Yibin ports, all located in the upper Yangtze, and Leshan port on the Minjiang. After that the relief goods could be transported by road to the disaster areas. By using water - road combined transport, the pressure on road transport was reduced and the comparative advantages of waterway transport for bulk goods fully exploited.

The Sichuan Provincial Communications Department and the Chongqing Communications Committee were well prepared for the possible need to utilize water - road transport for the delivery of relief aid, ensuring the link to the disaster area could be maintained. The shipping management departments in Chongqing, Luzhou, Yibin and Leshan devised contingency plans to ensure the ships which carried relief goods would have preferential passage, berthing priority and primary unloading and cargo transfer. The Changjiang River Administration of Navigational Affairs organized related departments in such a way as to make the deployment of resources clear and rapid. This ensured the efficient, smooth flow of relief supplies with priority lock and goods transport security in the main stream of the Yangtze River (China Communication News Net, 2008).

The data used here were primarily published by the Ministry of Transport, with some data derived from published sources and on-the-ground reporting. From this, it was estimated to take an average of one day to load the relief the goods; the transfer time was about one day; the ship speed was approximately 15 knots and the truck speed was 90 kilometers per hour. These speeds, while ambitious, were achievable due to the urgency of the situation and the priority attributed to the emergency consignments. Four main routes emerged the delivery of aid into the disaster zone in order to ensure continuous supply as far as possible. These are summarized in Table 2 and analyzed below.

Shanghai – Chongqing – Chengdu. On this route, the relief goods were shipped via waterway i.e. the Yangtze River, from Shanghai to Chongqing port. After transfer in Chongqing, the relief goods were moved on by truck along the Yusui highway, Chengyu highway or Yulin highway. The Yusui speedway was the main channel for the transport of the relief goods and is the shortest one of the three routes. Chongqing is the furthest inland river port on the Yangtze, servicing Shanghai and abroad and as such, it serves as the port for water transport of all cargo originating west of Chongqing. The distance between Shanghai port and Chongqing port is 2336 kilometers

Table 2. Emergency transport routes, Shanghai to Chengdu

	Stage 1				Stage 2				Total
	Route	River	Dist. (Km)	Time (hrs)	Route	Highway	Dist (Km)	Time (hrs)	Time* (hours)
1	Shanghai - Chongqing	Yangtze	2335	85	Chongqing – Chengdu	Yusui	300	3.5	135
2	Shanghai - Luzhou	Yangtze	2600	95	Luzhou - Chengdu	Chengdu – Longchang - Luzhou	265	3	145
3	Shanghai - Yibin	Yangtze	2720	100	Yibin - Chengdu	Chengdu- Nei-jiang - Yibin	280	3.	150
4	Shanghai - Leshan	Yangtze/ Minjiang	2880	105	Leshan - Chengdu	Chengle	130	2	155

* All totals include transfer times of approx. 45 – 48 hours

and Chonqing is thus further inland than almost any other port in the world. A 2000 ton ship can navigate the route in any season. As the assumed speed is 15 knots (28 kph) so the shipping time with no delay is 84 hours. The second transport stage was the 295 kilometers of the Yusui highway, taking three to four hours to move goods to Chengdu from Chongqing at an average speed of 90 kph. Preparation work by the Chinese Government meant that many processes were simplified; for example, paper work was much reduced during the emergency and the time to pass through security gates was shortened. Including the initial loading time and intermediate cargo transfer time which were each one day, the total transport time from Shanghai to Chonqing was 135 hours (China Communication News Net, 2008).

Shanghai- Luzhou- Chengdu. On this route, the relief goods were again moved by waterway, using the Yangtze River from Shanghai to Luzhou port via Chongqing. After transfer at Luzhou, the relief goods were transported by truck to Chengdu along the Luzhou – Chengdu highway via Longchang, a distance of 265 kilometers (Zhongguangnet, 2005). The distance from Chongqing port to Luzhou port is 270 kilometers, taking 9.7 hours to get Luzhou port by ship. The relief goods were then transported by road along the Chengdu – Longchang - Luzhou highway, taking 2.9 hours. Including loading times and transfer times, the total time for relief goods transport along this route was typically 145 hours (China Communication News Net, 2008).

Shanghai- Yibin- Chengdu. On this route, the relief goods were shipped along the Yangtze River from Shanghai, but this time to Yibin port. After transfer at Yibin, the relief goods were transported by truck along the Chengdu – Neijiang - Yibin highway, a distance of 280 kilometers. There are several railways to Chengdu, Chongqing, Neijiang, Xunchang etc. from Yibin and four highways to Shuifu, Leshan, Luzhou and Chengdu. Yibin relies on the Yangtze and Jinsha rivers as the two

main water transport corridors. The distance from Chongqing to Yibin is 385 kilometers. It takes around 14 hours to get to Yibin by ship; and the relief goods were then transported for 3 hours along the Chengdu - Neijiang - Yibin highway. Including loading and transfer times, the total time for the relief goods to be transported along this route averaged 150 hours (China Communication News Net, 2008).

Shanghai - Leshan- Chengdu. To reach Leshan from Shanghai, the relief goods were shipped along the Yangtze and Minjiang rivers, but then the goods had to be transferred to smaller vessels (less than 500 tons in the dry season and 800 - 1000 tons in the wet season. There are two highways to Chengdu from Leshan; one is the Chengle highway, a distance of 130 kilometers; the other a special 160 kilometer highway for large-scale equipment (China Communication News Net, 2008). The length of the route between Yibin and Leshan is 162 kilometers; with a journey time of around six hours. However equipment for night sailing does not exist on the route, so sometimes the ships had to wait until the morning, wasting several hours. No matter which route the trucks took, it was a two hour transit from Chengdu to Leshan. Provided that vessels could navigate without waiting at night, the total transport time on this route was around 155 hours.

Summarizing the four routes, the Shanghai – Chongqing - Chengdu route proved the most suitable compared to the other 3 routes; the reason being that it takes the shortest time to arrive at Chengdu from Shanghai; the other advantage was that a greater number of ships could reach Chongqing by using the C compared to the other three ports, so that more relief supplies could be transported on the leg from Shanghai to Chongqing.

International Aid Supply

Foreign countries also contributed substantially to the relief effort following an initial needs assessment by the Chinese government during the first few days after the earthquake. Most relief goods were delivery by military aircraft to Chengdu Shuangliu airport. At 15.00 on May 14th the first batch of post-earthquake international aid supplies, flown from Russia, arrived at Chengdu. The total weight was 24.4 tons and included 95 sets of winter tents, 1,500 quilts and 500 mattresses (Chinanews, 2008a). Subsequently, the Russian government dispatched six transport planes to help deliver relief supplies to Chengdu on May 14th, 16th and 17th (PRC, 2008). The Government of Pakistan delivered relief supplies by two C-130 military aircraft to Chengdu on May 16th. The aid included 300 tents, 4,000 blankets, mineral water, moisture-proof mattresses and other drugs (Chinanews, 2008a). On May 21st, a truck convoy carrying 6,000 tents departed from Islamabad, Pakistan to China by road with the goods transferred to the Chinese government control at Hongqilapu. Relief supplies arrived by air at Chengdu from Singapore, France, the Philippines and Spain on the afternoon and night of May 17th; all the relief supplies were transported by military transport planes and Boeing 747 (PRC, 2008). The USA dispatched C-17 strategic transport aircraft to deliver relief supplies worth $US 1.6 million to Sichuanon on May 18th (Xinhuanet, 2008c). Further relief cargo from different countries was delivered to Chengdu over the next few days. Other transport methods were also used; Japan, for example, sent goods to Zhanjiang port by sea (Chinanews, 2008b). The withdrawal of the troops and military hardware started on July 21st and ended on August 17th 2008.

It was clear that, once the Chinese government had carried out an initial needs assessment, the role of the international community was vital in bringing in a substantial proportion of supplies and in terms of logistics capability.

THE HAITI EARTHQUAKE: JANUARY 2010

Summary of Impact

On January 12th 2010 Haiti was struck by its worst earthquake in two centuries. The earthquake measured 7.0 on the Richter Scale with the epicenter only 25 km from the capital Port-au-Prince. The impact was therefore extremely severe and the proximity of the epicenter to such a highly populated area, combined with the problem of sub-standard building construction, resulted in an estimated 70% to 90% of buildings in the capital either destroyed or seriously damaged including government and public buildings such as the Palace of Justice, the National Assembly, and the Supreme Court (Guillon, 2010). A similar situation occurred in Jacmel where 70% - 90% of the buildings were destroyed (Bacon, 2010). Further damage was caused in the two months as at least 59 aftershocks measuring 4.5 or greater and sixteen over 5.0 occurred (Simmins, 2010). Initial estimates of the death toll were in the 30,000 to 100,000 range but this estimate quickly climbed to well in excess of 200,000 and by the middle of February 2010 confirmed deaths stood at between 217,000 and 230,000 (BBC, 2010a).

Emergency Phase: 12th January 2010 to 23rd January 2010

Figure 2 summarizes the main phases of response to the Haitian earthquake. The three periods of intense activity immediately following the disaster, emergency – restoration – reconstruction gave way to a less intensive rebuilding program after about six weeks.

The response to the earthquake, as in most crises which attract global attention, included national governments, non-governmental organizations (NGO) and for-profit organizations. Some countries sent relief and rescue workers and humanitarian supplies directly to the earthquake

Figure 2. Indicative timeline for Haitian earthquake emergency response phases

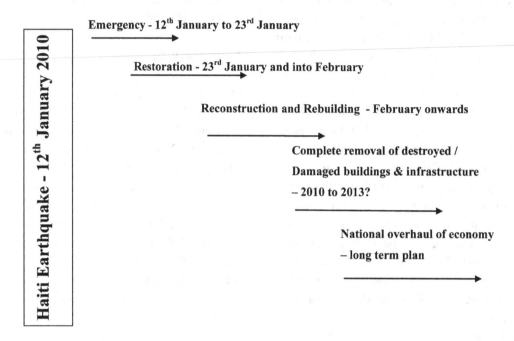

damage zones, while others raised funds to support the NGOs working in Haiti. The Dominican Republic was the first country to provide aid to Haiti, sending water, food and heavy-lifting machinery (Diaz, 2010). Hospitals in Dominican Republic were made available, and the airport opened to receive aid that would then be distributed to Haiti. The first international teams to arrive in Port-au-Prince within 24 hours of the earthquake were from Iceland, China and Qatar, the latter sending 50 tons of relief materials (ICE SAR, 2010; CRIENGLISH.com, 2010).

During the 16th and 17th January 130,000 food packets and 70,000 water containers were distributed and the airport had handled 250 tons of relief supplies. British search and rescue teams were the first to arrive in Léogane, the town nearest to the earthquake's epicenter, on 17th January, these were followed by around 300 Canadian relief support personnel (BBC, 2010b). Meanwhile efforts to reopen the Port-au-Prince seaport continued and by 21st January one pier

was functional enabling humanitarian aid to be offloaded. During the same period, a key road into the city from the port had been repaired, allowing more effective inland transport. By this time the US Navy had committed substantial military hardware to the area (US Navy, 2010a) and by 20th January had completed 336 air deliveries, including 123,000 litres of water, 111,082 meals and 4,100 kg of medical supplies. The first stage in the establishment of the US military sea-base had also been completed (CNNa, 2010).

In spite all the problems facing the emergency relief effort, the scale of the response can be judged by the fact that by the 23rd January, 4 million emergency food packs had been distributed (Dolmetsch, 2010). Similarly impressive progress had been made regarding the delivery of medical equipment, tents and other relief supplies. Thus, on 23rd January, the Haitian government declared an end to rescue efforts and the second phase with greater emphasis on restoration and reconstruction began (BBC, 2010c).

Second Phase: 23rd January 2010 to 28th February 2010

Early in the restoration phase it emerged that the severity of damage to the south pier of the Port-au-Prince port meant that it could not be used effectively, although low capacity use by military landing craft was feasible (Slevin, 2010). Meanwhile, the US military began to move some flights from the main airport at Port-au-Prince to Jacmel Airport in order to allow civilian flights to commence. This required Canadian Forces to manage the increase in air movements, as well as carrying out running repairs to the airstrip itself which was deteriorating due to increased usage (Lett, 2010).

Around 30th January the UN launched a major distribution program which aimed to deliver food and water to two million people within two weeks. The mission, with the World Food Programme (WFP) leading the logistics cluster, began at Port-au-Prince with the US military maintaining security (CNNb, 2010) During this period the US military started to draw down its resources, with four carriers redeployed to other duties (King, 2010; Steele, 2010; Anderson and Katz, 2010). At the same time, Cuba dispatched a fifth field hospital and South Korea sent 250 peacekeepers to Leogane (Xinhua, 2010). Restoration of electricity was an ongoing process involving specialists from several countries such as Japan (Anderson and Katz, 2010). In mid February, for instance, 350 Japanese troops were allocated to demolition and reconstruction with some specialist structural engineers engaged in building stability and safety.

A fresh contingent of 240 Korean peacekeepers (later supplemented by a further 190) was tasked with rebuilding, medical duties and equipment distribution from mid-February onwards. At the same time, a barge with 100,000 tons of relief supplies departed from New Orleans, to arrive at the port of Jacmel on March 1st (Times-Picayune, 2010). Meanwhile, the drawdown of US military personnel continued with the relief force being reduced from around 20,000 to 13,000 (Anderson and Katz, 2010). Likewise, Canadian commitments reduced with the withdrawal of HMCS Halifax on 19th February, the cessation of Canadian evacuation flights, and the switching of the movement of Canadian nationals leaving Haiti from military to commercial flights; by now these were departing from Port-au-Prince International Airport where partial commercial operations had recommenced (Canadian Press, 2010). During the latter part of February the capacity of Port-au-Prince seaport was increased to around 600 containers a day, roughly 350 per day more than that it had been handling before the earthquake. The port thus once again became the major freight gateway for Haiti (Siegler, 2010; BBC, 2010d).

Table 3 summarizes some of the international support that was provided to the Haitian government during the emergency and restoration phases following the earthquake. It is notable that a very large proportion of the response was in the form of logistics or transport capability, which in the early stages of the emergency overwhelmed both the very limited port capacity and the airport. The organization of a sea-air bridge thus became paramount as discussed above.

A measure of the length of time that it will take for the reconstruction of the national infrastructure to be completed is that the Haitian president suggested that at least three years would be required just to clear damaged buildings, bridges, roads and other infrastructure (Associated Press, 2010). The Haitian government also estimated that it will take around $US 11.5 billion to rebuild and completely overhaul Haiti and the rebuilding effort will require commitments from many governments, as well as the involvement of international aid agencies. In this context, an example of contingency planning is that a Canadian support group is currently preparing for a five to ten year commitment to contribute to reconstruction in Haiti (Williams, 2010).

Table 3. *Examples of international support to Haiti, January to March 2010*

| | | Form of Support | | | | | | | | | |
| | | Non Transport | | | | Transport | | | | | |
Country	Dominant Access Method	Water, Food Kitchens	Medical Supplies / Medics	Tents Shelters Mixed Cargo	Other	Large Freighters	Helicopters / Small Planes	Ships	Cranes / Machinery	Trucks	Communication Equipment
USA	Sea/Air	•	•	•	Security/Rescue, Survey/Salvage	•	•	•	•		•
Canada	Air	•		•	Personnel	•					
Dominican Republic	Road	•			Use of Roads, Port, Airport				•	•	•
Cuba	Sea		•		Personnel						
Jamaica					Personnel						
Columbia	Air		•	•	Personnel	•	•				
Brazil	Air		•		Personnel						
Argentina	In situ		•		Security		•				
Iceland	Air			•		•					
Switzerland	Air		•	•							
UK	Air		•	•	Rescue Teams, Personnel						
Italy	Air		•		Personnel						
France	Sea		•		Dry Dock, Personnel			•	•	•	
Netherlands	Sea			•	Personnel			•			
Qatar	Air			•	Personnel	•					
Israel	Air		•		Rescue Teams						
S. Korea	Air	•	•	•	Peace Keepers, Personnel						
China	Air		•	•		•					
Japan	Air	•	•	•	Peace Keepers, Personnel						

Transport Infrastructure

Access Constraints and Organization. From the point of view of access and supply chain organization, what immediately became apparent was the barrier presented by the destruction of the local port which had the inevitable effect of diverting virtually all the first phase response through the main airport of Port-au-Prince which itself did not entirely escape damage. In particular, the control tower was damaged affecting flight management for several days immediately after the earthquake. A further hindrance to the relief effort was the destruction of the headquarters of the United Nations Stabilisation Mission in Haiti, which severely hampered the UN's ability to respond and coordinate activity in the immediate aftermath of the earthquake. The earthquake damaged almost every part of the country's transport infrastructure with air, land, and sea transport facilities all affected. The overall picture was one of large scale, chaotic delivery of aid via the airport, where cargo rapidly accumulated for need of local distribution capability, and by sea where damage to both the port infrastructure and superstructure necessitated sea-basing using, for example, military carriers from which aircraft and helicopters fulfilled final delivery requirements.

The emergency phase, which lasted around 14 days, therefore saw an abnormally high involvement of air and air – sea aid delivery; during this time a second access route via the port was forged. The land routes from the Dominican Republic also became operational on a limited scale as they were hindered by severe damage to the roads themselves and blockage by debris. On 22nd January the UN and the United States formalized the coordination of relief efforts by signing an agreement giving the US responsibility for the ports, airports and roads, and making the UN and Haitian authorities responsible for law and order, symbolically this marked the end of the emergency phase, and the beginning of the re-orientations of resources towards restoration, reconstruction and rebuilding

(BBCc, 2010). The key transport activities were all severely compromised in the early period immediately following the earthquake. It is therefore pertinent to consider how these problems impacted on the logistics response in the emergency phase.

Air Transport. As the marine terminal in Port-au-Prince was severely damaged, initially air became the predominant means for the import of aid cargo. Air traffic congestion, and problems with prioritization of flights complicated the early stages of the relief effort, but the airstrip at Toussaint L'Ouverture International Airport itself fortunately remained useable (US Navy, 2010b). Damage to the air traffic control systems took several days to resolve, however, limiting the airport's freight handling capacity. The airport apron only had the capacity to accommodate 18 aircraft which further constrained the inward flow of aid. By the 17th January, 600 emergency flights had landed, and congestion was reduced further on 18th January when the United Nations and the US military began prioritizing humanitarian aid flights over security operations (Boadle, 2010; MSNBC, 2010). Thus, after about a week, the airport was able to support around 100 landings a day, three times its non-emergency operational capability, although this was still far below the capacity needed. On the 18th January the Haitian government formally placed the airport under the control of the US military, thus streamlining the control of inbound flights. Initially incoming flights into Port-au-Prince were stacked whenever there was a requirement for US airplanes to transport injured Haitians to the US (Mackey, 2010; Cowell *et al.*, 2010). There was some movement of aid in via Santo Domingo airport in the Dominican Republic.

As an additional buffer, an airfield near Jacmel was repaired and prepared to take incoming flights to increase aid to more isolated towns west of Port-au-Prince. This airstrip had suffered damage which rendered it unusable until 22nd January (Padgett, 2010). Once it had been reopened, Canadian Forces installed runway lighting, enabling aircraft to land at night, significantly increasing its capacity to around 160 military and civilian fixed-wing and helicopter flights per day (AFP, 2010). The airstrip was later extended to accommodate heavy-lift aircraft.

Delivery of aid through the use of airdrops was also carried out on a substantial scale but took some time to commence due to security concerns and the possibility of rioting. By the 18th January air drops had commenced and these were concentrated on three sites in less densely populated rural areas where security could be maintained. This approach was also used to test whether the system could be successful and expanded across Haiti. These drops were facilitated through the use of US military helicopters and were supported by the US 82nd airborne division (Jelinek and Burns, 2010).

Roads, Ports and Sea Basing. The earthquake had a dramatic impact on the early stages of the relief operation as the seaport at Port-au-Prince was rendered unusable for several days due to the severe damage caused to both the infrastructure and superstructure. When a bridgehead at the port was eventually forged, it took the form of a limited capability, fragile, roll-on roll-off (Ro-Ro) facility rather than a container crane based lift-on, lift off (Lo-Lo) operation as the terminal had been running prior to the earthquake. The Ro-Ro system was sub lower capacity. Gonaives seaport, in the northern part of Haiti, however, did remain operational (Brannigan, 2010). Relief logistics was also severely compromised due to the damage caused to the road transport infrastructure. Not only were roads blocked with debris but in many locations the road surface was also severely damaged. The main road linking Port-au-Prince with Jacmel was blocked for ten days after the earthquake, making the movement of aid to Jacmel extremely difficult. Transport by road from the airport was especially problematic, and had an immediate impact as the bottleneck created by the restricted use of the airport blocked arriving aid (Jelinek and Burns, 2010).

The immediate solution to the problems created by both the lack of port facilities and damage to the road infrastructure was for the US military to use a sea-base which took the form mainly of carrier-based helicopter operations. Other US Navy vessels were used to assess the damage and assist in the repair of the seaport (Garamone, 2010; Fuentes, 2010; Washington Post, 2010); they were also used for the unloading of shipments from larger vessels during the period before the port was reopened. By 18th January, the Bataan Amphibious Ready Group (ARG) had begun to arrive in Haiti as part of the US Humanitarian-Assistance Disaster-Relief (HADR) mission (JD News, 2010; US Navy 2010c). By the following day, most of the key elements of the sea-base were in place, including 17 ships, 48 helicopters and 12 fixed-wing aircraft; supporting 11,600 US military personnel. A variety of other vessels from other countries such as France, Canada, Spain, Italy and the Netherlands were dispatched with appropriate personnel to contribute to the aid mission (Expatica, 2010; Trenton *et al.*, 2010).

Non-Transport Infrastructure

Non-transport infrastructure was also severely damaged affect the ability to the agencies on the ground to respond. The most high profile of these was the UN's Mission building which collapsed, killing many UN staff. Many hospitals were also severely affected: three Médecins Sans Frontières (MSF) medical facilities around Port-au-Prince were damaged, one collapsing completely as did a hospital in Pétionville and the St. Michel District Hospital in Jacmel. By 16th January the first of several Red Cross - Red Crescent Emergency Response Units had been established to provide basic health care for up to 30,000 people. The Red Cross was able to respond quickly with pre-positioned stocks of emergency kits released from warehouses locally and the wider region (Red Cross, 2010). A further medical facility was established near to the United Nations building by the Israeli Military (JTA, 2010).

Communications infrastructure and electricity distribution networks suffered considerable damage and both the landline and cellular phone systems were affected. Radio transmissions were severely hindered and only 20 of the 50 stations in Port-au-Prince were back on air one week after the earthquake (India PRWire, 2010; Rhoads, 2010). This widespread damage severely hampered rescue and aid efforts; especially important was the lack of clarity concerning the chain of command and the relationships between NGOs, other support groups and the Haitian government. On 14th January the Dominican and Haitian presidents established an emergency plan for assistance which, among other aspects, included re-establishing communications, electricity and water supplies. The Dominican Institute of Telecommunications assisted with the restoration of some telephone services, but the communication infrastructure remained fragile for several days (DiarioLibre. com, 2010; Global Voice Online, 2010).

Understandably, as the scale of the disaster became clear, the need for law enforcement was a significant part of the aid effort as inland distribution of supplies was very slow and the securing of transport corridors and aid distribution centers became central to the effectiveness of the recovery effort. International police and military units were therefore widely employed in key areas as part of the peacekeeping process.

FUTURE RESEARCH DIRECTIONS

The scale and gravity of the two earthquakes at Wenchuan and Haiti highlight the scope for further research in several areas. There is clearly a need for contingency planning, covering a range of 'what if…' scenarios prior to the occurrence of large-scale emergencies. At the implementation stage, a flexible approach to the engagement of external agencies, especially where international help is required, should be encouraged. The unique nature of major earthquakes such as those discussed above highlights the considerable need for flexibility. In addition, a review of relationships between military and non-military bodies, government and non-government agencies and other relevant parties would be fruitful in shedding light on the possible options for future similar emergencies. With regard to specific supply chain issues, there is need for further research in the area of strategic stockholding for cargo allocated to emergencies, embracing locational issues and suitability of the cargo itself. This extends into an examination of the link between initial needs assessment, appeals launch and cargo mobilization. Finally, with specific reference to supply chain dynamics, there is considerable scope for examining in greater detail the relationships between route, method, mode and carrier in a given emergency to ascertain whether time, cost or risk should be minimized or whether a compromise solution should be reached.

CONCLUSION

If the two earthquakes are compared, the contrast between capabilities of the two countries and their indigenous response mechanisms becomes clear. Especially notable is the differing capability in terms of political structures which, in the first instance, determines the form of response, military capability to support the devised strategy and the available hardware to implement the response on the ground. Thus in China the mechanisms allowed an effective internal response supplemented by some external aid, whilst in Haiti external support was fundamental and vital to the aid effort. The main characteristics of the logistics environments in the respective countries are shown in Table 4.

In the case of the Wenchuan Earthquake, a form of freight transport relay system was deployed. But in Haiti, the destruction of port infrastructure and roads running inland from Port-au-Prince constituted the main barriers to distribution;

the main airport was almost undamaged so inland distribution from large-scale marine or air supply rapidly emerged as the primary bottleneck. The disaster response model proposed by Haas *et al.* (1977), which was subsequently refined by Pettit and Beresford (2005), thus represents an accurate portrayal of the main phases of the response process and of the allocation of military and non-military resources to a crisis hit area. The model fits the pattern of build-up and draw-down of Chinese military capacity especially well. Most military operatives who participated in the disaster relief operation arrived within four days making the best possible use of the critical first 72 hours. In addition, various vehicles, especially helicopters, boats and trucks were used in the relief operations in imaginative combinations; helicopters especially proved crucial as the affected region was so remote and mountainous. Two months after the earthquake effort in the disaster area was focused on reconstruction.

Unlike standard commercial supply chains in which cost versus time is normally the key trade-off speed is the crucial factor rather than cost in the context of emergency relief logistic operations. With the passage of time, however, the element of cost will be increasingly considered especially during the stages of restoration, reconstruction and rebuilding (Pettit and Beresford, 2005). In order to reduce the pressure on the roads, the Chinese government utilized four multimodal transport routes to Chengdu. This strategy also minimized the risk of emergency consignments failing to reach the disaster hit area, or emergency evacuation procedures breaking down. The value of contingency planning, rapid deployment of resources and operation of several supply routes simultaneously was clearly demonstrated.

In Haiti, in terms of human casualties, the impact of the earthquake was almost unprecedented. Unusually, gaining access to the affected area was far from simple despite the availability of marine routes and the close proximity of a number of other countries. The largely urban population, concentrated in and around Port-au-Prince fell victim to collapsed buildings and the destruction of the local infrastructure. The pattern of damage meant that access via the adjacent seaport was virtually impossible for several days following the event and access via the airport was constrained by inland distribution difficulties. What emerged was a sea-air bridge solution run on an unprecedented

Table 4. Humanitarian aid logistics: Wenchuan vs. Haiti

	Wenchuan	Haiti
Distance from Coast	Very long	Coastal
Access Options	Air - Road	Air – Road
	Sea – Waterway	Sea – Air
	Sea – Waterway – Lake – Foot	Sea – Road
Sea-basing	None	Extensive
Military Involvement	Very high	Very high
Main Military Resources	Soldiers, road vehicles, helicopters, landing craft, heavy equipment	Military personnel, aircraft carriers, other naval vessels, helicopters, aircraft
Security Level: distribution corridors and warehouses	Extremely High	Variable
International Aid	Very important	Extremely important
Countries Involved in Aid Effort	Fewer than 20	40 +
Overall Control	Internal	Joint Internal - External

scale. This was supplemented by land access from the Dominican Republic, but the land routes were of limited capacity; only when the seaport had been rehabilitated and the airport decongested could several supply lines run in parallel.

REFERENCES

Agence France Presse. (2010). *Haïti le Canada va rouvrir l'aéroport de Jacmel.* Retrieved 10th March, 2010, from http://www.romandie.com / infos/news2/100120160031 .a9e10wyg.asp

Aldworth, P. (Ed.). (2009). *Lloyd's maritime atlas of world ports and shipping places* (25th ed.). London, UK: Informa Professional.

Anderson, J., & Katz, J. M. (2010). *US forces scale back Haiti earthquake relief role.* Retrieved 10th March, 2010, from http://www.breitbart.com/ article.php?id=D9DS9B880 &show_article=1

Associated Press. (2010). *Post-earthquake rubble removal to take three years: Haitian president.* Retrieved 10th March, 2010, from http://dcnonl. com/ article/id37658

Bacon, L. M. (2010). *Carl Vinson, other ships headed to Haiti.* Retrieved 10th March, 2010, from http://www.navytimes.com/ news/2010/01/ navy_vinson _haiti_update_011310w/

Barbarosoglu, G., & Arda, Y. (2004). A two-stage stochastic programming framework for transportation planning in disaster response. *The Journal of the Operational Research Society, 55*(1), 43–53. doi:10.1057/palgrave.jors.2601652

Barbarosoglu, G., Ozdamar, L., & Cevik, A. (2002). An interactive approach for hierarchical analysis of helicopter logistics in disaster relief operations. *European Journal of Operational Research, 140*(1), 118–133. doi:10.1016/S0377-2217(01)00222-3

BBC. (2010a). Haiti will not die, President Rene Preval insists. *BBC News.* Retrieved 10th March, 2010, from http://news.bbc.co.uk/1/hi /world/ americas/8511997.stm

BBC. (2010b). Haiti quake victims' bodies 'piled up by roads. *BBC News.*Retrieved 18th January, 2010, from http://news.bbc.co.uk/2/hi /uk_news/ england/ devon/8465916.stm.

BBC. (2010c). Haiti quake victim rescue operation declared over. *BBC News.* Retrieved 23rd January, 2010, from http://news.bbc.co.uk/1/hi / world/americas/8476474.stm.

BBC. (2010d). Haiti aid effort one month after earthquake. *BBC News.* Retrieved 30th March, 2010, from http://news.bbc.co.uk/1/hi /world/ americas/8509333.stm

Boadle, A. (2010). *U.S. military says Haiti airport jam easing.* Retrieved 30th March, 2010, from http://www.reuters.com/article/ idUS-TRE60H00020100118

Brannigan, M. (2010). *Haiti seaport damage complicates relief efforts.* Retrieved 10th March, 2010, from http://www.miamiherald.com/ news/ breaking-news/ story/1426067.html

Brazier, D. (2009). Heavy lifting for United Nations peacekeeping: Strategic deployment stocks . *Logistics and Transport Focus, 11*(11), 37–40.

Canadian Press. (2010). *Canada stops Haitian evacuation flights, death toll set to jump.* Retrieved 28th February, 2010, from http://www.google. com/ hostednews/canadianpress /article/ALeqM-5jMrV3 QsuxEtZVBiLmFwVcL _XgMnQ

China Communication News Net. (2008). *The Ministry of Transport issued 4 water-road transport routes.* Retrieved 20th June, 2010, from http:// www.zgjtb.com/101179/101182/101215/32931. html

China.com. (2008). *Up to July 20 2008, the Sichuan Wenchuan earthquake has caused 69,197 deaths, 374,176 people injured, and 18,222 people were missing.* Retrieved 20th June, 2008, from http://www.china.com.cn/news/zhuanti/wxdz/2008-07/20/content_16038392.htm

Chinanews. (2008a). *The first international aid relief supplies has arrived in Chengdu.* Retrieved 20th June, 2008, from http://www.chinanews.com.cn /gn/news/2008/05-14/1250176.shtml

Chinanews. (2008b). *Japan military will send the relief supplies to the disaster area.* Retrieved 20th June, 2008, from http://bjyouth.ynet.com/ view.jsp?oid=41090285

CNN. (2010a). *Haiti pier opens, road laid into Port-au-Prince.* Retrieved 20th March, 2010, from http://www.cnn.com/2010/ WORLD/americas/01/21/ haiti.earthquake/index.html?hpt=T2

CNN. (2010b). *Massive food distribution begins in quake-ravaged Haitian capital.* Retrieved 20th March, 2010, from http://edition.cnn.com/2010/WORLD/americas/01/31/ haiti.food.aid/

Cowell, A., & Otterman, S. (2010). *Relief groups seek alternative routes to get aid moving.* Retrieved 20th March, 2010, from http://www.nytimes.com/2010/01/16/world/ americas/ 16relief.html?hp

DiarioLibre.com. (2010). *LF viaja a Haití, acuerda con Préval plan para mitigar daños.* Retrieved 18th March, 2010, from http://www.diariolibre.com/ noticias_det.php?id=230910

Diaz, R. (2010). *Dominican Republic: Helping neighboring Haiti after earthquake.* Retrieved 16th March, 2010, from http://globalvoicesonline.org/ 2010/01/14/dominican-republic -helping-neighboring- haiti -after-earthquake/

Dolmetsch, C. (2010). *UN urges Haiti coordination as supplies flood airport.* Retrieved 16th March, 2010, from http://www.businessweek.com/ news/2010-01-22/un-urges- haiti-relief-coordination-as- supplies- flood-airport.html.

English, C. R. I. (2010). *Chinese team offers aid in Haiti.* Retrieved 20th March, 2010, from http://english.cri.cn/6909/ 2010/01/15/45s542729.htm

Expatica. (2010). *Expatica, Dutch aid ship arrives in Haiti.* Retrieved 16th March, 2010, from http://www.expatica.com/nl/ news/dutch-rss-news/ dutch-aid- ship-arrives-in-haiti_20254.html

Fawcett, P., McLeish, R., & Ogden, I. (1992). *Logistics management.* London, UK: Pitman Publishing.

Fuentes, G. (2010). *Bunker Hill en route to help Haiti mission.* Retrieved 20th February, 2010, from http://www.navytimes.com/ news/2010/01/ navy_bunkerhill_011610/

Gaoyan, C. (2008). *Strategy assessments of China army in Sichuan earthquake relief.* Retrieved 20th June, 2008, from http://military.china.com/zh_cn /critical3/27/20080604/ 14886960.html

Garamone, J. (2010). *Top navy doc predicts long USNS comfort deployment.* Retrieved 20th March, 2010, from http://www.defense.gov/news /news-article.aspx?id=57565

Guillon, J. (2010). *In Haiti, the Jacmel cathedral clock stopped at 5:37 pm.* Retrieved 20th March, 2010, from http://www.mysinchew.com /node/34251

Haas, J. E., Kates, R. W., & Bowden, M. (1977). *Reconstruction following disaster.* Cambridge, MA: MIT press.

Haghani, A., & Oh, S. C. (1996). Formulation and solution of a multi-commodity, multi-modal network flow model for disaster relief operations. *Transportation Research Part A, Policy and Practice, 30*(3), 231–250. doi:10.1016/0965-8564(95)00020-8

ICESAR. (2010). *The Icelandic urban SAR team has landed at Haiti*. Retrieved 20th March, 2010, from http://www.icesar.com/

ICRC. (2010). *Haiti earthquake: Reaching victims outside the capital*. Retrieved 20th March, 2010, from http://www.icrc.org/web/eng/ siteeng0.nsf/ html/haiti- earthquake-update-190110

India PRWire. (2010). *Statement from Digicel on Haiti earthquake*. Retrieved on 10th March, 2010, from http://www.indiaprwire.com/ pressrelease/ telecommunications /2010011441347.htm

Jelinek, P., & Burns, R. (2010). 10,000 troops on scene by Monday. *Navy Times*. Retrieved 30th March, 2010, from http://www.navytimes. com/news/ 2010/01/ap_military_haiti _update_011510/

Jennings, E., Beresford, A. K. C., & Pettit, S. J. (2002). Emergency relief logistics: A disaster response model. In *Proceedings of the Logistics Research Network Conference* (pp. 121–128).

JTA. (2010). *Israeli medical, rescue workers help Haitians*. Retrieved 30th March, 2010, from http:// www.jta.org/news/article /2010/01/17/1010200/ israeli- medical-rescue-workers-help-haitians

King, L. (2010). *Hampton roads, The Carl Vinson departs Haiti*. Retrieved 30th March, 2010, from http://hamptonroads.com/2010/02 /carl-vinson-departs-haiti

Ku, B. (2008). *Relief operations in Wenchuan*. Retrieved 10th June, 2008, from http://blog.sina. com.cn/s/ blog_4bb4c26301009e9h.html

Lett, D. (2010). *Canada earns its wings*, Retrieved 30th March, 2010, from http://www.winnipegfree-press.com/ opinion/columnists/Canada-earns-its-wings-83147562.html

Long, D. C., & Wood, D. F. (1995). The logistics of famine relief. *Journal of Business Logistics, 16*, 213–229.

Mackey, R. (2010). *Latest updates on rescue and recovery in Haiti*. Retrieved 30th March, 2010, from http://thelede.blogs.nytimes.com/ 2010/01/15/latest-updates-on- rescue-and-recovery-in-haiti/?hp

McClintock, A. (1997). Global cases in logistics and supply chain management . In Taylor, D. H. (Ed.), *The logistics of third-world relief operations* (pp. 354–369). London, UK: International Thomson Business Press.

MSNBC. (2010). *Haiti aid bottleneck is easing up*. Retrieved 30th March, 2010, from http://www. msnbc.msn.com/id /34915151/ns/world _news-haiti_earthquake/

Navy, U. S. (2010a). *Maritime force serves as cornerstone of relief operations in Haiti*. Retrieved 20th March, 2010, from http://www.navy.mil/ search/ display.asp?story_id=50696

Navy, U. S. (2010b). *Vinson helicopters perform medical evacuations*: *Sea Base on the way*. Retrieved 20th March, 2010, from http://www.navy. mil/search/ display.asp?story_id=50582

Navy, U. S. (2010c). *USS Normandy arrives off coast of Port-Au-Prince*. Retrieved 20th March, 2010, from http://www.navy.mil/search/ display. asp?story_id=50593

News, J. D. (2010). *22nd MEU departs for Haiti*. Retrieved 20th March, 2010, from http://www. jdnews.com/news/ uss-71828-equipment-leave. html

Padgett, T. (2010). *With the military in Haiti: Breaking the supply logjam.* Retrieved 20th March, 2010, from http://www.time.com/time/specials/ packages/article/0,28804,1953379 _1953494,00.html

Pettit, S. J., & Beresford, A. K. C. (2005). Emergency relief logistics: An evaluation of military, non-military and composite response models. *International Journal of Logistics Research and Applications, 8*(4), 313–331.

PRC. (2008). The relief supplies arrived in Chengdu successfully and will be delivered to the disaster areas as soon as possible. Retrieved 15th July, 2008, from http://www.gov.cn/jrzg/2008-05 /17/content_980034.htm

Red Cross. (2010). *Haiti earthquake appeal.* Retrieved 30th March, 2010, from http://www. redcross.org.uk/ donatesection.asp?id=102168

Rhoads, C. (2010). *Earthquake sets back Haiti's efforts to improve telecommunications.* Retrieved 30th March, 2010, from http://online.wsj.com/ article/ SB10001424052748703657604575005453223257096.html

Siegler, M. (2010). *Twitter strikes deal to bring free SMS tweets to Haiti.* Retrieved 20th March, 2010, from http://www.washingtonpost.com / wp-dyn/content/article/2010/02/ 23/AR2010022 300234.html

Simmins, C. (2010). *Two months after the Haitian earthquake.* Retrieved 20th March, 2010, from, http://northshorejournal.org/ two-months-after-the-haitian-earthquake

Slevin, P. (2010). *Quake-damaged main port in Port-au-Prince, Haiti, worse off than realized.* Retrieved 30th March, 2010, from http://www.washingtonpost.com /wp-dyn/content/article/2010/01 /27/ AR2010012705250.html

Sohunet. (2008). *Premier Wen answered the questions from pressmen in Yingxiu County.* Retrieved 15th July, 2008, from http://news.sohu.com/ 20080903/ n259341067.shtml

Steele, J. (2010). *Navy destroyer to return after helping out in Haiti.* Retrieved 30th March, 2010, from http://www.signonsandiego.com/ news/2010/feb/03/navy-destroyer -return-after-helping-out-haiti/

Thomas, A., & Mizushima, M. (2005). Logistics training: Necessity or luxury? *Forced Migration Review, 22,* 60–61.

Times-Picayune. (2010). *New Orleans to Haiti barge initiative seeks donations of cash, goods.* Retrieved 30th March, 2010, from http://www. nola.com/news/ index.ssf/2010/02/new_orleans _to_haiti_ barge_ini.html

Trenton, D., Clark, L., & Rosenberg, C. (2010). *New airfield, more troops to increase delivery of aid, security.* Retrieved 20th March, 2010, from http://www.miamiherald.com/ 2010/01/19/1433097/us- pledges -aid-security-will-improve.html

United Nations ISDR. (2004). *Terminology of disaster risk reduction.* Retrieved 25th July, 2010, from http://www.unisdr.org/ eng/library/ lib-terminology -eng%20home.htm

Washington Post. (2010). *US Navy en route to make Haiti seaport usable.* Retrieved 20th March, 2010, from http://www.washingtonpost.com /wp-dyn/ content/article/2010/01 /16/AR2010011601601. html

Wijkman, A., & Timberlake, L. (1988). *Natural disasters: Acts of God or acts of man?* Earthscan Publications.

Williams, P. (2010). *Emergency architects of Canada to aid in Haiti reconstruction effort.* Retrieved 20th March, 2010, from http://dcnonl. com/article/id37551

Xinhua. (2010). *S.Korea to dispatch peacekeepers to Haiti next week if parliament approves.* Retrieved 20th March, 2010, from http://www. istockanalyst.com/ article/viewiStockNews/ articleid/ 3841113

Xinhuanet. (2008a). *China publicized the deployment of the military in the disaster area first time.* Retrieved from http://www.chinaelections.org/ NewsInfo.asp?NewsID=127999

Xinhuanet. (2008b). *The military mission focuses on the relief operation of remote villages.* Retrieved 15th July, 2008, from http://www.ce.cn/xwzx/gnsz /gdxw/200805/23/t20080523_ 15598398.shtml

Xinhuanet. (2008c). *1.6 million dollars relief supplies provided by U.S. military has arrived in Chengdu.* Retrieved 15th July, 2008, from http:// china.zjol.com.cn/05china /system/2008/05/18/ 009525011.shtml

Xinhuanet. (2008d). *Military relief operation record.* Retrieved 15th July, 2008, from http:// news.xinhuanet.com/ mil/2008-05/19/ content_8205680.htm

Xinhuanet. (2008e). *113080 soldiers have been dispatched to the earthquake relief mission.* Retrieved 15th July, 2008, from http://news.sina.com. cn/c/ 2008-05-18/163215566124.shtml

Xinhuanet. (2008f). *People will never forget.* Retrieved 15th July, 2008, from http://www. ce.cn/xwzx/gnsz/ gdxw/200808/27/t20080827 _16635783.shtml

Xinhuanet. (2008g). *The earthquake relief military continue retracing.* Retrieved 15th July, 2008, from http://news.sohu.com/ 20080814/ n258843155.shtml

Zhongguangnet. (2008). *Helicopters and assault boats are dispatched urgently into the relief operation.* Retrieved 15th July, 2008, from http://www. sznews.com/news /content/2008-05/17/ content_ 2052018.htm

Zhongguangnet. (2008). *Port of Luzhou.* Retrieved 15th July, 2008, from http://www.cnr.cn/zhuanti1/ gkwlx/zjgk/t20050628 _504078120.html

ADDITIONAL READING

Alexander, D. (2006). Globalization of disaster: Trends, problems and dilemmas. *Journal of International Affairs, 59*(2), 1–22.

Altay, N., Prasad, S., & Sounderpandian, J. (2009). Strategic planning for disaster relief logistics: Lessons from supply chain management. *International Journal of Services Sciences, 2*(2), 142–161. doi:10.1504/IJSSCI.2009.024937

Banomyong, R., Beresford, A., & Pettit, S. (2009). Logistics relief response model: the case of Thailand's tsunami affected areas. *International Journal of Services Technology and Management., 12*(4), 414–429. doi:10.1504/IJSTM.2009.025816

Beresford, A., & Pettit, S. (2009). Emergency logistics and risk mitigation in Thailand following the Asian tsunami. *International Journal of Risk Assessment and Management, 13*(1), 7–21. doi:10.1504/IJRAM.2009.026387

Carroll, A., & Neu, J. (2009). Volatility, predictability and asymmetry: An organising framework for humanitarian logistics operations? *Management Research News, 32*(11), 1024–1037. doi:10.1108/01409170910998264

Heaslip, G. (2008). Humanitarian aid supply chains . In Mangan, J., Lalwani, C., & Butcher, T. (Eds.), *Global Logistics and Supply Chain Management.* Chichester: John Wiley & Sons.

Hoehling, A. A. (1973). *Disaster – Major American Catastrophes.* New York: Hawthorne Books Inc.

Hofmann, C.-A., & Hudson, L. (2009). Military responses to natural disasters: last resort or inevitable trend. *Humanitarian Exchange Magazine*, *44*, 29–31.

Kaatrud, D. B., Samii, R., & Van Wassenhove, L. N. (2003). UN joint logistics centre: a coordinated response to common humanitarian logistics concerns . *Forced Migration Review*, *18*, 11–14.

Kelly, C. (1996). Limitations to the use of military resources for foreign disaster assistance. *Disaster Prevention and Management*, *5*(1), 22–29. doi:10.1108/09653569610109532

Kent, R. C. (2004). International humanitarian crises: two decades before and two decades beyond. *International Affairs*, *80*(5), 851–869. doi:10.1111/j.1468-2346.2004.00422.x

Kovács, G., & Spens, K. M. (2007). Humanitarian logistics in disaster relief operations. *International Journal of Physical Distribution and Logistics Management*, *36*(2), 99–114.

Kovács, G., & Spens, K. M. (2009). Identifying challenges in humanitarian logistics. *International Journal of Physical Distribution and Logistics Management*, *39*(6), 506–528. doi:10.1108/09600030910985848

Lane, F. W. (1966). *The Elements Rage, David and Charles*. Newton Abbott.

Long, D. C. (1997). Logistics for disaster relief: Engineering on the run. *IIE Solutions*, *29*(6), 26–29.

Oloruntoba, R., & Gray, R. (2006). Humanitarian aid: an agile supply chain? *Supply Chain Management – . International Journal (Toronto, Ont.)*, *11*(2), 115–120.

Pettit, S., & Beresford, A. (2009). Critical success factors in the context of humanitarian aid supply chains. *International Journal of Physical Distribution & Logistics Management*, *39*(6), 450–468. doi:10.1108/09600030910985811

Pujawan, I. N., Kurniati, N., & Wessiani, N. A. (2009). Supply chain management for disaster relief operations: principles and case studies. *International Journal of Logistics Systems and Management*, *5*(6), 679–692. doi:10.1504/IJLSM.2009.024797

Sheu, J. B. (2007). Challenges of emergency logistics management. *Transportation Research Part E, Logistics and Transportation Review*, *43*(6), 655–659. doi:10.1016/j.tre.2007.01.001

Sheu, J. B. (2007). An emergency logistics approach for quick response to urgent relief demand in disasters. *Transportation Research Part E, Logistics and Transportation Review*, *43*(6), 687–709. doi:10.1016/j.tre.2006.04.004

Taylor, D., & Pettit, S. (2009). A consideration of the relevance of lean supply chain concepts for humanitarian aid provision. *International Journal of Services Technology and Management*, *12*(4), 430–444. doi:10.1504/IJSTM.2009.025817

Van Wassenhove, L. N. (2006). Humanitarian Aid Logistics: supply chain management in high gear. *The Journal of the Operational Research Society*, *57*(5), 475–589. doi:10.1057/palgrave.jors.2602125

Weeks, M. R. (2007). Organizing for disaster: Lessons from the military. *Business Horizons*, *50*, 479–489. doi:10.1016/j.bushor.2007.07.003

Whiting, M. C. (2009). Enhanced Civil-Military Collaboration in Humanitarian Supply Chains. In: Gattorna, J. (2009) *Dynamic Supply Chain Alignment* (p. 107-121). Gower Publishing Ltd: Farnham, UK.

Whittow, J. (1980). *Disasters – the anatomy of environmental hazards*. London: Penguin Books Ltd.

KEY TERMS AND DEFINITIONS

A Natural Disaster: "A serious disruption of the functioning of a community or a society involving widespread human, material, economic or environmental losses and impacts, which exceeds the ability of the affected community or society to cope using its own resources" (UN/ISDR 2004).

Access: The possibility to reach a disaster area, and beneficiaries in a disaster area.

Disaster Response: Encompasses humanitarian activities before, during and after a major emergency. The response is normally subdivided into several phases embracing: preparedness, response and recovery. These phases can themselves be subdivided according to specific conditions prevailing at the time of the natural disaster.

Emergency Transport: The movement of people or goods from one location to another involving vehicles which may or may not be designed for the purpose of emergency conditions.

Humanitarian Aid: All cargo required in response to a major emergency or crisis. Such aid includes: emergency food rations, water and water purification facilities, sanitation equipment, tents and shelters, equipment necessary to aid the construction and maintenance of temporary shelters, medical supplies, clothing, blankets, and all other materials required to support a population left without access to normal living facilities.

Humanitarian Logistics: "The process of planning, implementing and controlling the efficient, cost-effective flow and storage of goods and materials, as well as related information, from point of origin to point of consumption for the purpose of meeting the end beneficiary's requirements" (Thomas and Mizushima, 2005).

Sea Basing: The use of a vessel as a facility such as a warehouse or clinic.

Supply Line: The logistics pipeline of an individual humanitarian organization.

Transport Infrastructure: Points and ways of access for different transportation modes that facilitate local distribution capability. Points of access include ports, airports, border crossings, bridges etc. Ways of access include waterways, railways, roads, etc.

Transportation Mode: The means by which people and freight achieve mobility. These include road, rail, air, water transportation as well as pipelines, intermodal transportation and a telecommunications superstructure.

Transportation Route: The selected way of access to serve a disaster area and its population.

Chapter 5

The Application of Value Chain Analysis for the Evaluation of Alternative Supply Chain Strategies for the Provision of Humanitarian Aid to Africa

David H. Taylor
Sheffield, UK

ABSTRACT

The study reported in this chapter was commissioned in 2009 by the charity Advance Aid in order to provide an independent evaluation to compare conventional methods of supplying humanitarian aid products to Africa from outside the continent, with a proposed model of locally manufactured and pre-positioned stocks. The evaluation was carried out using "value chain analysis" techniques based on "lean" concepts to provide a strategic evaluation of alternative supply models. The findings show that a system of local manufacture and pre-positioned stockholding would offer significant advantages over conventional humanitarian supply chains in terms of responsiveness, risk of disruption and carbon footprint, and that delivered costs would be similar to or significantly better than current non-African supply options. Local manufacture would also have important benefits in terms of creating employment and economic growth, which in the long run would help African states to mitigate and/or respond to future disasters and thus become less dependent on external aid.

The chapter also gives a more general consideration to the potential of value chain analysis concepts and techniques to the measurement, evaluation, and improvement of humanitarian supply chain operations in locations and scenarios beyond that described in the current case study.

DOI: 10.4018/978-1-60960-824-8.ch005

INTRODUCTION

Currently it is estimated that over 90% of products supplied in Africa for humanitarian purposes are sourced outside the continent primarily from Europe, North America and South East Asia. In 2007, a new charity called Advance Aid was established in the UK with the objective:

to implement more effective humanitarian response and bring economic benefits to Africa through a lower cost, more timely supply of basic relief items, which have been manufactured, stored and pre-positioned in strategic locations around Africa.[1]

In many respects this objective has a strong intuitive appeal not only because it reduces the time, cost and risks involved in lengthy international aid supply pipelines, but also because of the potential to stimulate employment and economic growth in Africa. Indeed at a more general level the need for Africa to become less dependent on external aid has been powerfully advocated in the book 'Dead Aid' by Dambisa Moyo (2009).

The study reported in this paper was commissioned in 2009 by Advance Aid in order to provide an independent and systematic evaluation to compare typical conventional international supply models with an alternative strategy of African manufacture and local stockholding. Subsequently the research was used by Advance Aid in building a case to obtain financial support for the establishment of a pilot project in Africa.

Whether aid is required in response to rapid onset disasters or ongoing humanitarian development work, there are a variety of supply chain strategies that can be adopted by aid agencies. As in other industry sectors, humanitarian supply chains have evolved in response to customer needs (beneficiaries in this case) and the characteristics and objectives of the provider organizations (aid agencies). With a growing number of humanitarian needs to be met and an ever increasing number of aid agencies responding, it is not surprising that

many different supply chain solutions have been developed. It is also understandable that once particular agencies have established a supply chain strategy and a modus operandi that works, they are inclined to persist with it and perhaps make incremental changes and improvements as time progresses. However, an issue that seems to have been less well developed in many agencies both large and small is the development of appropriate methods to measure the performance of their supply chains. Without such Key Performance Indicators (KPI's), it is difficult to comprehensively evaluate alternative supply strategies or indeed to evaluate the performance of existing chains with a view to improving their performance.

The approach adopted in the study was to use 'Value Chain Analysis' (VCA) techniques to provide a top level comparison of alternative supply strategies. One of the most important features of VCA is that it provides a consistent and holistic set of performance measures for supply chain activity. KPI's are developed both for individual elements of the supply chain (e.g. transport, warehousing, manufacturing) and importantly for the supply chain as a whole. In recent years there has been an increasing recognition of the need for appropriate KPI's to monitor and evaluate humanitarian aid supply chains. (Moxham *et al.* 2007; Beamon *et al.* 2008, Whiting *et al.* 2009). It is suggested that VCA provides metrics that not only facilitate comparison of alternative supply strategies as reported here, but importantly, also provide a rigorous method with which to evaluate the performance of existing supply chains as a starting point for their systematic improvement.

The objective of the chapter is thus twofold. Firstly to highlight the key features and comparative performance of some typical supply chains that have developed for supplying aid to Africa and compare these to the local-for-local supply model proposed by Advance Aid. Secondly to demonstrate the applicability and potential of VCA in evaluating and improving humanitarian supply chains be they in Africa or elsewhere.

THE 'LEAN' CONTEXT

The theoretical framework and practical tools used in the analysis are based on the approach known as 'Lean Thinking.' (Womack and Jones, 1996). Lean Thinking provides a methodology for the analysis and improvement of business and is based on the system of business management originally developed by Toyota the Japanese automotive manufacturer. In recent years lean approaches have been adopted by many companies, and organizations throughout the world in both commercial and non-commercial environments. Lean applications were originally focused on improving manufacturing operations, but as time has gone by, they have been successful applied to other functions such as warehousing, procurement, transport and administration and over the last decade, there have been increasing examples of their application to supply chain operations. (e.g. Hines *et al.* 2004; Taylor, 2005; Taylor, 2009). At the same time lean has spread from its origins in manufacturing industry to many other sectors including healthcare, defense, financial services, local government and agri-foods to name but a few. Lean not only provides a set of tried and tested methods with which to analyze and benchmark business processes, but importantly also provides a framework for their radical and continuous improvement. Leading companies that have adopted lean such as Tesco, Boeing, Porsche and Wal-Mart have achieved startling improvements in performance and reported significant reductions in costs and waste, whilst at the same time achieving dramatic improvements in lead-times, quality and delivery performance.

Value Chain Analysis

Value chain analysis (VCA) is the approach to supply chain improvement based on lean principles and methodologies. In all sectors supply networks are complex, involving many organizations many products, many markets and customers. VCA focuses on the analysis of specific value chains, where a value chain is identified as a particular 'target product' or product group, sourced from a particular supplier and destined to a particular end-user or market group. This tightly focused and disaggregated approach has two important benefits: firstly it provides an analytical mechanism with which to break into the complexity of supply networks in a manageable and effective manner; secondly experience has shown that analysis of particular target or pilot value chains almost always highlights features, problems and opportunities that are common across the wider supply network and as such provides a basis for broader decision making

The overall the aim of value chain analysis is to improve the efficiency and effectiveness of supply chains, with efficiency focused on reducing waste and eliminating 'non-value-adding' activities and effectiveness being conceptualized as providing 'value' to the customer particularly in terms of quality, cost and delivery. A critical aspect of VCA is to develop an appropriate set of key performance metrics with which to quantify the current performance of a supply chain, highlight opportunities for improvement and subsequently monitor improvement initiatives. These KPI's also permit the benchmarking and comparison of alternative supply chain operations and strategies and it is in this respect they have been used within this study.

THE ADVANCE AID CASE STUDY

Advance Aid proposed to provide an 'Emergency Kit' which would enable a displaced family to survive by providing the six basic items needed: a shelter, blankets, a stove, a kitchen set, a water carrier and a hygiene kit (Figure 1). The plan was to assemble kits at three key locations in Africa: Nairobi, Accra and Johannesburg with products sourced from manufacturers clustered around each of these locations. In addition to stocks held at these primary locations, a number of satellite stocks would be pre-positioned close to known disaster-prone areas. Advance Aid would not undertake to distribute kits within the areas of need, but instead would sell the product to established NGO's and other aid agencies for distribution to beneficiaries through their networks.

The specific product on which this analysis is based is tarpaulins (plastic sheet). This product was selected for two reasons. Firstly it is a primary means of shelter used in many emergencies and in consequence is a high demand humanitarian product. Secondly one of the world's leading suppliers of tarpaulins into the humanitarian sector, agreed to cooperate in the study by provision of data. For reasons of commercial confidentiality this company will be referred to as 'Tarp Co.

Figures 2 and 3 give an overview of the current worldwide supply network for the manufacture, stock holding and distribution of tarpaulins used in humanitarian aid.

Manufacturing. Production of tarpaulins is concentrated in two main areas, South East Asia (80%) with India and China as the major producers and the USA (18%). Details of the main tarpaulin manufacturers are shown in the table on Figure 2. Over half the output from India is used within the subcontinent; similarly much of the Pakistan output is used internally. It should also be noted that product from a number of the Indian manufacturers does not meet current international quality standards for humanitarian goods, although in times of high demand such products will be used. US output is primarily channeled through US aid organizations. In consequence China is a main source of tarpaulins for many of the world's aid agencies. Product supplied by Tarp Co. is manufactured in China. At the time the study was undertaken there was no manufacture of tarpaulin in Africa, although Tarp Co. were considering establishing a factory in Nairobi.

Stockholding. Stockholding locations for tarpaulins can be classified as follows:

Figure 1. The advance aid emergency kit

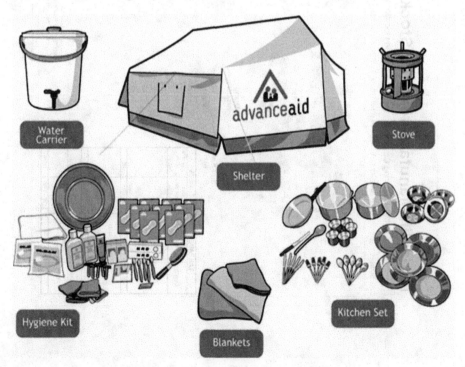

Figure 2.Tarpaulin manufacturing and stock locations

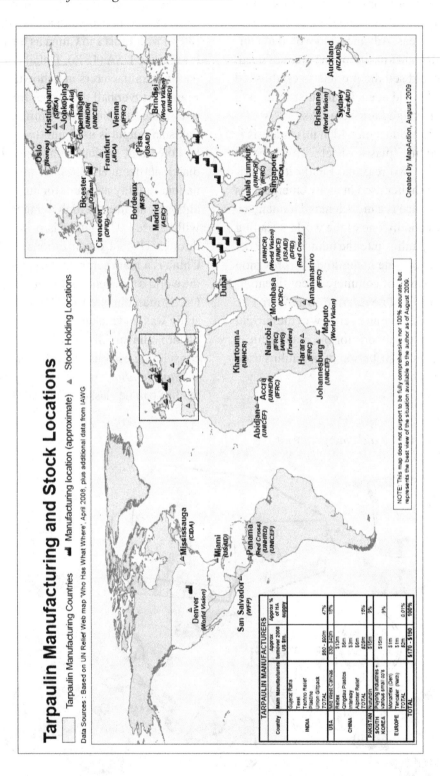

Figure 3. Tarpaulin supply routes to Africa

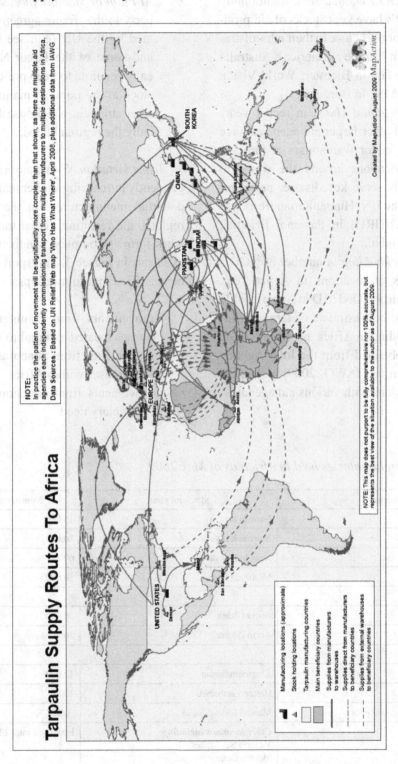

- *Close to NGO Headquarters*: traditionally many NGO's have kept stocks of aid product in warehouses close to their administrative HQ's in Europe / America / Australia etc (e.g. Oxfam in Bicester; World Vision in Denver; MSF in Bordeaux)
- *Strategic Regional Hubs*: in recent years the UN and other larger aid agencies have established major warehouses at three or four key locations around the world to more readily serve key disaster prone regions, e.g. the UN 'Humanitarian Response Depots' (UNHRD) in Panama, Brindisi, Malaysia, Dubai.
- *African Stockholding*: a number of NGO's hold stocks at locations in Africa (e.g. IFRC in Nairobi; UNHRD in Accra). Table 1 shows the best estimate of tarpaulin inventory holding in Africa as of May 2009, with data obtained from the Inter Agency Working Group (IAWG, 2009) and informal discussions with various aid agencies.

- *At Point of Manufacture*: suppliers that are party to the 'frame agreements' now operated by the UN (e.g. see UNHCR, 2010) and some of the larger NGO's, are typically required to keep pre-defined levels of stock at the point of manufacture, as well as at strategic warehouse locations - typically the regional hubs.

Transportation. With many aid agencies independently purchasing shelter materials and many potential manufacturers in the various supply regions, the resulting global transportation network forms a complex 'spaghetti' of movement as shown in Figure 3. Transport movements can be classified into four groups:-

- movements from manufacturer to warehouses located external to Africa
- movements from manufacturer direct to warehouses within Africa
- movements from warehouses to areas of beneficiary need

Table 1. Tarpaulin inventories held in Africa as of May 2009

Agency	Stock location	Number of tarpaulins
IFRC	Nairobi Kenya	15,000
IAWG	Nairobi Kenya	20,000
Independent traders	Nairobi Kenya	6000
ICRC	Mombasa Kenya	50,000
UN HRD	Accra Ghana	20,000
IFRC	Accra Ghana	3000
UNHCR	Khartoum Sudan	43,000
IFRC	Harare Zimbabwe	3000
World Vision / Advance aid pilot	Maputo Mozambique	5000
French Red Cross French Red Cross French Red Cross French Red Cross	*Other locations including:* Victoria Seychelles Moroni Comoros Antananarivo Madagascar Sainte Marie Reunion	Estimated total 20,000

Note: This table has been based on best estimates of stockholding as of May 2009 with data provided by IAWG and other agencies.

- movements from manufacturer direct to areas of beneficiary need

Figure 4 shows a simplified schematic diagram of the structure and movement patterns of the supply network.

Target Value Chains. Four value chains were selected to exemplify the main supply chain options found within the supply network. Figure 5 shows the selected value chains as well as the African supply model proposed by Advance Aid. In each case, a region in northern Uganda was taken as the assumed area of beneficiary need as this is an area that has a continuing demand for humanitarian aid due to natural disasters, refugee issues and civil war.

Classification of Relief Scenarios. Relief operations were classified into three categories depending upon the type of response:

- *Blue Scenario*: situations where there is an on-going requirement for relief. Goods are taken from warehouses either within or outside the African continent utilizing surface-based transport and manufacturers located external to Africa.
- *Red Scenario*: more extreme, rapid-onset disasters, where demand exceeds the capacity of normal supply chains and goods are supplied under emergency conditions from sources outside Africa. Air freight is typically used to achieve rapid response, but if air freight is unavailable or too ex-

pensive, protracted supply lead times inevitably result.

- *Green Scenario*: the proposed Advance Aid model with product manufactured in Africa, combined with pre-positioned stockpiles at key locations, permitting rapid response without the need to resort to expensive air transport.

Data Sources

The primary data source used in the analysis of value chain lead times and costs was operational records held by Tarp Co. the tarpaulin manufacturer. Information was collected to show the lead time for each element of the supply chain of 25 orders required in Africa and placed on the manufacturer in China in the period between October 2007 and March 2009. In 14 of these cases, Tarp Co. was responsible for delivery of product to the 'port of entry' in Africa whilst in 11 shipments it was responsible for delivery beyond the port to an inland destination. Detailed data was gathered to show actual inland transportation times for these 11 shipments. Additional data regarding scheduled international transportation times and routes was provided by Scan Global Logistics (2009). Data for the proposed 'African manufacture model was based on a feasibility study carried out in 2008 by Tarp Co. for a proposed factory in Nairobi. Carbon footprint calculations were based on the emission standards set out in the '2008 Guidelines

Table 2. Classification of relief scenarios

Relief Scenario	Value chain	Agency	Source of supply	Transport modes	Stockholding points	Destination
Blue	VC 1	Typical NGO	China	Sea / road	Nairobi	Uganda
Blue	VC 2	Typical UN	China	Sea / road	Dubai	Uganda
Blue	VC 3	Typical NGO	India	Sea / road	Nairobi	Uganda
Red	VC 4	Typical UN	China	Sea / Air / road	Copenhagen	Uganda
Green	VC 5	Advance Aid	Nairobi factory	Road	Entebbe	Uganda

Figure 4. Schematic diagram of the structure of Tarpaulin Supply Networks

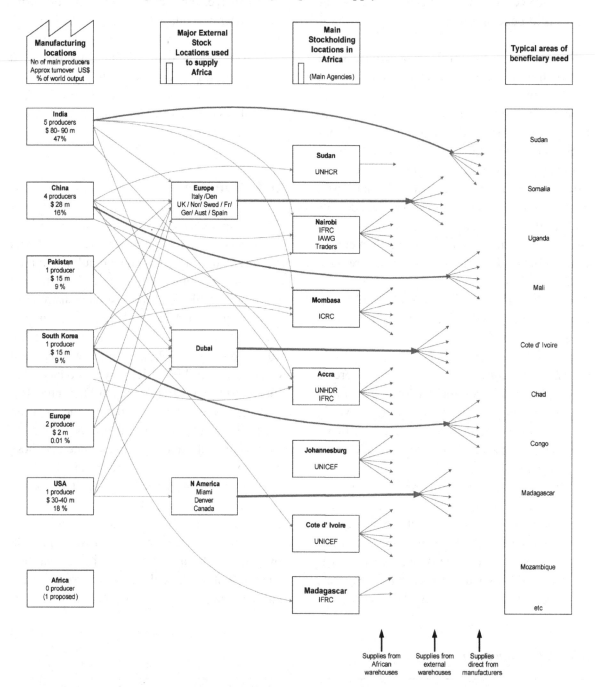

Notes
This chart presents an overview of the worldwide supply network.
In practice the pattern of movements will be significantly more complex than that shown because at each node in the chain there is typically more than one organisation involved (i.e. multiple manufacturers, multiple NGOs) each arranging independent transportation movements using different shipping agents and different routes through the network. In addition there are a significant number of smaller NGOs independently arranging stock holding and transportation

Figure 5. Value chains selected for analysis

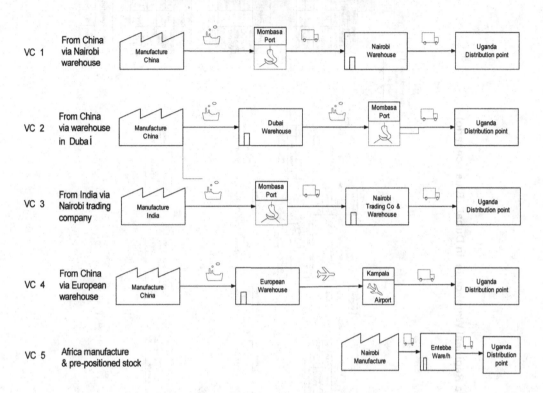

Value Chains Selected for Analysis

to DEFRA's Greenhouse Gas Conversion Factors' (DEFRA, 2009), together with inter-port shipping distances derived from the 'Port World Distance Calculator' (Port World, 2009).

VALUE CHAIN MAPPING

Value chain mapping has a number of benefits:

- It provides a succinct visual representation of value chain structures and operations.
- It provides quantified key performance indicators (KPI's) both for the individual elements in a value chain e.g. factories, warehouses, transport links and for the chain as a whole.

- In adopting a standardized methodology for analysis it provides a basis for comparison of alternative supply chain models.
- It identifies opportunities for improvement in supply chain performance and prioritizes these opportunities in terms of their impact on supply chain KPI's. (although this aspect is not included in this paper)

A 'current state' map was constructed for each of the five value chains selected for analysis. By way of example, the map for Value Chain 2: Manufacture in China and Supply Via Regional Hub in Dubai Using Surface Transport is shown in Figure 6.2 Each of the value chain maps has five distinct elements:

Value Chain Structure. The top level of the map shows the physical elements that make up the supply chain including factories, ports, ware-

Figure 6. Example Current State Map

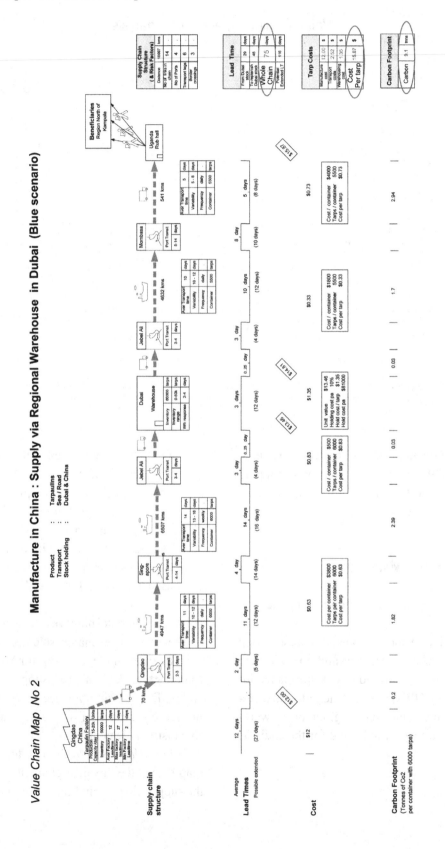

Manufacture in China : Supply via Regional Warehouse in Dubai (Blue scenario)

houses etc and the methods of linking these nodes. Specific KPI's for each activity are shown in the data boxes beneath each element of the chain.

Lead Times: shown on the second level of the map. The time-line has two levels: the top level indicates times when the product is moving through a facility e.g. factory or port, whilst the bottom level of the line indicates the time the product is in transit. Average times for each of these elements are shown above the time line, whilst the variability in time at each point (minimum and maximum) is shown below the line.

Cost: the third level of the map shows the cumulative cost for one tarpaulin from factory production through to delivery to the area of need.

Carbon footprint: the fourth level of the map indicates the carbon footprint of transportation elements of the chain. No attempt has been made to assess carbon impact of production, warehousing or port activities. The figures indicate the carbon footprint of a batch of 6000 tarps, which is the typical shipment quantity.

Summary KPI's for the entire value chain: the data boxes on the right hand side of the map, summaries the overall performance of each value chain in terms of four key performance characteristics.

COMPARATIVE ANALYSIS OF ALTERNATIVE SUPPLY MODELS

Assessment of the alternative supply chain models was made on the basis of four parameters:

- Lead time- in terms of the ability to provide the required goods to beneficiaries in a timely manner
- Supply chain vulnerability - in terms of the likelihood of disruption or delay
- Cost - in terms of the total cost of supply chain operations including manufacturing cost, transport costs and stockholding costs
- Environmental impact - in terms of the carbon footprint of transportation.

Supply Chain Lead Time

Once a requirement for tarpaulins (or any other product) has been identified by an aid agency and an order placed, three aspects of lead time need to be considered:

- The lead time from the nearest stockholding point to delivery to beneficiaries
- Lead time to replenish stocks
- Total lead time for the supply chain as a whole[3]

The charts in Figure 7 show the lead times for the five value chains in respect of these three criteria.

Lead Time From Stock: 'The Initial Response Time'. Figure 7a, showing the delivery lead time from the nearest stocking point, indicates the initial response time for each supply chain, on the assumption that stock is available at that relevant stock point. Short initial response times are of paramount importance in the early phases of a disaster, where the speed of response can be a matter of life or death. Rapid response can be achieved either by having stocks close to the area of need or by use of air-freight from more distant locations. Some of the larger aid agencies already hold stock in Africa, albeit at a limited number of locations (Figure 2). Lead times to areas of need using surface transport can be significant, given the large distances in the continent and the often poor state of the transport infrastructure.

If stocks are not available close to the disaster area, the only option to achieve rapid response is by use of air freight from distant stocking points. Value Chain 4, centered on a European warehouse, typifies the air freight supply strategy adopted by a number of major agencies. When rapid response is required, airfreight is also used from the regional hub at Dubai (Value Chain 2), as a response time averaging 28 days by surface transport is too slow in emergency situations.

Figure 7. Supply chain lead time charts

7a.

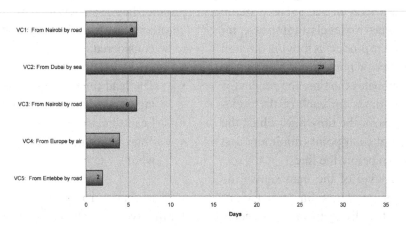

7b.

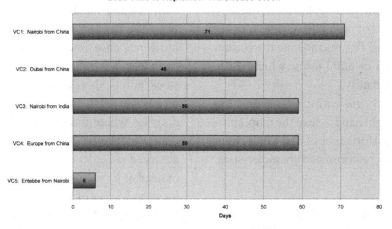

7c.

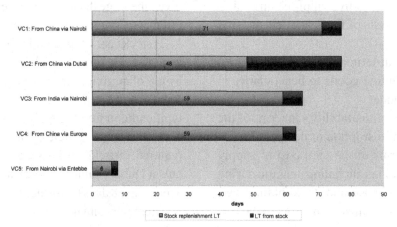

Use of air freight clearly has a high cost penalty, with the cost of chartering a plane typically in the order of US$190,000. There is also another potential drawback. In times of major disasters, the demand for aircraft increases and the lead times for air transport can be extended because of difficulties in obtaining charter aircraft. Anecdotal evidence suggests that it could take anything up to one to two weeks to obtain a charter aircraft, whilst charter costs can double.

Holding pre-positioned stock at satellite locations close to known disaster-prone areas would provide a vital first line of response in rapid-onset disasters. In the regions that are served by the satellite stocks, response times would be less than from the current African stockholding hubs and similar, if not better, than those achieved by air freight.

Stock Replenishment Lead Time. Stock replenishment lead time influences both the amount of stock required to meet anticipated demands. Longer replenishment times require higher stocks and increase the potential for a warehouse to have insufficient stock when demands arise. If an unexpected demand has drained the warehouse and subsequent demands arise in a shorter space of time than the replenishment lead time, then there will be a stock-out resulting in unfulfilled demand or a requirement to resort to more expensive emergency supply systems. The shorter the replenishment lead time, the lower the probability of being out-of-stock. Figure 7b shows that Value Chain 5 has a significant advantage over the other supply options, with a replenishment lead time at a pre-positioned stock location in Entebbe of six days, compared to 71, 48, 59 and 59 days respectively for the other chains considered.

Total Supply Chain Lead Time. If as a result of sustained demand, the various stock points in a supply chain become completely drained, the lead time to meet beneficiary requirements will be the total supply chain lead time from order placement to delivery at the point of use (or the need to resort to air freight, which may be pro-

hibitively expensive). In a number of the recent major disasters e.g. Pakistan Earthquake 2007, Myanmar 2008, supply chains were completely drained of stock, meaning that 'Total Supply Chain Lead Time' became the pertinent measure. For example during the Myanmar disaster, Tarp Co. reported supply lead times in excess of 60 days to supply tarpaulins from China to Burma. Figure 7c shows that in such circumstances, the local manufacturing and stockholding model has again a very significant advantage with a total lead time of eight days, compared to average total lead times for the other supply models ranging from 63 to 77 days. This advantage accrues primarily because of the location of manufacturing in Africa.

Supply Chain Vulnerability / Risk

Growing reliance on long and complex global supply chains in many sectors of industry has increasingly brought the issue of supply chain vulnerability to the forefront. Many commercial companies are starting to appreciate their vulnerability to disruption of their global supply chains and of taking steps to reduce the risks. Intuitively it would be expected that longer supply chains have greatest risk of disruption and in practice there are a number of reasons why this may be the case. Longer supply chains have more links, including transport legs, ports, warehouses etc., which creates increased potential for disruption and delay both within each operation and more particularly at the interchange between operations. At the same time complex global supply chains also present a greater challenge in terms of coordination, monitoring and management because of the many, often independent agencies involved such as transport operators, freight agents and customs authorities.

When considering international chains, and particularly those supplying humanitarian needs, three elements present particular risks of disruption and delay:

Ports: Ports often represent a significant bottleneck for a variety of reasons including inefficient port handling systems, congestion (both land-side and sea-side) and delays due to coordination of vessels for trans-shipment. These issues are particularly apparent at seaports, but also occur at airports. Furthermore port congestion is frequently exacerbated in disaster situations when large amounts of un-coordinated aid arrive at a particular port which may have only limited capacity.

Border crossings: Considerable delays frequently occur at border crossings due to import/export regulations, customs procedures, bureaucratic incompetence, corruption or political obstruction. Furthermore in some humanitarian disaster situations these obstacles appear to increase rather than diminish.

Transport legs: The greater the number of transport legs and the greater the distance of those legs, the greater the potential for delay in a supply chain. All transport operations are vulnerable to a variety of disruptions ranging from weather and other natural events, availability of vehicles, strikes and breakdowns and again in emergency situations some of these vulnerabilities may be exacerbated.

The charts in Figure 8 demonstrate the extent to which a local supply model is less vulnerable to disruption than conventional chains, not only because of the shorter distances involved, but also because of reduced exposure to the elements which create the potential for disruption or delay. Recent research into supply chain vulnerability (Peck, 2006) highlights three key causes: disruption to transportation (as described above), disruption to power supplies and loss of labor. The latter two factors are particularly relevant to manufacturing and to a lesser extent warehousing. In this analysis it has been assumed that manufacture in Africa would be as reliable as manufacture in other parts of the world which in practice is unlikely to be the case. However a key element in the Advance Aid development proposal was to work with African

manufactures to ensure reliability of supply from a production point of view. Furthermore the proposed location of a tarpaulin factory in Nairobi was selected in part because of its reliable power supply and the availability of local labor.

Supply Chain Cost

Total supply chain costs were assessed in terms of manufacturing cost, transportation costs and warehousing costs. No allowance was made for information handling and other administration costs involved in supply chain management.

The manufacturing costs shown in Figure 9 were based on known factory-gate prices for one tarpaulin as of May 2009. Transportation costs were based on actual transportation charges for shipments made during the period between 2007 and May 2009. Warehousing costs were based on an annual holding cost at a nominal 10% of the capital value of one tarpaulin. The holding cost includes allowance for warehouse operating costs such as rent, labor, insurance etc as well as an allowance for the cost of capital tied up in inventory. In the case of Value Chain 3, inventory was held in Nairobi by a trading company and the cost of $6.02 includes both the inventory holding costs and the company's profit requirement.

In terms of total delivered cost, the local supply model was competitive with all the options of supply from manufacturers external to Africa when surface transport was employed and had a threefold cost advantage compared to the use of air freight. Thus in situations where pre-positioned stocks can provide an acceptable alternative to air freight in terms of speed of response, use of such a strategy would allow approximately three times as much product to be made available for the same capital outlay.

Environmental Impact

Environmental impact was assessed solely in terms of the carbon footprint of the transportation legs of

Figure 8. Supply chain vulnerability charts

8a.

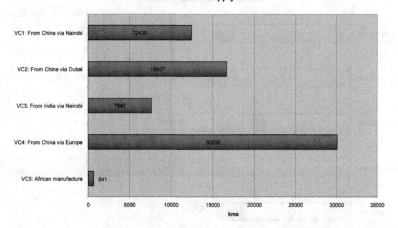

8b.

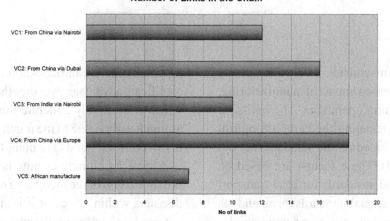

8c.

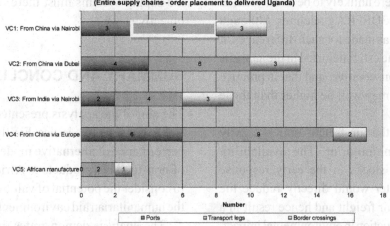

Figure 9. Supply chain vulnerability charts

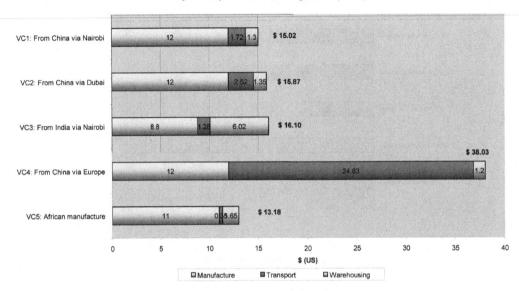

Cost per Tarp - delivered Uganda ($ US)

the five supply chains considered. No attempt was made to evaluate carbon impact of manufacturing, ports or warehousing operations. The emission factors used in calculating the carbon footprint of transport were based on the standards set out by DEFRA (2009). These factors are based on current UK vehicle standards for road transport and modern international standards for air and sea transport. It was recognized that some vehicles used in aid operations (e.g. trucks in Africa, older charter aircraft) were unlikely to be as efficient as those detailed in the DEFRA guidelines. Although some allowance was made for such differences, it is suggested that the estimates used in the analysis are likely to be 'conservative' and that in practice carbon emissions may well be higher than those indicated.

In Figure 10 the air freight option clearly dominates carbon footprint. The availability of pre-positioned stocks in the early response phases of a disaster would directly reduce the requirements for air freight and hence result in a commensurate reduction in environmental impact. However, leaving aside air freight, the model of

local manufacture and stockholding still has a very significant advantage over the other supply options which respectively generate three or four times as much carbon due to the much longer distances involved when goods are manufactured external to Africa. As climate change is recognized as a possible cause of the increasing number of natural disasters within Africa, it is ironic that supply of aid goods is itself a significant contributor to CO_2 levels. Strategies to reduce the carbon footprint of aid supply systems must therefore be a sensible objective.

SUMMARY AND CONCLUSION

The aim of the analysis presented in this chapter was to consider two issues. Firstly to compare the performance of alternative models for the supply of non-food aid products to Africa and secondly to consider the potential of value chain analysis in the humanitarian aid environment more generally.

The analysis demonstrates that if a model of local manufacture and local stockholding could

Figure 10. Environmental impact

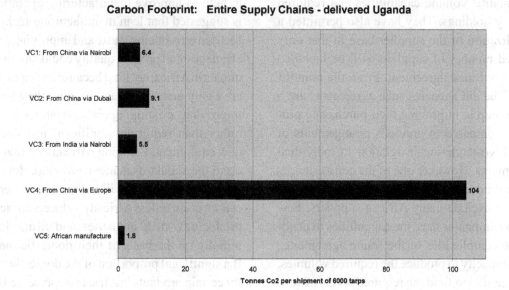

be established it would either equal or improve on existing methods of supply in all of the dimensions considered. Supply lead times would be significantly reduced compared to currently used international sources. Establishment of strategically located pre-positioned stocks would permit rapid response at the onset of a disaster and reduce the need for expensive air freight in the initial response phase. Supply chains entirely contained within Africa would not be subject to the risks of disruption and delay encountered in the international movements. Local chains would still be vulnerable to disruption of inland transportation within Africa, however existing supply models are also exposed to these vulnerabilities in the final distribution legs of the chains wherever these involve surface transport. In terms of the tarpaulin product analyzed, the delivered cost associated with the African-based supply chain is comparable to that of non-African supply models using surface transport, but offers a threefold cost advantage compared to a supply strategy involving air freight. Tarp Co. was confident that manufacturing in Africa could be achieved at a similar cost to other parts of the world. With

other products this may not always be the case and the manufacturing cost model would need to be evaluated on a product by product basis. Local African supply would greatly reduce the environmental impact of transport compared to the much longer international supply routes currently used and where ever pre-positioned stocks could avoid the use of air freight there would be a massive carbon saving.

Overall therefore the analysis confirms the intuitive view that local manufacture and stock-holding of aid supplies in Africa, if it could be effectively achieved, would be advantageous from a logistical point of view and would also have the benefit of stimulating economic activity in Africa. However, a number of questions arise as to whether local African manufacture and pre-positioning of stocks could provide a realistic alternative to international sourcing.

In recent years in an endeavor to improve the efficiency of their purchasing and acquisition procedures, a number of the major aid agencies such as UNICEF, UNHCR, IFRC have introduced 'frame agreements' for supply. These agreements set out clear criteria that the aid agencies require from suppliers in terms of price guarantees, quality standards, volume capabilities and required inventory holdings. They have also permitted a rationalization of the supplier base in that only a limited number of suppliers will be included within any frame agreement. From the point of view of the aid agencies such agreements are a positive step in improving their purchasing procedures compared to previous arrangements of competitive tendering in relation to individual requirements. However one of the consequences of the introduction of frame agreements is that they may exclude many African suppliers. Few African companies have the capabilities to negotiate the complexities of the frame agreements, nor the capacity to produce the required volumes, nor the capital to fund the required levels of consignment stock and may also have difficulties, at least initially, in meeting the quality requirements that are usually established in terms of stringent western criteria. It is thus hard to see how small or medium-sized manufacturers in Africa could supply to the aid goods under such agreements. Indeed there are examples of African producers that previously supplied products to international aid agencies that lost the business as a result of the introduction of frame agreements. One of the objectives of Advanced Aid is to act as an intermediary between groups of African producers and the major aid agencies in order to establish a system whereby locally produced goods can be included within the frame agreements. However, it surely behoves the aid agencies themselves to review their purchasing strategies to enable local African manufacturers to participate albeit, if in the short run, this means relaxing some of their procedures and requirements.

A second and closely related issue with a local supply model is whether African manufacturers can achieve the levels of reliability and quality achieved by producers elsewhere in the world. In practice this may not be the case, at least in the short run. Advance Aid recognized this as a problem and part of their proposed strategy was to

work closely with their chosen African suppliers to help to improve manufacturing operations. It is suggested that lean manufacturing techniques focused on reducing waste and improving process efficiency and product quality could have a major impact in Africa, not least because lean approaches are easily understood and because they focus on improving existing operations and equipment rather than requiring significant investment in new equipment. Bringing African manufacturers up to the standards achieved in more developed industrial economies will not be easy, but certainly cannot occur unless African producers are actually producing! Most industries and firms develop initially on the basis of their domestic markets. If a significant proportion of the domestic market for certain products in Africa is represented by aid requirements, it would seem perverse that local manufacturers are excluded from these markets by the policies of non-African aid agencies!

Leaving aside manufacturing efficiency, a further difficulty in achieving reliable supply from within Africa relates to the significant challenges created by bureaucracy and corruption. Solutions to these issues are clearly beyond the scope of logistics and supply chain management, yet if there is a fundamental conviction on the part of the aid community that local supply is better from both a logistical point of view and in terms of stimulating economic development, these barriers will have to be addressed.

In May 2010, Tarp Co. commenced production at a new factory in Nairobi, albeit that it had taken some two years of negotiations with local and national authorities to establish an operable legal, financial and trading framework. Once fully operational this factory will employ 300 local people and will supply tarps to Advance Aid as well as to other customers in Africa.

A further challenge with the proposed Advanced Aid model relates to pre-positioning of supplies in proximity to areas of need. It is sometimes argued that it is impossible to predict where aid should be pre-positioned within Africa, as demands arise from an ever-changing landscape of natural and man-made disasters. Further research is being carried out to analyze historical patterns of natural disasters, together with the locations of refugees and internally displaced persons and early results suggests that there is some predictability to the geographical demand for humanitarian assistance. In the meantime Advance Aid established a pilot pre-positioning of 5000 kits in Mozambique. These were used in the initial response phase for the Malawi earthquake in February 2010 and were delivered at significantly lower total cost than had air freight been used. At the time of writing, Advance Aid had made an agreement with a major African logistics company to provide warehousing space to pre-position some 15,000 kits at two locations in East Africa.

The second objective of this chapter was to illustrate the potential of lean concepts and value chain analysis techniques in the humanitarian sector. It is suggested that these approaches have significant potential to assist in improving humanitarian supply chain operations. Value chain analysis provides a systematic and holistic method of identifying and evaluating supply chain processes. Value chain maps provide a visual and concise method of presenting and communicating supply chain characteristics. However, perhaps the most important benefit of VCA is the development of a comprehensive and consistent set of key performance indicators which measure both the performance of individual operations within the chain and of the chain as a whole. Such KPI's have hitherto been absent in many humanitarian supply operations.

In the study outlined in this chapter, VCA was used to compare alternative supply models at a strategic level. It is more common however for it to be used as a basis for identifying opportunities for improvement in existing supply chains followed by the systematic application of well-established lean improvement tools and techniques. In spite of the best efforts of charities such as Advanced Aid to develop local supply within Africa, international supply chains will undoubtedly continue to be used

for many years to come. It is suggested that aid agencies both large and small could benefit by the application of VCA in order to reduce waste and inefficiency in their existing supply chains and at the same time improve service to the beneficiaries.

The KPI's used in this project were developed in response to the requirements for a strategic comparative evaluation of alternative supply models. They were also strongly influenced by the data that was available within the confines of the study and under ideal circumstances some additional KPI's would have been included. The metrics reported here largely focus on the efficiency aspects of the chain. There is a requirement for appropriate measures to show the effectiveness of the chain in terms of actual delivery of aid to beneficiaries. The ultimate measure would be number of lives saved or lives improved as a result of an aid effort. However this is subject to multiple variables, many of which lie beyond the influence of supply chain management. Nevertheless it is suggested that one supply chain effectiveness measure that ought to be included in any humanitarian VCA is an indication of the amount of aid reaching intended recipients. In practice in most humanitarian chains less than 100% of the aid that enters the pipeline will get through to the beneficiaries. Losses occur for a variety of reasons including damage, deterioration, theft, corruption and misappropriation. VCA could provide a good framework with which to track where aid seeps from the supply chain and also to understand the causes of the losses. Such a measure would be a good starting point in trying to increase the proportion of aid getting through.

The development of an appropriate set of KPI's through VCA would also have a benefit to aid agencies in reporting to donors. Increasingly donors, be they governments or individuals are concerned as to whether the money provided is well spent. Estimates vary as to the proportion of aid budgets consumed by logistics and supply chain activity ranging from 40% to 80% (Whiting *et al.*, 2009; Van Wassenhove 2006) Whatever the actual figure, logistics undoubtedly accounts for a high proportion of the costs of a relief effort. If aid agencies can demonstrate through rigorous KPI's that they are making improvements to the efficiency and effectiveness of their operations it will increase the confidence of donors to continue to provide funds.

As far as is known the work carried out for Advance Aid in Africa was the first to apply value chain analysis techniques to the humanitarian supply chain environment and as such is acknowledged as preliminary and somewhat limited analysis. However it is suggested that the concepts and methodologies of lean and VCA offer very significant potential for the measurement, evaluation and importantly the improvement of humanitarian supply operations wherever they occur. There is a now a need for more pilot projects and applications in order to refine and adapt the lean /VCA methodologies, measures and tools so as to be most appropriate for the humanitarian environment. More and more value chain improvement initiatives are occurring in commercial industry sectors resulting in significant improvements in operational efficiency, better customer service and increased profits. The bottom line of value chain improvement in the humanitarian sector would hopefully be an increase in the number of lives saved.

ACKNOWLEDGMENT

The author would like to acknowledge the contribution made by the following people in providing assistance and information for the research described in this chapter:

- David Dickie and Howard Sharman: Advanced Aid
- Simon Lucas and Sally Prosser: Reltex Ltd
- George Fenton: World Vision International
- Emese Csete: Map Action
- Maud Duchemin: A willing volunteer

REFERENCES

Beamon, B. M., & Balcik, B. (2008). Performance measurement in humanitarian relief chains. *International Journal of Public Sector Management, 21*(1), 4–25. doi:10.1108/09513550810846087

DEFRA. (2009). *2008 guidelines to Defra's greenhouse gas conversion factors- Methodology paper for transport emission factors*. Retrieved from www.defra.gov.uk

Hines, P., Holweg, M., & Rich, N. (2004). Learning to evolve: A review of contemporary lean thinking. *International Journal of Operations & Production Management, 24*(9/10).

IAWG. (2009). Retrieved from http://www.iawg.gov/

Moxham, C., & Boaden, R. (2007). The impact of performance measurement in the voluntary sector. *International Journal of Operations & Production Management, 27*(8), 826–845. doi:10.1108/01443570710763796

Moyo, D. (2009). *Dead aid*. London, UK: Allen Lane.

Peck, H. (2006). Reconciling supply chain vulnerability with risk and supply chain management. *International Journal of Logistics Research and Applications, 9*(2), 127–142.

Port World. (2009). *Distance calculator*. Retrieved May 2009, from http://www.portworld.com/map

Scan Global Logistics. (2009). *Distance calculator*. Retrieved May 2009, from www.scangl.com

Taylor, D. H. (2005). Value chain analysis: An approach to supply chain improvement in agri-food chains. *The International Journal of Physical Distribution & Logistics Management, 35*(10), 744–761. doi:10.1108/09600030510634599

Taylor, D. H. (2009). An application of value stream management to the improvement of a global supply chain: A case study in the footwear industry. *International Journal of Logistics Research and Applications, 12*(1). doi:10.1080/13675560802141812

UNHCR. (2010). Retrieved from http://www.unhcr.org/ pages/49f6d3d26.html

Van Wassenhove, L. N. (2006). Humanitarian aid logistics: supply chain management in high gear. *The Journal of the Operational Research Society*, *57*, 475–489. doi:10.1057/palgrave.jors.2602125

Whiting, M., & Ayala-Ostrom, B. (2009). Advocacy to promote logistics in humanitarian aid. *Management Research News*, *32*(11), 1081–1089. doi:10.1108/01409170910998309

Womack, J., & Jones, D. (1996). *Lean thinking*. New York, NY: Simon and Schuster.

ADDITIONAL READING

Taylor, D. H., & Pettit, S. (2009). Consideration of the relevance of lean supply chain concepts to humanitarian aid provision. *International Journal of Services Technology and Management*, *12*(4), 430–444. doi:10.1504/IJSTM.2009.025817

KEY TERMS AND DEFINITIONS

Carbon Footprint: CO_2 as greenhouse gas emissions resulting from manufacturing and transportation.

Key Performance Indicator (KPI): Methods and metrics to measure the performance of the supply chain. KPIs can be used to evaluate the performance of the existing supply chain or to evaluate alternative strategies.

Lean Thinking: A methodology for the analysis and improvement of business and is based on the system of business management originally developed by Toyota the Japanese automotive manufacturer.

Local Sourcing: A supply model that prefers manufacture from the disaster area.

Pre-Positioning: Stock-holding for purposes of preparedness.

Supply Chain Cost: Total supply chain costs in this chapter were assessed in terms of manufacturing cost, transportation costs and warehousing costs. Other costs include information and other administrative costs.

Supply Chain Risk: The possibility of disruption and delay in the supply chain.

Value Chain Analysis: VCA is the approach to supply chain improvement based on lean principles and methodologies. The overall the aim of value chain analysis is to improve the efficiency and effectiveness of supply chains, with efficiency focused on reducing waste and eliminating 'non-value-adding' activities and effectiveness being conceptualized as providing 'value' to the customer particularly in terms of quality, cost and delivery.

ENDNOTES

[1] www.advanceaid.org
[2] Value chain maps are specifically designed to be produced on A3 sized sheets of paper. Constraining the map to this size is an effective filter in selecting relevant information and still remaining legible. Unfortunately when reproduced at a smaller size legibility suffers.
[3] This analysis has not included any allowance for the lead time required for aid agencies to undertake needs assessment, nor to complete internal procedures required to place an order. In a rapid onset disaster needs assessment could typically take 1 to 2 weeks, whilst evidence from the VCA in commercial organizations indicates that administration time for order processing is often longer than might be expected.

Chapter 6
Designing Post–Disaster Supply Chains:
Learning from Housing Reconstruction Projects

Gyöngyi Kovács
HUMLOG Institute, Hanken School of Economics, Finland

Aristides Matopoulos
University of Macedonia, Greece

Odran Hayes
European Agency for Reconstruction, Ireland

ABSTRACT

Post-disaster housing reconstruction projects face several challenges. Resources and material supplies are often scarce; several and different types of organizations are involved, while projects must be completed as quickly as possible to foster recovery. Within this context, the chapter aims to increase the understanding of relief supply chain design in reconstruction. In addition, the chapter is introducing a community based and beneficiary perspective to relief supply chains by evaluating the implications of local components for supply chain design in reconstruction. This is achieved through the means of secondary data analysis based on the evaluation reports of two major housing reconstruction projects that took place in Europe the last decade. A comparative analysis of the organizational designs of these projects highlights the ways in which users can be involved. The performance of reconstruction supply chains seems to depend to a large extent on the way beneficiaries are integrated in supply chain design impacting positively on the effectiveness of reconstruction supply chains.

INTRODUCTION

In contrast to the developments in increasing the accuracy of forecasting a number of natural disasters, the aftermath of these events, particularly the part related to disaster relief operations, often remains very problematic. The increased frequency of both human and manmade disasters which implies that more resources have to be allocated more efficiently, more frequently and sometimes more unexpectedly, has resulted in increased complexity in the delivery of humanitarian

DOI: 10.4018/978-1-60960-824-8.ch006

assistance (USAID, 2002; EM-DAT, 2008). Complexity is further increased by the large growing number of organizations, both governmental, and non-governmental, which are nowadays devoted to providing humanitarian assistance. Another novelty is that emergency relief efforts rarely remain within the boundaries of single countries. In most cases, multi-country collaboration is required, adding thus global implications in the development of relief efforts.

Given that logistical efforts account for a very significant portion of the humanitarian aid spending (van Wassenhove, 2006), many researchers are pointing out the crucial importance of having an efficient and effective logistics system. But as Kovács and Spens (2007) argue when it comes to humanitarian aid, there is an important distinction to be made between logistical activities that pertain to 'continuous aid work' vs. 'disaster relief'; or, as van Wassenhove (2006) points out, slow-onset vs. sudden-onset disasters. Yet, distinct phases can also be seen within disaster relief, such as preparation, immediate response and reconstruction (Kovács and Spens, 2007). Whilst the focus in the immediate response phase is one of time efficiencies, the later reconstruction phase has a longer-term focus and thus, deals with more predictable demand and the possibility to plan for constant schedules (Maon *et al.*, 2009; Taylor and Pettit, 2009).

The reconstruction phase of disaster relief operations is at the heart of this chapter. In particular, the chapter sheds light on two major European-based reconstruction housing programs with the aim of increasing the understanding of the overall supply chain design. The chapter starts with a review of relief supply chain literature, with particular emphasis on supply chain design and performance in post-disaster reconstruction. Next, the research methods of the study are presented, followed by empirical evidence from the housing reconstruction programs. The chapter ends with the key findings and conclusions.

EMPIRICAL BACKGROUND

The chapter reports the findings of a comparative analysis of two studies of reconstruction housing programs. Study 1 is based on a European Housing Reconstruction Programme in the Kosovo, while study 2 sheds light on a similar Housing Reconstruction Programme in the Former Yugoslav Republic of Macedonia (FYROM).

Study 1: Housing Reconstruction in Kosovo

Kosovo is located in the central Balkan peninsula in Southern-eastern Europe. It is a landlocked region and borders the FYROM to the south, Albania to the west and Montenegro to the northwest. For many decades it was an autonomous part of Yugoslavia, but after 1989 conflicts between Kosovo Albanians Serbians started which were continued until 1999 when NATO forces bombed Serbia. The end of the Kosovo conflict revealed a typical complex emergency situation characterized by refugees and a large-scale destruction of houses. An estimated 120,000 houses out of a total of over 250,000 were damaged or destroyed. The European Union played an important and multifaceted role in Kosovo's reconstruction particularly through the European Agency of Reconstruction (EAR). According to the EAR (EAR, 2002), 41,000 were less badly damaged, 32,000 were seriously damaged (41-60% of the house damaged) and 47,000 were very seriously damaged (61-100% of the house damaged) - most of these houses were effectively destroyed, with often not even a sound foundation remaining. This large-scale destruction, as well as the need to rapidly re-house families in Kosovo urged for increased efficiency in the reconstruction effort. Without the return of families from temporary accommodation to their homes, normal life could not have resumed in Kosovo. Several issues added to the problem. For example, large-scale refugee returns (mostly Kosovo Albanians) from late spring 2000

(100,000 estimated by UNHCR until the end of the 2000) added to the complexity of the situation. In other words, while demand is rather predictable in reconstruction, reconstruction supply chains that deal with post-military conflicts need to take the potential of renewed hostilities into account (Taylor and Pettit, 2009). Problems related to property rights also appeared. Many families who have had their homes damaged or destroyed were not in the most vulnerable category of beneficiaries, but lacked the resources to fully pre-finance their speedy reconstruction. In addition, given that the Housing Reconstruction Programme 2000-2001 targeted approximately 12,000 homes, the houses that would be assisted in reconstruction needed to be selected carefully. Finally, the damage assessment conducted by International Management Group (IMG) in 1999, revealed problems in the supply of housing materials. On the one hand there was an urgent need for the procurement of timber, roof tiling and other materials for the rehabilitation of private dwellings and some public buildings. On the other hand several problems were reported with respect to supply imports, such as embargo problems, closed borders and delays in deliveries.

Study 2: Housing Reconstruction in FYROM

The FYROM is a landlocked country located in the central Balkan Peninsula in South-eastern Europe. It declared independence in 1991 after the disintegration of the former Yugoslavia. The country is bordered by Kosovo to the northwest, Serbia to the north, Bulgaria to the east, Greece to the south and Albania to the west. FYROM's Housing Repair and Reconstruction Programme started in June 2001, a few months after the conflict between Ethnic Albanian armed groups and Government forces which took place in the Northwest (Tetovo) and North-East of Skopje (Skopska Crna Gora). The conflict caused extensive damage to buildings in former conflict areas, including buildings of particular religious and historical significance,

as well as housing infrastructure, in particular in the north regions of the country and also in other parts of the country (Kumanovo, Arachinovo, Bitola). The levels of damage to individual houses varied considerably, with many being in the more lightly damaged categories. After the end of the conflict there was an urgent need to start quickly on repairing / reconstructing the houses damaged by the conflict thereby facilitating the return of the displaced persons, to re-establish normal living conditions and to rebuild confidence between the ethnic groups. An initial assessment carried out in the Tetovo area (April 2001) indicated about 190 houses of the first phase of conflict to be repaired/reconstructed. The assessment on the Northeast of Skopje (Skopska Crna Gora) was delayed due to the need to clear the area of mines. The 190 damaged houses in the Tetovo area accommodated about 1,500 people. The estimated 250 houses damaged in Skopska Crna Gora accommodated about 2,000 people. Implementation of the EC/EAR House Repair and Reconstruction Programme started in September 2001 and was undertaken in different phases and under different budget lines. In total, 1150 houses were reconstructed (or scheduled to be reconstructed) with a cost of approximately €7.5 million. The program was initially implemented by the Commission Services with support from the existing operational centers of the European Agency for Reconstruction and the relevant national and/or local authorities. In addition, "Grant contracts" with NGO implementing partners were signed with the selected NGOs being responsible in managing assistance allocations to beneficiaries and also undertaking technical assessments and materials' voucher allocation. Moreover, NGOs were involved in the provision of technical advice and labor support quality control and monitoring of reconstruction work; and management of any works and supply sub-contracts.

SUPPLY CHAINS IN RECONSTRUCTION

Humanitarian logistics and relief supply chain management distinguishes between disaster relief with all its complexities (upon man-made or natural disasters or a combination of both) and development aid. Yet also within disaster relief, several phases are set apart: (a) the preparedness phase with its measures to prevent disasters or to prepare populations and international humanitarian organizations for an effective response to them, (b) the immediate response phase, from search and rescue operations to actual disaster relief, i.e. any activities related to providing for beneficiaries, i.e. the population affected by a disaster, and (c) the reconstruction phase. Often neglected in humanitarian logistics literature (to the extent that authors such as Long, 1997, and van Wassenhove, 2006, do not even mention it), reconstruction is the time when infrastructure and housing in the disaster area is rebuilt, people resettled etc. Reconstruction and restoration thus concludes immediate "emergency" response in a cycle of reaction and recovery (cf. Maon *et al.*, 2009). But as Pettit and Beresford (2005) pinpoint, reconstruction does not only indicate recovery and rehabilitation but is intrinsically linked with preparedness activities. This is especially the case in disaster-prone areas such as earthquake zones due to tectonic faults, or areas with cyclical disasters such as cyclones, hurricanes and annual floods. Yet while literature has considered post-disaster prevention since the Indian Ocean tsunami in 2006 (Beresford and Pettit, 2007; Banomyong *et al.*, 2009), research on reconstruction has remained scant. What is more, reconstruction suffers from a lack of funding, as donors tend to emphasize immediate relief.

Construction supply chains have been characterized as *converging* (several supply lines coming together at site), *temporary* (set up on a project basis, though project as well as supply chain members can come together for several projects

in a row), and following a *make-to-order* principle (Vrijhoef and Koskela, 2000). These converging supply chains may be better described as an extremely complex construction supply network with a main contractor at the construction site (logistically to be seen as a hub), with links to the client, main supply agencies as well as design and specialist management services (Dainty *et al.*, 2001). Vrijhoef and Koskela (2000) further distinguish four focal areas of supply chain management in construction: (a) on-site activities, i.e. project management and the coordination of all supply lines at the construction site, (b) supply chain design with a focus on cost efficiencies in setting up the supply chain, (c) a transfer of activities away from the construction site to more prefabrication of materials and components in earlier echelons, and (d) integrated management of the site and the converging supply chains. Saad *et al.* (2002) add the focus of relationship management and partnering, i.e. a move away from traditional arms-length and short-term relationships in construction as a result of a new supply chain orientation even of public sector clients (such as the UK's Ministry of Defence), though even follow-up studies found little evidence of this being put in practice (cf. Briscoe and Dainty, 2005). Reconstruction supply chains observe similar focal areas and related challenges. They are in effect converging temporary supply chains that follow a make-to order principle. These issues are of importance in the design of reconstruction supply chains. At the same time, the convergence on site is a matter of not only bringing together different construction companies (and their related supply chains) but also, different humanitarian organizations involved in a reconstruction program.

Relief supply chain design needs to be flexible enough to "evolve from an initial emergency response to an ongoing reconstruction operation" (Maon *et al.*, 2009). Yet reconstruction poses new questions for relief supply chains. Contrary to the agility maxim of immediate relief (cf. Oloruntoba and Gray, 2006), the reconstruction phase can indeed be planned more in advance (Taylor and Pettit, 2009) and thus, focus more on cost as well as time efficiencies. Rather comparable to other construction projects (see Fearne and Fowler, 2006), reconstruction supply chains are designed for temporary purposes, though without a potential reassembly of the same supply chain members for further projects. In the humanitarian context, the cost efficiency focus alongside long-term goals of reconstruction is related to the development side of humanitarian aid, filling what Oloruntoba and Gray (2006) call the transitional stage of a relief to development continuum. Measuring the performance of reconstruction supply chains thus, differs from performance measurement in immediate relief that has focuses on short-term activities (cf. Beamon and Balcik, 2008; Maon *et al.*, 2009). Having said so, literature on performance measurement in any phase of disaster relief is scant (Kovács and Tatham, 2009). The few exceptions include van Wassenhove's (2006) general assessment of at least 80% of costs of aid to be attributed to logistics, two case studies on measuring performance in immediate relief (van der Laan *et al.*, 2009, on Médecins Sans Frontières, MSF and Schulz and Heigh, 2009 on the International Federation of Red Cross and Red Crescent Societies, IFRC), and, probably most importantly, Beamon and Balcik's (2008) evaluation of the effectiveness of a relief mission. As they suggest, "the challenges identified for performance measurement in the non-profit sector include the intangibility of the services offered, immeasurability of the missions, unknowable outcomes, and the variety, interests and standards of stakeholders" (Beamon and Balcik, 2008, p.8). Yet performance measurement in relief supply chains, including reconstruction, is particularly important from the perspective of accountability to beneficiaries as well as donors.

Supply chain performance measurement traditionally focuses on the dimensions of efficiency and effectiveness (cf. Fearne and Fowler, 2006). As Kovács and Tatham (2009) discuss, breaking down these two results in debates on product and process quality, on-time deliveries, flexibility, time and cost efficiencies, and customer service levels. Beamon and Balcik (2008) suggest a tripartite measurement in terms of (a) resource performance metrics (resource utilization, quantities, output) such as inventory holding costs to man-hours, (b) output performance metrics (i.e. looking at effectiveness) such as lead times, back-orders and stock-outs, product quantities and qualities, all in accordance with the strategy of an organization, and (c) flexibility metrics such as shortest delivery lead times etc. The latter is the only key performance indicator cited in Maon *et al.* (2009), pinpointing its importance. Similarly, Fearne and Fowler (2006) emphasize the importance of delivering construction projects on time – and within budget. Furthermore, Balcik *et al.* (forthcoming) discuss equity considerations as performance metrics in the not-for-profit and public sectors that are equally applicable to relief supply chains. Equitable aid distribution targets the most vulnerable people without discrimination and according to their needs. Equity can be seen as a stand-alone measure, or incorporated in the concept of aid effectiveness.

Thus one of the most interesting dimensions related to relief supply chains is that of effectiveness, as it is far from unclear whether it is the effectiveness of an organization, a mission, or aid effectiveness en large that should be measured. What is more, while there is a call to look at all stakeholders of a "mission", it is still organizational (or program) effectiveness that is typically under evaluation (such as in Schulz and Heigh, 2009, van der Laan *et al.*, 2009). It can be argued that to measure effectiveness in a humanitarian context, the concept needs to be approached from both the beneficiary perspective (not unlike a customer focus in "commercial" supply chain management, though including the equity aspect), from the perspective of the supporting supply chain, as well as from a stakeholder perspective. In this paper, the focus is on the beneficiary perspective on the performance and design of reconstruction supply chains.

EMPIRICAL RESEARCH

This paper is based on the analysis of secondary data as reported in the European Agency of Reconstruction (EAR) evaluation studies of two Housing Reconstruction Programmes in Kosovo and FYROM (former Yugoslavian Republic of Macedonia). Both studies were conducted for the purposes of an internal analysis of the EAR (EAR, 2002, EAR, 2003) and thus did not have the aims of this research in mind. Nonetheless, they are unique studies in that they investigate the effect of the aid program from the perspective of beneficiaries. In other words, in contrast to other surveys (e.g. Long and Wood, 1995; Oloruntoba and Gray, 2006; Pardasani, 2006) the role of beneficiaries is not only conceptually recognized, but empirical investigations are also provided and documented in the EAR (2002) report. In addition, one of the authors was part of the original evaluation studies and was involved in primary data collection. Secondary data analysis was employed as a research strategy in this chapter due to the major practical constraints in accessing the research object. Not surprisingly, several articles in humanitarian logistics are based on secondary data analysis (e.g. Pardasani, 2006; Beresford and Pettit, 2007) as this method allows for the analyses of events in what would otherwise be inaccessible settings, due to practical weaknesses in accessing the research object. Data collection involved desk research with access to files and relevant documents, as well as structured interviews (by the Evaluators) with different stakeholders, including task and program managers, NGOs, contractors, suppliers, Housing Reconstruction committees, etc. Case study data were collected from the members of the reconstruction supply chain as listed above, including a random sample of village committees, in in-depth interviews. The research tools developed covered a wide range of issues, such as: weaknesses and strengths during the selection process (targeting, participation, time-consumption, guidelines, etc.), comparison

of the three reconstruction approaches (self-help, assisted self-help, contractors) and also coordination issues. This data was complemented with a mail survey sent to other NGOs in the area (of which 12 were returned from Kosovo and 3 from FYROM). Considering a potential bias in village reconstruction committees that made approval decisions as well as represented beneficiaries, these data were complemented (and triangulated) with focus group interviews in some areas. Furthermore, the two Housing Reconstruction Progammes (in Kosovo and FYROM) were analyzed comparatively before arriving to common findings from the studies.

Findings

Reconstruction relief supply chains display the features of construction supply chains such as convergence, temporariness and observing a make-to-order principle. Yet *convergence* here starts from beneficiaries as "main clients", and with needs assessments of these beneficiaries. As beneficiaries are usually not attributed any purchasing power (Kovács and Spens, 2008), needs assessment processes replace the function of placing orders in the relief supply chain. In fact, humanitarian organizations often act as proxies for beneficiaries when placing orders in the supply chain. Thus humanitarian organizations involved in the needs assessment process become part of the already complex (re-)construction supply chain (see Figure 1). An important aspect of convergence in the reconstruction supply chain is though the involvement of beneficiaries as active supply chain members – something that is unusual given their (otherwise) lack of purchasing power in relief supply chains.

Reconstruction supply chains are also designed for a given time, following the *temporariness* of the aid programme that serves as their background. In fact, relief supply chains in general obey principles of temporariness especially in the field or disaster area (cf. Tatham and Kovács, 2010).

Figure 1. The reconstruction supply chain

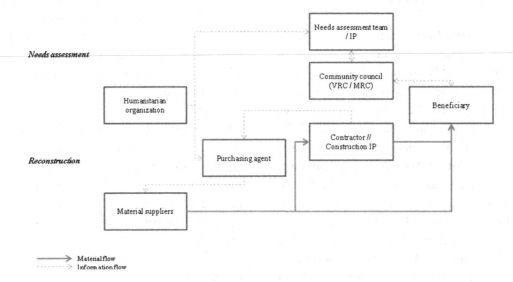

However, more permanent supply chain design can be employed on the global, strategic levels – bearing in mind that local sourcing is commonly preferred due to its positive impact on the local economy (Long and Wood, 1995; Jahre and Spens, 2007).

Vrijhof and Koskela's (2000) third aspect of construction supply chains is that they follow a make-to-order principle. This allows for customization, and here, for meeting the actual needs of beneficiaries. Nonetheless, design principles of prefabrication as well as modularization can still apply to reconstruction supply chains as long as they contribute positively to supply chain performance.

Involving Beneficiaries in Reconstruction Supply Chains

Beneficiaries are the end customers of the relief supply chain (Oloruntoba and Gray, 2006; Maon et al., 2009) and as such, the main clients of the reconstruction supply chain Equitable aid distribution is based on the actual beneficiary needs, observing the scarcity of available resources (cf. Balcik et al., forthcoming). Given a scarcity of

funds for housing reconstruction, both Housing Programmes needed to establish criteria for the selection of the beneficiaries most in need of their assistance. Anderson and Woodrow's (1998) Capacities and Vulnerability Analysis (CVA) was employed to match people, vulnerabilities and their capacities with the programme. This analysis is based on a matrix that evaluates the vulnerabilities as well as capacities of beneficiaries in three dimensions: the physical/material, social/organizational, and motivational/attitudinal. As a result of this analysis, most vulnerable households were deemed the ones least able to access the necessary resources to rebuild. Different organizations were involved in the identification of the most vulnerable households. Implementing partners (IPs) of NGOs brought in international as well as local social assessment experts to carry out the capacities and vulnerability analysis. Yet the identification of beneficiaries started at village level. Local partners in the form of community-based organizations that link implementing NGOs to beneficiaries are typical for relief supply chains (Oloruntoba and Gray, 2006). In this case, village reconstruction committees (VRCs, see Figure 1) were formed through a bottom-up approach, their members elected from and by the community. This approach ensured beneficiary participation as well as empowerment, as well as ensures the precise articulation of needs (see also Pardasani, 2006, for such an approach used in post-tsunami reconstruction). The aim of the VRCs was to ensure the transparency of the beneficiary selection process, rendered accountability to both selected and non-selected beneficiaries and informed the community about the Housing Programme. Still, the assessment of the selection criterion of the income situation of beneficiaries proved more difficult. Thus the capacity and vulnerability analysis had to be adapted in that wealth ranking was replaced with social assessors of the IPs received information from VRCs but essentially, triangulated this with indicators such as visible disposable assets and general living conditions to judge the income/asset situation of beneficiaries.

The Use of Local Resources

Such a community-based approach to reconstruction also ensured access to local suppliers and capacities. As Long and Wood (1995) point out, humanitarian logistics should always priorities information from local personnel as well as use local expertise and labor as much as possible, so that local leaders would take personal interest in

the success of operations. Local sourcing, where possible, has a positive impact on the economic situation in the region, as well as ensures the cultural and regional applicability of solutions as well as the potential to maintain local lifestyles (cf. Long and Wood, 1995). Not surprisingly, thus, there is a trend towards local sourcing in relief supply chains (Jahre and Spens, 2007). Important regional conditions for reconstruction programmes include meteorological conditions of a region as well as assessments of potential natural hazards (Pande and Pande, 2007) and emphasizing the need of local knowledge in reconstruction (Pardasani, 2006). The implementing partners (IPs) of both Housing Reconstruction Programmes thus adopted construction labor techniques that were based on the community as well: self-help, assisted self-help, and contractors (see Table 1). These could be mixed, so for example a nominally self-help house could have a contractor for the roof. Assisted self-help could comprise unpaid village labor teams as well as the more typical paid mobile teams of craftspeople. Table 1 not only summarizes the different degrees of beneficiary involvement in reconstruction but also assesses their strengths and weaknesses as reported by beneficiaries in the household survey.

IPs had the flexibility to decide which method was the most appropriate in the individual case.

Table 1. Beneficiary participation in reconstruction (Source: EAR, 2002)

Labor assistance	Strengths	Weaknesses
Self-help	• Encourages beneficiary participation/ownership • Generates local income and maximizes involvement of local labor • Moderates envy of non-selected neighbors • Cost-efficient	• The most vulnerable families cannot benefit from the self-help approach because of lack of expertise and economic means • Need for more supervision by IP • More time-consuming
Assisted self-help	• Encourages beneficiary participation/ownership • Ensures a more timely delivery if used to supplement "slow" beneficiaries	• Higher pressure for timely delivery of material • More supervision (i.e. clarifying all of the obligations the beneficiaries have to meet in order to be problem-free and to finish on time)
Contractors	• The only feasible approach for the most vulnerable • Time-efficient • Quality control	• Higher costs (i.e. fewer houses possible within the same overall budget)

Some IPs adopted the direct financial assistance to the beneficiaries in order to utilize the skilled labor present within the assisted family (extended) or the community and, as a means of ensuring that the cash flow was absorbed by the local economy. In the beneficiary household survey, beneficiaries showed great satisfaction and reported very few problems with all construction techniques employed.

Whilst it was possible to use local labor and contractors, local sourcing of construction materials proved more difficult. In the case of Kosovo, the existing market could not cope with the rapidly increasing demand for construction materials. Therefore, most of the building materials had to be imported. For example, Kosovo did not have a functioning brick factory hence bricks were imported. Timber also needed to be imported from Bulgaria. The resulting estimate of local input, mostly of sand and gravel, sets its rate of housing materials at 25%. The supply chain design of the Kosovo Housing Reconstruction Programme changed over time. At the beginning, materials were centrally procured through an agent contracted directly by the donor, and the agents subcontracted suppliers. In the later phases of the programme, materials supply was organized through international open tenders for each municipality. Contracted material suppliers were also in charge of all logistical activities including last mile deliveries and inventory management. Implementing partners (IPs) employed, however, a controller for warehouses, while procurement specialist teams were responsible for quality control.

The main challenge in the supply of housing materials was delays in deliveries. These were caused by a lack of experience of suppliers in trading with housing materials, as well as by the sheer scale of the Housing Reconstruction Programme. Further delays were instigated by the closing of the FYROM border and the prior destabilization of transport infrastructure such as railways. The main problem, however, remained the scarcity of supply of housing materials facing such a surge in demand.

In the case of FYROM, the use of local labor was also an important decision. A link to a higher utilization of local labor without violation of the local tendering procedures could have been an important condition for subcontractors, which was not the case in all villages visited.

Local suppliers also experienced cash flow problems as no payment in advance was allowed. This led to the introduction of a voucher system in 2000 in both housing programmes. Beneficiary households were given vouchers (value corresponding to the assessed damage category) to be exchanged for specified reconstruction materials at nominated supply locations, whether these are private trader's premises, or warehouses managed by IPs. Emphasis was placed on flexibility and maximum control of the beneficiaries. Problems due to corruption have not been seen and the use of voucher system countered partly this possibility.

FUTURE RESEARCH DIRECTIONS

Community-based supply chain design in reconstruction empowers beneficiaries as well as seemingly improves the effectiveness in meeting their needs. However, community-based approaches can go beyond mere village reconstruction committees in beneficiary identification and selection and even beyond incorporating beneficiaries as active members of the supply chain if adding a cash component to aid. Cash components have been used as early as in 1998 by the International Federation of Red Cross and Red Crescent Societies (IFRC) in response to Hurricane Mitch. They restore the purchasing power of beneficiaries and give them the opportunity to decide actively on their most urgent needs. At the same time, relief supply chains benefit from a reduced need to organize the purchasing and transportation of materials. Also, as long as materials are available on the local market, a cash component counteracts a sudden surge in imports. What is more, in the reconstruction supply chain, they can mitigate cash flow problems of upstream suppliers as well.

Further research is though needed on the aspects of direct cash components versus various voucher systems (as in the Housing Reconstruction cases in Kosovo and FYROM) in disaster relief.

Construction supply chains have long embraced the topic of prefabrication (Vrijhoef and Koskela, 2000). Even though newer developments in shelter and reconstruction have initiated projects in development architecture that include aspects of modularization and prefabrication, this phenomenon needs further investigation also in research. Arguably, as in construction supply chains, prefabrication can reduce the need for the management of convergence at site as well as contribute to purchasing economies. What is more, sustainable construction embraces e.g. prefabricated concrete as a less polluting variant of concrete materials. Sustainable construction is yet to embrace both the ecological and social dimensions of construction projects. Community-based approaches to construction are rarely combined with aspects of energy efficiency, not to speak of passive housing to plus-energy housing projects.

CONCLUSION

This chapter has explored the nature and scope of the post war reconstruction efforts required in Kosovo and in the FYROM. This research was based on secondary data from two studies and their related reports that were developed for a different purpose than the research itself. Notwithstanding a potential bias of the EAR (2002) and EAR (2003) reports that were analyzed, there are a number of conclusions to be drawn. Both programs due to budget and time restrictions targeted the most vulnerable families, therefore the selection process and procedures were of high importance. In both programs the involvement of local communities were considered highly appropriate and successful by beneficiaries, the IPs and the Agency.

Regarding construction procedures three delivery mechanisms were applied namely: self-help, assisted self-help and cooperative forms and they

were proved to be highly appropriate. In both programs problems with reference to the speed of delivery of building materials, or supply of labor were encountered. By incorporating local labor, small construction firms and local suppliers of construction materials, all these problems were confronted. In addition, this gave a positive impulse to the local economies. Indirect and induced effects were seen in the field of social cohesion, gender equity and democratic procedures. An interesting aspect of supply chain design was raised in the introduction of a voucher system in the Housing Reconstruction Programme. Not only did the voucher system empower beneficiaries to take their own decisions related to housing materials but the de-coupling of financial fro material flows actually reduced delivery times in the reconstruction supply chain. The practical implication of this case is to further emphasize financial instruments (vouchers and micro-credits) as well as direct cash donations as a means to increase supply chain effectiveness. Moreover, it was proved that the introduction of a voucher system for material worked satisfactorily and safeguarded against risks of fraud.

ACKNOWLEDGMENT

The authors would like to thank the European Agency for Reconstruction for access to the data and their kind support.

REFERENCES

Anderson, M. B., & Woodrow, P. J. (1998). *Rising from the ashes. Development strategies in times of disaster*. Boulder, CO: Lynne Rienner Publishers.

Balcik, B., Iravani, S., & Smilowitz, K. (2010in press). A review of equity in nonprofit and public sector: A vehicle routing perspective. In Cochran, J. J. (Ed.), *Wiley encyclopedia of operations research and management science*. John Wiley & Sons.

Banomyong, R., Beresford, A., & Pettit, S. (2009). Logistics relief response model: The case of Thailand's tsunami affected area. *International Journal of Services Technology and Management, 12*(4), 414–429. doi:10.1504/IJSTM.2009.025816

Beamon, B., & Balcik, B. (2008). Performance measurement in humanitarian relief chains. *International Journal of Public Sector Management, 21*(1), 4–25. doi:10.1108/09513550810846087

Beresford, A., & Pettit, S. (2007). Disaster management and mitigation: A case study of logistics problems in Thailand following the Asian Tsunami. In Á. Halldórsson & G. Stefánsson, (Eds.), *Proceedings of the 19th Annual Conference for Nordic Researchers in Logistics, NOFOMA 2007, Reykjavík, Iceland* (pp.121-136).

Briscoe, G., & Dainty, A. (2005). Construction supply chain integration: An elusive goal? *Supply Chain Management: An International Journal, 10*(4), 319–326. doi:10.1108/13598540510612794

Dainty, A. R. J., Millett, S. J., & Briscoe, G. H. (2001). New perspectives on construction supply chain integration. *Supply Chain Management: an International Journal, 6*(4), 163–173. doi:10.1108/13598540110402700

EAR. (2002). *Kosovo housing reconstruction programme* 2000-2001. Evaluation Report, Programming, Coordination and Evaluation Division Evaluation Unit, September 2002.

EAR. (2003). *FYROM housing reconstruction programme* 2001-2003. Evaluation Report, Programming, Coordination and Evaluation Division Evaluation Unit, August 2003.

EM-DAT. (2008). *Emergency events database-Université Catholique de Louvain*. Retrieved January 29, 2009, from http://www.emdat.be/Database/terms.html

Fearne, A., & Fowler, N. (2006). Efficiency versus effectiveness in construction supply chains: The dangers of lean thinking in isolation. *Supply Chain Management: An International Journal, 11*(4), 283–287. doi:10.1108/13598540610671725

Jahre, M., & Spens, K. (2007). Buy global or go local – That's the question. In P. Tatham, (Ed.), *Proceedings of the International Humanitarian Logistics Symposium*, Faringdon, UK.

Kovács, G., & Spens, K. (2007). Humanitarian logistics in disaster relief operations. *International Journal of Physical Distribution and Logistics Management, 37*(2), 99–114. doi:10.1108/09600030710734820

Kovács, G., & Spens, K. (2008). Humanitarian logistics revisited. In Arlbjørn, J. S., Halldórsson, A., Jahre, M., & Spens, K. (Eds.), *Northern lights in logistics and supply chain management* (pp. 217–232). Copenhagen, Denmark: CBS Press.

Kovács, G., & Tatham, P. (2009). Humanitarian logistics performance in the light of gender. *International Journal of Productivity and Performance Management, 58*(2), 174–187. doi:10.1108/17410400910928752

Long, D. (1997). Logistics for disaster relief. *IIE Solutions, 29*(6), 26–29.

Long, D. C., & Wood, D. F. (1995). The logistics of famine relief. *Journal of Business Logistics, 16*(1), 213–227.

Maon, F., Lindgreen, A., & Vanhamme, J. (2009). Developing supply chains in disaster relief operations through cross-sector socially oriented collaborations. *Supply Chain Management: An International Journal, 14*(2), 149–164. doi:10.1108/13598540910942019

Oloruntoba, R., & Gray, R. (2006). Humanitarian aid: An agile supply chain? *Supply Chain Management: An International Journal, 11*(2), 115–120. doi:10.1108/13598540610652492

Pande, R. K., & Pande, R. (2007). Resettlement and rehabilitation issues in Uttaranchal (India) with reference to natural disasters. *Disaster Prevention and Management, 16*(3), 361–369. doi:10.1108/09653560710758314

Pardasani, M. (2006). Tsunami reconstruction and redevelopment in the Maldives. A case study of community participation and social action. *Disaster Prevention and Management, 15*(1), 79–91. doi:10.1108/09653560610654257

Pettit, S. J., & Beresford, A. K. C. (2006). Emergency relief logistics: An evaluation of military, non-military, and composite response models. *International Journal of Logistics: Research and Applications, 8*(4), 313–331.

Saad, M., Jones, M., & James, P. (2002). A review of the progress towards the adoption of supply chain management (SCM) relationships in construction. *European Journal of Purchasing and Supply Management, 8*(3), 173–183. doi:10.1016/S0969-7012(02)00007-2

Schulz, S. F., & Heigh, I. (2009). Logistics performance management in action within a humanitarian organization. *Management Research News, 32*(11), 1038–1049. doi:10.1108/01409170910998273

Tatham, P., & Kovács, G. (2010). The application of swift trust to humanitarian logistics. *International Journal of Production Economics, 126*(1), 35–45. doi:10.1016/j.ijpe.2009.10.006

Taylor, D., & Pettit, S. (2009). A consideration of the relevance of lean supply chain concepts for humanitarian aid provision. *International Journal of Services Technology and Management, 12*(4), 430–444. doi:10.1504/IJSTM.2009.025817

USAID. (2002). *Foreign aid in the national interest: Promoting freedom, security and opportunity* (pp. 1–149). Washington, DC: U.S. Agency for International Development.

van der Laan, E. A., de Brito, M. P., & Vergunst, D. A. (2009). Performance measurement in humanitarian supply chains. *International Journal of Risk Assessment and Management, 13*(1), 22–45. doi:10.1504/IJRAM.2009.026388

van Wassenhove, L. N. (2006). Humanitarian aid logistics: Supply chain management in high gear. *The Journal of the Operational Research Society, 57*(5), 475–489. doi:10.1057/palgrave.jors.2602125

Vrijhoef, R., & Koskela, L. (2000). The four roles of supply chain management in construction. *European Journal of Purchasing and Supply Management, 6*(3-4), 169–178. doi:10.1016/S0969-7012(00)00013-7

ADDITIONAL READING

Cox, A., Ireland, P. & Townsend, M. (2006). *Managing in Construction Supply Chains and Markets. Reactive and Proactive Options for Improving Performance and Relationship Management.* Thomas Telford Publishing, London, UK.

Lee, C. W., Kwon, I.-W. G., & Severance, D. (2007). Relationship between supply chain performance and degree of linkage among supplier, internal integration, and customer. *Supply Chain Management: An International Journal, 12*(6), 444–452. doi:10.1108/13598540710826371

Pryke, S. (2009). In Construction, S. C. M., Ed.). Chichester, UK: Blackwell Publishing Ltd.

Safran, P. (2003, January). *A strategic approach for disaster and emergency assistance,* Paper presented at the meeting of the 5th Asian Disaster Reduction Center International and the 2nd UN-ISDR Asian Meeting, 15-17 January 2003, Kobe, Japan. Retrieved June 26, 2006, from http://www.adb.org/Documents/Policies?Disaster_Emergency/disaster_emergency.pdf.

Sheffi, Y. (2005). *The Resilient Enterprise: Overcoming Vulnerability for Competitive Advantage.* Cambridge, MA: MIT Press.

Thomas, A., & Kopczak, L. (2005). From logistics to supply chain management. The path forward in the humanitarian sector. *Fritz Institute*, Retrieved January 25, 2009, from http://www.fritzinstitute. org/PDFs/WhitePaper/FromLogisticsto.pdf.

KEY TERMS AND DEFINITIONS

Beneficiary Empowerment: Actions and processes that recover the decision-making power of beneficiaries.

Community Based Approach: An approach in disaster relief that emphasizes socially sustainable development through a focus on the involvement of beneficiaries and their social environment in relief activities.

Local Sourcing: Purchasing materials and services in, and close to the disaster area.

Reconstruction Supply Chain: A construction supply chain in the reconstruction phase of disaster relief. It converges construction supply chain(s) with needs assessment teams, donors and beneficiaries.

Relief Supply Chain Performance: How a supply chain operates and achieves its aims can be measured in terms of (cost and time) efficiency and effectiveness. Relief supply chains performance distinguishes between supply chain effectiveness related to the completion of a mission, and aid effectiveness overall. A further performance measure in relief supply chains is that of equity.

Supply Chain Design: Determination of the structure and configuration of the supply chain in terms of its members, degree of collaboration, geographical (facility) locations, and supporting systems.

Chapter 7
Local Sourcing in Peacekeeping:
A Case Study of Swedish Military Sourcing

Per Skoglund
Jönköping International Business School, Sweden

Susanne Hertz
Jönköping International Business School, Sweden

ABSTRACT

This case study explores the Swedish armed forces' sourcing from local suppliers in the area of the peacekeeping operation in Liberia. The chapter discusses why, what, and how the Swedish armed forces develop local sourcing. For the study, a theoretical framework was developed with an industrial network perspective based on three cornerstones: supplier buyer relation development, internationalisation, and finally, souring and business development in a war-torn country. The results of the study show that both implicit and explicit reasons to source locally exist. Every operation is unique, and therefore the sourcing needs to be tailored for each operation. Local sourcing was developed in the country based on existing needs and when opportunities arise. Theoretically, new insights of differences between business relations in military operations and normal business to business relations were gained. Practically, this study illustrates the importance to develop and diversify sourcing in international operations.

INTRODUCTION

The Last 20 years have seen a significant change in European military engagement. During the Cold War era most European countries focused their military efforts on homeland defence. Today participation in international peacekeeping operations have become an important factor for the military

planning. During the Cold War, local sourcing was a natural part of the homeland defence planning. Due to the change towards international engagements in peacekeeping, sourcing have become much more complex and difficult. Sourcing and especially local sourcing under these conditions requires a different approach and knowledge. Military conflicts requiring peacekeeping efforts

DOI: 10.4018/978-1-60960-824-8.ch007

are international. They mainly occur in regions where poverty is severe and small changes or disturbances can cause starvation and death. In conditions like these, the most urgent need is to save human lives and, in a more long-term perspective, extend aid to help the society/nation to build/rebuild the infrastructure and revitalise the businesses.

Most peacekeeping operations take a long time. Stable conditions must be reinstated and the parties in the conflict must realize that they will all gain from the prevalence of peace. The war economy has to convert to peace-economy. Crime incidence has to drop and provision of protection decrease in importance. Furthermore business development and job creation are important parts of the stabilization process. Military peacekeeping operations in foreign countries have demands and needs for goods and services. Therefore they should be able to make a difference to the local business as well as for the future business development. However, the forces mainly source their needs from their home country or from international networks.

There are few articles on military logistics in peer reviewed journals. The bulk of logistics literature describes, theorizes, or provides handbook information on large scale operations (e.g. Pagonis 1992; Kane 2001; Tuttle Jr. 2005). Studies on how small nations handle their military logistics are rare (e.g. Tysseland 2007, Markowski, Hall and Wylie 2010). Important work has been done about the growing private military industry (e.g. Singer 2008, Verkuil 2007). Previous literature focuses almost exclusively on international sourcing or sourcing from the home country. There is very little on local sourcing in military operations.

This study aims to explore the Swedish armed forces' sourcing from local suppliers in the area of a peacekeeping operation. The paper will discuss why, what and how the Swedish armed forces develop local sourcing. The study is based on a case concerning the Swedish participation in the UN-mission in Liberia. Data was collected during a field visit in September 2005 and from

interviews in Sweden. Further comparative data has been collected about Swedish operations in Afghanistan and Kosovo.

After the introduction, the paper presents a theoretical framework on post conflict development and relevant commercial theory, applicable to non-profit military environments. In section three, the paper gives some basic description on issues important for local sourcing of military operations. In section four the methodology and the empirical material from the Liberia-case are presented. Thereafter the empirical material is discussed in the light of the presented theory. Finally, conclusions and thoughts for future research are presented.

THEORETICAL FRAMEWORK

The main area to be discussed covers local sourcing in an international environment. We see the supply chains as a network of suppliers and discuss supplier development from that perspective (Lambert, Cooper and Pagh, 1998). The existing theories about sourcing in an international environment and supply chains in the form of networks come from the area of private business (Agndal 2004; Lambert et al. 1998). The intention is to adopt and adjust private commercial theories so they can be used to describe the sourcing of military operations.

The meaning of the term local sourcing requires some clarification. Based on international theory, the engagement of a military unit can be viewed as a temporary Greenfield venture or investment. The military unit starts up an operation in a foreign area where their services are requested, and they stay as long as their services are asked for by the international society. Their knowledge about the area is limited in advance to their arrival, but they know that that it exist a "business opportunity" in the area. Their services are requested by local organisations and/or the international society and it will stay this way until the conflict is solved and a stable peace is created. In the end the services will no longer be requested ant the military unit will close down and withdraw from the area. The sourcing of the venture can be divided into three types. Two of these are represented by the parent company i.e. national sourcing in the parent company's homeland and international sourcing. The third type is the sourcing performed by the Greenfield venture in the specific region where they are operating. This third type is what we call local sourcing in this paper. Local sourcing is linked to spatial proximity and synchronous deliveries which are applied in supply chain management and economic geography (Gullander and Larsson 2000). These aspects are also reflected by the requirements on a minimized military logistic footprint in the operational area (Piggee 2002).

Sourcing and Creation of Supply Relationships

Sourcing is mostly described in the literature as the forming of external relations with suppliers in order to promote and develop effective supply chains (Van Weele 2005). This view is too limited for describing all issues that are important for sourcing. Axelsson, Rozemeijer and Wynstra (2005) define sourcing as:

Sourcing essentially is a cross-functional process, aimed at managing, developing and integrating with supplier capabilities to achieve a competitive advantage. (Axelsson et al. 2005 p.7)

In their opinion sourcing includes both internal and external activities. Internal activities, they say, include organisational development while external activities comprise market research. It is necessary to take the internal considerations of both organisational and process character into account. This is considered to be important for military operations where other aspects can be more important than just the commercial parts of a supplier relation. In a public non-profit envi-

ronment, competitive advantage is exchanged by the ambition to get the best value for money. The definition may thus be read as follows: Sourcing is essentially a cross-functional process, aimed at managing, developing and integrating with supplier capabilities to achieve the highest possible cost benefit solution.

As for sourcing in peacekeeping operation, Kaldor (1999) points out that to achieve value for money it is important to achieve supplier relations that support the development of political legitimacy of the operation and the new government. Supporting the grassroots economy and creating new jobs are keys to lasting peace, supplier relations with small businesses are important for peacekeeping units (Korac 2006). The behaviour aspect of local sourcing is important in peacekeeping operations.

International sourcing requires proactively coordination and integration with suppliers (Trent and Monczka 2005). In their view, international sourcing requires internationally located organisations, which also include local sourcing in foreign operations. Local sourcing decisions (Steinle and Schiele 2008) and strategic service sourcing decisions (Nordin 2008) creates competitive advantage, which in a military operation could mean more value for the money. Sourcing requires development of supplier relations, which is an important ongoing research area (e.g. Pagano 2009; Wagner, Fillis and Johansson 2007; Wouters, van Jarwaarde and Groen 2007). We have chosen to focus on two models (Dwyer, Schurr and Oh 1987 and Gadde and Snehota 2000) as the basis for our study of creation and development of local supplier relations in peacekeeping operations. We argue that, since military operations have not been studied with a business sourcing perspective, we need to investigate if the military setting creates new questions to the "classic" models. It is also believed to be important to use a network perspective. Peacekeeping operations can by supporting the development of the supplier networks also support the peace-building process.

Dwyer et al. (1987) present a five-step development process for the buyer-seller relation. They hold that relations evolve in the following phases:

1. Awareness - seller and buyer identify each other and internally evaluate if the other party could be of interest;
2. Exploration – the phase of bilateral contact where the parties try out and present expectations to each other;
3. Expansion – the phase of increasing interdependence, where trust and increased risk-taking are typical signs;
4. Commitment – long-term relations in a deeply interdependent partnership; and
5. Dissolution – disengagement or withdrawal from the relation.

The model is not directly adoptable to create business relations as in the type of environment destroyed by war. It will require some modification on detail level to make it suitable. In the peacekeeping operations, the political dimension is very important. It affects roughly all phases in the relation model. Especially the commitment phase will require aspects that are partially different from what a regular long-term business relation requires. Impartiality and integrity of the military forces must be taken into consideration (Boutros-Ghali 1992). It is important for the military unit not to favour one side of the conflict and not to accept any type of corruption, since as mentioned above the building of political institutions and public authorities is important for the de-escalation of the conflict. Handling relations the wrong way can hinder this development. Long-lasting deep relations with one single supplier can be seen as partial and might create problems (Ebersole 1995).

According to Gadde and Snehota (2000) sourcing can also be discussed in terms of managing a supply network. They see relationship involvement (high or low) and number of suppliers (single or multiple) as key parameters for the sourcing policy. The choice between the different parameters depends on the situation of the buying organization and what types of suppliers are available. Single sourcing with high involvement is regularly recommended for long-term business relations. There are reasons to have more than one relation with high involvement, though. The reason can be related to customer requirements or limitations in supply capacity. Single source policy with low involvement is a possible solution to reduce administrative cost, achieve economy of scale, or in some cases it may be the only possible solution. Finally, multiple sourcing with low involvement is the "traditional" or transactional way of purchasing. There are arguments that this solution may entail risk reduction or promote a competitive situation.

Does this model apply and if so how can this model be applied for sourcing in an environment where peacekeeping operations are taking place and under what circumstances? This will depend on the buying organisation in the peacekeeping unit, available local businesses and how war thorn the area is. Peacekeeping operations mostly take place in severely war-thorn areas. This situation will influence the number and size of businesses available on the local market. In countries where peace recently has been established, many new small businesses have shown are likely to appear.

The model does not explicitly take the size of the suppliers into consideration. A high involvement solution in the model is recommended, and is linked to the rule that deep relations require management resources and are costly. In the case of small firms (1-10 employees) this is not so apparent, and small suppliers may not be able to supply all the requested products.

Sourcing in an International Environment

With local sourcing we mean the sourcing in the area of the military peacekeeping operational theatre, delineated by e.g. UN Security Council decision. In terms of business relations, local sourcing by the deployed units can be regarded as an act of internationalization. In international operations, the Swedish armed forces are required to be able to act, up to 5000 km from Sweden. The commitment to undertake an international operation by the armed forces is believed to be relative to homeland security, the higher the importance for homeland security the higher commitment. We argue that the Uppsala-model and the concept of psychic distance can be used to describe these activities. Johanson and Vahlne (2009) demonstrate that the model still is valid for the internationalisation process, and also is useful concerning professional services, which are important in a military context. The Uppsala-model is often viewed as behavioural, compared with other models which have an economic focus on the internationalization (Buckley and Casson 1976) or the eclectic theory on internationalization (Dunning 1980). The Uppsala-model is a process description of internationalization. The model is based on four cornerstones, market knowledge, market commitment, commitment decisions, and current activities, where the two first are looked upon as state aspects and the second two are change aspects (Johanson and Vahlne 1977). Market commitment is the amount of resources used for a certain market and the difficulty to move them to another market. Market knowledge is based on Penrose (1959), where she divides knowledge into objective and experiential. The experiential knowledge is to a large degree country specific, and needs to be obtained at the location. Johanson and Vahlne (1990, 2003) integrate the Uppsala-model with the industrial network approach, explaining the relations with other firms in the current activities as network interaction. In general terms the Uppsala-model explains two patterns; first how the engagement in a new market develops, and second entrance to new markets with successively higher psychic distance. The focus of the Uppsala-model is on marketing and not purchasing, but we think it is relevant to have both these patterns in mind when the armed forces are ordered to join an operation

in a foreign country. Research investigating sourcing or upstream activities as a part of the internationalisation process (e.g. Agndal 2004; Ghezán Mateos and Vileri 2002) is gathering momentum. Since participation in international operations is a political decision, the armed forces can not choose where to go. Anyhow, we believe that the Uppsala-model can explain how the armed forces act when going in to a new area of operation and consider local sourcing.

Psychic distance on the organizational level is defined as the difference in perceptions between buyer and seller regarding either needs or offers. Three determinants form the concept of psychic distance; trust, experience, and cultural affinity (Hallén and Wiedersheim-Paul 1982). Experience on the individual level causes preconceptions regarding the suppliers in the foreign country. This will affect both, executive level when sourcing directive decisions are taken for an operation and the operational level when local supplier contacts are taken. Cultural affinity expresses the effects of differences in different aspects as legal environment, business habits, language, and cultural environment. The relevance of psychic distance as a concept has been questioned (e.g. Stöttinger and Schlegelmilch 1998). It can not alone explain differences in performance, but it has shown to be a helpful predictor of corporate performance (Evans, Treadgold and Mavondo 2000). Swift (1999) chooses to look at the issue in the opposite direction, and talks instead about psychic or cultural closeness. He points at finding personnel with cultural similarity, pre-training should focus on similarities, and finally the personnel should be made aware of important cultural differences. O'Grady and Lane (1996) argue that psychic distance can be of much larger magnitude than one could expect, and it is therefore always of importance to analyze the parameters before entering a new market. The psychic distance and its three determinants of trust, experience and cultural affinity is believed to be important for armed forces' willingness to source locally.

Post Conflict Development

When researchers talk about peace-building, they discuss how to rebuild a sustainable society. There are many different aspects to include, building state institutions, develop local governance, re-establish rule of law, reconstruct infrastructure, create business etc. (Kaldor 2003; Gotthand and Humle 1999). To create a lasting peace, it is important to identify which dominating factors could stabilise a weak peace and, Korac says, underscore: " ...the need for long-term peace-building initiatives that should critically involve identifying and supporting local capacities for peace" (Korac 2006, p. 518). Privatisation of public property is often done to generate resources to an under funded post-conflict government. The problem with privatisation is that in most cases the only parties that have capital to buy state owned property are the people that have prospered on the war-economy (Kamphuis 2005), the war-economy dominated by theft, drug production and trade, sex trade, and security services. Criticism of the neo-liberal economic ideas dominant in peace-building operations is common among researchers (e.g. Kaldor 2003).

Kaldor (1999) argues that the reconstruction of a community must begin with organizing political institutions and governmental authorities, and these issues must come hand in hand with the development of local businesses. The problems in the post conflict society are many: weak government, high unemployment rates e.g. 65-75% in Bosnia and Herzegovina (UNHCR, 1997), demilitarisation of combats, destroyed infrastructure, and a war-economy. There is no single solution to all problems but the basic needs of humanity do not change and are important to take into consideration.

Most important in post conflict development is to secure peoples' needs to get jobs, to support themselves with food, clothes and shelter as well as protection against criminals (Maslow, 2000). The final step in Maslow's hierarchy of needs is self-actualisation, which is vital factor in a developing economy and post-war society. Creating business relations, giving micro-loans to support new businesses is by many believed to be the sustainable development that will hinder future conflicts. These activities are commonly regarded to support the grassroots economy (e.g. Korac 2006).

It is also commonly held that foreign aid organisations, Non-Governmental Organisations (NGOs) and peacekeepers can hinder peace-development. Groups of the society get their income from the support they offer to these organizations and therefore they will strive to keep them in the former conflict area (Kamphuis 2005; Kaldor 1999; Macrae and Leader, 2000). Without knowledge on how to do local sourcing the possibilities to do wrong are many. The problem appears to be of importance when an NGO or peacekeeper becomes a dominant customer in the area of operation. It is essential for local sourcing to avoid disturbing the peace-process. Done in a circumspect way, it could instead contribute to reduced unemployment by means of the development and growth of local businesses, and thus to the rebuilding of a sustainable society.

Type of Products

The variation of products needed for a peacekeeping operation is huge. US (10 classes) use more detailed product categorisation compared with NATO (5 classes) (Foxton 1994). In Sweden the NATO product definition is used (Försvarsmakten 2005). The NATO definition of categories is too broad and brings together very different types of goods and does not meet the needs for a detailed discussion. The US 10 classes are internationally used and give a usable categorisation to discuss what to source locally. The ten classes are: Class 1: Sustenance and Water, Class 2: Individual Clothing and Equipment, Class 3: Petroleum, Oils and Lubricants, Class 4: Field Fortifications and Supplies, Class 5: Ammunition, Mines and

Explosives, Class 6: Re-sale and Personal Items, Class 7: Major Equipments, Class 8: Medical and Dental, Class 9: Spare parts, Class 10: Aid to Civilians.

In military applications the classes have previously only been used for physical products. To get a complete picture about local sourcing it is necessary to use the Porter (1980) definition of products which includes both physical products and services. These classes differ from each other not just in their physical context but also if they are strategically important, if they are standard products or commodities, or bottleneck products (Skjøtt-Larsen 1999; Weele 2005).

Implications for the Empirical Parts

Building supplier relations are positive for the peace-building. Dwyer et al. (1987) and Gadde and Snehota (2000) give a theoretical baseline for studying the development of relations. The Uppsala-model is believed to be able to explain how the engagement of the armed forces in local sourcing in a new operational area develops based on the four operational parameters; market knowledge, market commitment, commitment decisions, and current activities. The psychic distance affects the willingness to supply in a new area of operation.

To rebuild a society, local businesses should not have aid organisations as their dominating customers. Important for the rebuilding of a society is the support of local businesses. Close supplier relations will ease the possibilities to give suppliers management and technical assistance.

THE CASE STUDY

Method

This paper presents an exploratory case study. The empirical material used in the study derives from 16 interviews in Sweden during the period from 2004 to 2007, documents concerning Swedish international operations, and a one-week field visit to Liberia September 2005. The field visit included 6 interviews of personnel from the unit and 3 interviews of suppliers, and one week of observations. The observations covered internal meetings, supplier meetings, reconnaissance activities and corridor chats. The field visit was at the Swedish National Support Element (NSE), the organization responsible for supporting the operating unit with maintenance, supply, services and a base camp. A limited amount of comparative data has been collected about other operations, mainly Kosovo and Afghanistan. For this type of study the construct validity is of special interest. Important for construct validity is the use of multiple data sources, chain of evidence and review of results (Ellram 1996). The main source of data for this paper is the field visit in Liberia. The interviews in Sweden had two purposes, first to get general data about sourcing for international operations and secondly to be able, through triangulation, to verify and validate the data from the Liberia visit. Documents have also been analysed to complement or verify the results. Reviewers are more adverse in their reaction when reviewing individual data compared with aggregated data (Yin 1981). To review the results of the study, the empirical content of the paper has been used instead of results from individual interviews. The review has been performed by experienced logisticians from the Swedish armed forces.

The Swedish Armed Forces Sourcing and Peacekeeping Operations

The Swedish Armed forces do not have a tradition to source locally. Until late 1990's the Swedish Armed Forces were organized for homeland defence. Supplying international operations were handled mainly by circulating some of the enormous stock held to equip a million man-strong conscript army. Changes in the doctrine during the late 1990s have reduced the volume of stored material and new ways of sourcing international operations have been discussed since then. Another reason to discuss new ways of sourcing is that the costs of international operations are soaring. Previously, when a portion of the material in stock would have to be circulated the additional costs were limited. Today much of the materiel used in an international operation has to be purchased.

Local Sourcing of the Swedish Armed Forces in International Operations

What is the volume and type of material and services the Swedish armed forces source? This depends on the type of operation, whether it is an UN-mission or an operation that is undertaken under another organization's command, such as NATO. UN-missions differ from other operations in the sense that UN supplies the participating units with some consumables, e.g. food and fuel. Other operations differ with some getting most of their supplies from another unit or being supplier for all units in an area. The normal situation for the participating nations is supply their own troops. Variances in sourcing where units from different nations get supplies from each other, is normally an important part for the logistic concept for the deployed units. The three operations mentioned in this paper differ in the quantities that are being sourced. For example in the case of the Liberia-mission, under UN-command, some of the sourcing was done together with the Irish battalion which the Swedish unit was a part of. The operations in Kosovo and Afghanistan are NATO-lead, in which the Swedish unit is responsible for its own sourcing.

One of the main issues for the logisticians planning the operation is to identify which resources are critical for the operation. In this sense critical means that without these resources the unit can not do the tasks it is assigned to perform within the area of its operation. A unit of 250 men in a ground operation has normally more than 1000 different unique types of consumables and equipments, spare parts excluded. It is easy to picture that some of these equipments, though few, are important for the operation. There are also consumables, like drinking water, which also are critical for the soldiers' capability to do their job. The main part of the equipments and consumables are not that important for the short-term capability of the unit. All non-critical items will though be critical in a long-term perspective. The time span that passes for an item to become critical differs. The criticality issue must be handled when planning sourcing for the operation.

Politics is an important factor for peacekeeping operations. Both the host nation and nations in the transit line for goods and personnel can have political reasons to hinder the operation. The capacity, possibility or capability to deliver to the unit in the area of operation is an evaluation criterion for the sourcing solution. When sourced from the home country there might be limitations in transportation capacity or problems to transit the goods through customs in transit countries or to get the resources into the area of operation.

When sourced from a local supplier there might be problems to guarantee supplies due to the fact that the suppliers themselves have problems with regularities of supply. There might also be issues of political character affecting the deliveries from a local supplier. The Swedish armed forces expect the risk of not getting the supplies is generally larger when dealing with local suppliers than getting the supplies from outside the area of operation.

The Mission in Liberia

The Liberia mission was UN-lead, which affected some product areas. Fuel, water and food were supplied by the UN. These products can otherwise be of importance for local sourcing.

Reasons to Source Locally, Market Knowledge and Commitment

When the Swedish unit arrived in Liberia, the situation was relatively stable. The risk of political hindrance of Swedish logistics transports or local purchases was small. The tropical climate put special requirements on the equipments and consumables. It required higher quality, special packaging and just in time deliveries of some of the products. The distance between Sweden and Liberia was long; it took about one month to send goods by sea. The air freight was limited, in most cases requiring chartered flights. The international package delivery companies had not yet established business in Liberia. The new logistics doctrine (Försvarsmakten 2005), which focuses more on international operations, has not been totally applied yet. This document does though require an efficient supply base and a minimized logistics footprint in the area of operations.

The intention was to be fully equipped from Sweden. Gradually, plans to source from the local market were established as needs arose and the deployed unit began to understand the environment in Liberia. The Operational Head Quarters (OHQ) had a strong commitment to international operations, and the personnel had international experience. But they had no new regulation about how to source from the local market. The routine that the forces used was; the NSE would get a limited budget for local purchasing, and if purchasing passed that limit, a request for approval would have to be sent back Sweden. On the national level in Sweden no activities were done to support or encourage the NSE to supply locally. On the contrary details reveal that individual persons in Sweden acted to stop the NSE from sourcing locally. Motives for this could be fear of being made redundant in Sweden or lack of trust in the professionalism of the personnel in Liberia.

The knowledge about the situation in Liberia before arriving with the unit was limited. It was based on a visit to the Irish battalion in Liberia and

information received from the UN, complemented with internal intelligence reports. The situation developed also very quickly as the situation stabilised and the population returned to "normal" life.

The development of local support and supply of the camp was handled through some of the local retailers and locally employed personnel for unqualified work and through Swedish personnel on site.

Furthermore the Swedish armed forces had no requirements for the Swedish or international suppliers concerning subcontracting of local supplies or services. The day to day contact and requirements were handled by the NSE with all three categories. The NSE could decide on using local sourcing instead of using the contracted Swedish or international suppliers.

When doing business, the Swedish unit did not know how to act. On one hand they were afraid of disturbing the growth of the community by buying off essential items and emptying shop shelves and thereby causing shortages of consumables necessary for the community's growth and at the same time fuelling inflation, which in turn would slow down the restoration of the community. On the other hand they knew intuitively that they would help business growth by buying some items locally.

When discussing with the officers in Liberia one could understand that even if there were a number of underlying factors to purchase locally, they had a backbone feeling that what they were doing must have some value for the spent money. All the other factors in the relatively calm situation that existed in Liberia were subordinated to this requirement. The NSE was aware of the fact that they would be blamed if the operating unit did not get what they needed. Because of the blame game, the NSE argued that they needed safeguards if deliveries from Sweden failed to reach the area of operation. It could be late deliveries, blocked borders, custom delays or a clogged port or airfield. Except for the risk of late deliveries, all other problems were in this case given a very low probability to occur. Late deliveries could occur due to limited resources in stock in Sweden or lack of transportation capacity, within the timeframe needed. Local sourcing was one way of limiting this risk.

The personnel in Liberia found that many necessities could have been purchased locally. The unit's main problem was that it did not know when or how much it could source locally. The unit's personnel were afraid of disturbing the local economy, and they did not want to be blamed for the emergence of new conflicts and delayed growth. They tried to buy things that they thought had little or no negative effect on local business life. The reasons to source locally depended also on the type of product, which will be discussed below.

Development and Selection of Relationships

The NSE developed their local purchasing by combining patrol assignments with learning about the slowly growing local business life. During the patrols the soldiers visited firms, stores, retailers and restaurants. The business life changed quickly due to the fact that many businesses had closed down during the civil war. The patrolling was discussed in meetings in which new and old firms were identified, shops or restaurants were presented and evaluated and follow-up visits were decided on. Different officers were assigned to tour different businesses. For stores and repair shops that handled maintenance-related issues a group from the spare parts and maintenance team did the visit. For general stores and wholesalers the supply team did the visit. Travel agencies, restaurants, and other recreation suppliers were visited by the travel, recreation and healthcare managers. In other words, the unit had organized a way to follow the development of the local business, and had also organized some sort of evaluation criteria for whom to buy from. By visiting the stores and retailers, they slowly developed business relations and thereby formed a small network or suppliers

for consumables, spare parts for standard vehicles and air-conditioning equipment. To maintain the relations with the suppliers they regularly bought small amounts. The objective was not really to supply their operation but to have the possibility to buy in case of shortage of deliveries from Sweden. The only goods which had their main sourcing locally were air-conditioning equipment. The local restaurants were regularly visited by the unit, when having official guests and by soldiers on leave. The restaurants were assisted and educated to improve kitchen and water hygiene. The NSE kept informal control over the visits so that the recommended restaurants would get roughly the same number. Generally, they had an ambition diversify relations within each product area.

During interviews they said they had no form to fill in about the suppliers but their way of describing the visits showed that they had developed a way to evaluate a supplier and to develop the business relations over time.

Cultural Closeness and Trust

The soldiers going to Liberia had personal motives to engage in different activities in place in the area of operation. They had not received any training or education on how to handle local suppliers before going to Liberia. Neither had they any documented guidance on how to plan local supply. The drafting process in Sweden was based on volunteers. Motivated soldiers created a cultural closeness to the people in the area of operation. The unit used its own capabilities and ideas to supplement deliveries from Sweden with local possibilities to supply. Lack of knowledge on how their engagement with local business should be done was very obvious. The soldiers/officers had in most cases no commercial business background, and most soldiers were young and inexperienced. The officers in Liberia had great concern on the possibility of the local sourcing causing problems for the peace-development.

The unit's personnel cared for the population and its situation. During their off duty time and on leave vacation in Sweden, the Swedish personnel collected money and helped building an orphanage in one of Monrovia's suburbs. It was not possible to evaluate if this activity had an effect on business relations, but during the visit one could sense a kind of respect for the local population, and their will to change their own situation. The personnel believed that most suppliers were fair and they had also recognized that Swedes had a good reputation from old mining businesses before the civil war.

According to Swedish laws the unit was required to adhere to and local rules and realities of life, which in some cases were impossible to pursue, contradicted each other. The most outstanding issue was related to garbage where the Swedish unit was under obligation to treat waste as well as sewage in the same way as in Sweden. In Liberia the Swedish unit had a contract with a supplier that agreed to take care of waste and sewage according to the requirements of the Swedish law. The representatives from the unit had their doubts that it was handled according to requirements in the contract.

Type of Products Sourced in Liberia

What to source in Liberia was mainly decided by the NSE. The influence from the OHQ in Sweden was limited. The complexity of services and physical products was mentioned above. To discuss what to source locally it is therefore necessary to discuss different product classes. The ten classes presented above will be the framework to present how they did in Liberia.

1. The first class is sustenance and water. This class was not supplied locally by the Swedish unit. This class was supplied through UN and the Irish battalion. Rules about hygiene and quality are factors that make it complicated both to source locally and to source from operating partners, since requirements differ between organisations and countries. Especially in warm climate like Liberia, water is a product which requires much transport resources if not produced locally.

2. The second class is individual clothing and equipment. Much of the products in this class are special products and the soldiers were equipped when debarking from home country. These products were supplied from the military stores in Sweden. These were very rarely resupplied locally.

3. The third class is petroleum and lubricants. In Liberia UN delivered petroleum, while oil and lubricants was a national responsibility. The unit did look for local suppliers for oil and lubricants, but in most cases it was difficult to find the same or equivalent products with the right quality. In this area things improved every month with more possibilities to supply locally as new maintenance workshops opened.

4. The fourth class is field fortification supplies. In this group the possibilities were good to source locally and much of the materiel needed that was not brought in with the first shipment was sourced locally.

5. The fifth class is ammunition and explosives. All deliveries in this product group came in with special flights from Sweden.

6. The sixth class is re-sale and personal items. This area was handled in Liberia by a contracted company, running the px-shop in the camp. No requirement was put on that company on how and where to source its products.

7. The seventh class is major equipments. Nothing in this product group was sourced locally.

8. Class eight is medical and dental items. All products in this class were delivered from Sweden. This was a requirement by the OHQ.

9. The ninth class is spare parts. The ambition in Liberia was to buy spare parts locally. If

the spares were attainable locally, the lead times was considerably reduced.

10. The tenth class is aid to civilians. In Liberia the Swedish unit was not engaged in organized aid to the civilians. Among the soldiers they did though raise funds for building an orphanage. All resources to build the home were locally purchased.

In Liberia the infrastructure was totally destroyed after the civil war, but as the society recovered more and more was possible to source locally.

ANALYSIS

Three questions were raised in relation to local sourcing in the introduction. They are: why, what and how, peacekeeping units source locally.

Why Source Locally?

The theoretical framework on post conflict development stresses the need to support grassroots economy and new businesses. At the same time it sends warning signals with regard to the persistence of the war economy. Peacekeeping forces and aid organizations can hinder its transformation into a peace-economy (Kamphuis 2005; Kaldor 1999). Consequently, the first and possibly most interesting question is if there are any strategies about sourcing of the Swedish armed forces.

The new doctrine does not discuss how to supply the armed forces; it only puts requirements on being efficient, effective, and to meet the operational units' requirements. In peace-support operations, only the security aspect is discussed in the doctrine. The peace-building aspects of sourcing are not considered. Two reasons for the absence of strategies concerning local sourcing can be identified. First, the military focus is to cooperate with other military units in international operations to provide security. Other aspects of

peace-building are other authorities supposed to handle. Secondly, the interviews indicated a limited knowledge concerning local sourcing and how it can contribute to peace-building, among officers within the OHQ. This relates to psychic distance and a lack of coping with the psychic distance (Hallén and Wiedersheim-Paul 1982; Grady and Lane 1996).

In the autumn of 2005 the OHQ in Sweden had only a limited strategy for the local sourcing in Liberia. The NSE had some small funds for conducting limited sourcing locally. OHQ decisions were only made in the request of the NSE. The strategy for local sourcing was developed and decided by the NSE and in some cases confirmed by the OHQ in Sweden. In Liberia it was apparent that sourcing from the local market in the area of operation was only of concern for the NSE. Liberia is about 6000km from Sweden. The Swedish force contribution was not aimed to be a long-lasting engagement. The top management within logistics in FM HQ had little knowledge about the local market in Liberia and how it developed. Based on these facts, the Uppsala-model (Johanson and Vahlne 1990) hints at low management involvement to source locally, indicating its limited scope and that in fact has been the case. The only driving force according to the model was current activities in Liberia. In the NSE activities were continually ongoing to find new ways to supply some of their needs from local suppliers, and that is why the sourcing strategy was developed locally.

The Liberia case indicates several reasons to source locally. The reasons in the Liberia case were few but rather obvious: First, the NSE needed a safeguard if deliveries from Sweden failed to reach the area of operation. For example it could be late deliveries, blocked borders, custom delays or a clogged port and airfield. The second reason was to supply the unit with products which not was possible to get from Sweden, such as local services and locally adapted products. The theoretical aspect of proximity to suppliers is important to minimize the logistic footprint. The logistics doctrine supports local sourcing if efficient and also to minimize the logistics footprint. Local sourcing was a way to connect to the local community and thereby stabilize the situation, or in other words de-escalate the conflict. Peacekeepers can hinder peace-development through local sourcing if they support only one ethnical group or warranting fraction (Macrae and Leader 2000). In the Monrovia area, no important conflicts between different groups in the population existed, which made it less problematic to source locally. Still the unit had the opinion that it was important to buy from more than one supplier. The argument was, in line with the theoretical framework, to avoid becoming a dominant customer and not to favour just one supplier in a way that could cause negative effects on the society. If a peace support operation is successful, the security problems are reduced, which opens up for the possibility to increase the contacts with local business. Local sourcing can be seen as a second step for military units to stabilize the situation.

To sum up, there has been only a limited strategy concerning local sourcing on the Swedish national level. The lack of knowledge about different aspects of local sourcing is believed to be the most important factor for the short term strategy. On the unit level the strategies have mainly been based on the needs to safeguard supply and connect to the local society. Meanwhile, the possibilities for local sourcing have not been fully developed. Even if there have been activities on unit level to support the local business, there is little awareness about how to support the post conflict development. Two key challenges for the future local sourcing in peacekeeping operations are to get OHQ involvement and to improve the professionalism in the NSE.

What to Source Locally?

As mentioned above, the logistics doctrine's main concerns are effectiveness and efficiency. So when it comes to what to source locally, each operation will arrive at decision emanating from its unique situation. In the theoretical framework, ten product classes were presented as a way to structure the discussion about what to source locally (Foxton 1994). No guidelines exist in military logistics models on what to source locally.

The first consideration can be to turn the question around; what should not be sourced locally? The second class, personal clothing and equipment should normally not be sourced locally. First of all, as foreign military you should be identified as such, local sourcing can reduce this clarity. Secondly, for military personnel, the correct personal equipment can be a matter of life and death. The fifth class, ammunition and explosives, should never be sourced locally. Swedish regulations are strict concerning arms control and handling of ammunition, which makes it more or less impossible to purchase locally. One of the main aims with peacekeeping operations is to reduce the availability of these types of products on the market; local sourcing would have the opposite effect. The seventh group, major equipment is not possible to source locally, from a practical point of view. They do not exist on the local market. The units need to train on the equipment, before going to the operational area, why it is also very impractical to source locally. The eighth class, medical and dental items, is normally a scarce resource in a conflict area. Medical and dental care follows very high standard requirements in Swedish units which the local market normally cannot meet.

This leaves us with six classes possible to source locally.

- The first and third classes, food, water and fuel, are the most bulky supplies needed in military operations. Military units have always strived for sourcing these locally to reduce the needed transportation capacity. In modern time the requirements on quality have become more and more important. This reduces the possibilities to source locally. In Liberia the UN mainly handled this area, but in other operations, e.g. food quality and hygiene are major concerns when choosing suppliers. There are good possibilities though to combine development aid and sourcing in these areas, e.g. creation of water purification capability in an affected country can be both self-help and development aid in the peace-process.

- The sixth class, re-sale and personal items. Much of the supplies in this area are the small luxurious things that are important for the unit's morale. The availability of these products is limited on the local market, so putting requirements on the international provider of the service, running the px-shop, to source some of their products locally seems to be unreasonable.
- The ninth class is spare parts. Local sourcing in this area has only positive effects. If the spares were attainable locally, the lead times were considerably reduced. Newly started maintenance workshops would get some extra revenue from the handling of spare parts, which helps the development of the peace-economy.
- The fourth class, fortification, and the tenths class, aid to civilians, require normally no large volumes. These areas can normally be locally sourced. By sourcing construction material and services the local industry, the production and distribution are supported in the start-up phase after the war. Aid to civilians can differ a lot but generally it is limited tasks where the needed goods can be found on the local market. Larger aid to the population is handled by other civilian organizations.

As we can see, it is possible to define which areas are suitable to source locally. Naturally the possibility differs depending on how developed the society in the area or operation is. In Liberia the infrastructure was totally destroyed after the civil

war, but as the society recovered more and more was possible to source locally. The division into these ten classes and the experience from Liberia give a detailed indication about products suitable to source locally. The Swedish operations in Kosovo and Afghanistan also support this conclusion. It is reasonable to argue that it would be beneficial for the OHQ to produce a sourcing strategy based on the ten product categories.

From another perspective, in Liberia the local sourcing could be divided into goods and services that are difficult to get from Sweden, and safeguards in case of delivery disturbances. In the first group are air-condition equipment and restaurant visits typical examples of supplies. In the second group the most important supplies are general spare parts, hand tools, and travel agency support. Some goods were supplied by UN in Liberia, e.g. fuel, food and water. In comparison, in Kosovo and Afghanistan other types of products were sourced locally. A tailored sourcing solution is needed for each operation. It depends on what is available on the market, how the market develops, the possible disturbance of the peace-building and what is needed for the operation.

The local sourcing is dependent on the situation. This is in line with Gadde and Snehota (2000), reasoning about availability of suppliers. Our findings point to the fact that sometimes the relevant supplier only exists on the local market. In other cases the numbers of suppliers are limited or the size of the single supplier is too small to deliver the needed volumes.

How to Source Locally?

Sourcing could be discussed in terms of internal and external activities (Axelsson et al. 2005). Internal activities are described as human resource development and development of organizational mechanisms. In Liberia there was evidence for some internal activities. At the NSE they had developed routines for learning how to deal with the local market, through visits and internal meetings with concerned officers for briefing and decision-making. During internal meetings, the process solution to having the local market as a standing point on the agenda guaranteed that all involved officers got the information they needed and that they could discuss the handling of suppliers. There was a lack of Internal activities on the headquarter level in Sweden. The rules about how to source locally had not changed much since the time before the change of doctrine, and the personnel going to an operation did not get any education about how to handle local suppliers. The external orientation with market research and visits was handled by the NSE. The OHQ in Sweden did only marginal activities before the deployment of the unit.

Figure 1 describes how the relation between the NSE and the suppliers developed over time. The relation that developed between the NSE and the suppliers did follow the first step in the model suggested by Gadde and Snehota (2000). The relation normally started on a low level involvement and gradually grew into a relation with a higher involvement. In some cases though more suppliers were added and initially the involvement stayed on a low level. However in the third step, most relations became multiple relations with high involvement. The explanation of the three steps is fairly simple. Firstly, it depended on how the business developed in different areas; secondly for risk reduction reasons it was seen as necessary to have more than one supplier if possible: thirdly the high involvement was not only based on pure business reasons but also a way of learning about the security situation in the country, and finally there was a need to try to treat most suppliers equally to avoid getting accusations of favouring certain groups in the community. Gadde and Snehota's (2000) model suggests high involvement with a single source as the final step. The model was of course not originally discussed and developed for environment of a peacekeeping operation, which in the beginning involved very small businesses and also a small military unit (in Liberia around 250 soldiers). In this situation the meaning of high and low involvement becomes different, since most businesses are relatively small and newly established. Low involvement means that the purchaser has learnt enough to know about the supplier's business and what types of products are being sold. High involvement means that the purchaser has learnt to know a little bit more about the business and its owner and can, to some degree, engage in a relation with him/her, where at least the owner and the officer know each other by name. Important is also that the soldiers had time to socialise with the local society.

When it comes to development of relations, the theoretical model (Dwyer et al. 1987) adjusted in the way discussed in the theoretical framework above, fits very well with the empirical results from Liberia. Two phases differed from the original model. In the commitment phase, the forces aimed at, if possible, getting more than one close relation in the same business area. Since the local market was immature – often only one supplier was available – the NSE avoided getting a too close relation, since it was in their interest to get more than one supplier and that they tried to have this in mind when handling the suppliers. The reason could also be attributed to the fact that in the final phase, they did not want to get into close relationships. The final phase, dissolution, differed in two senses from the normal model. Firstly the local market was immature with newly started small firms, which could disappear very quickly. Secondly, the Swedish force had been prolonging its mission in Liberia on a yearly basis. The relatively short-termed and unstable relation had obvious effects on the final phase and hindered the development of the commitment phase.

The Uppsala-model and the concept of psychic distance seem to be useful in this specific environment. The model can be interpreted, that lack of corporate strategy will limit local sourcing, which was the case in the Liberia-mission. The model implies that firms can decide on which country

Figure 1. Local supplier relationships development in peacekeeping operations

Posture of relationship   Sourcing Policy

		Single		Multiple
High involvement		2	⟶	3
		⇧		⇧
Low involvement		1	⟶	(2)

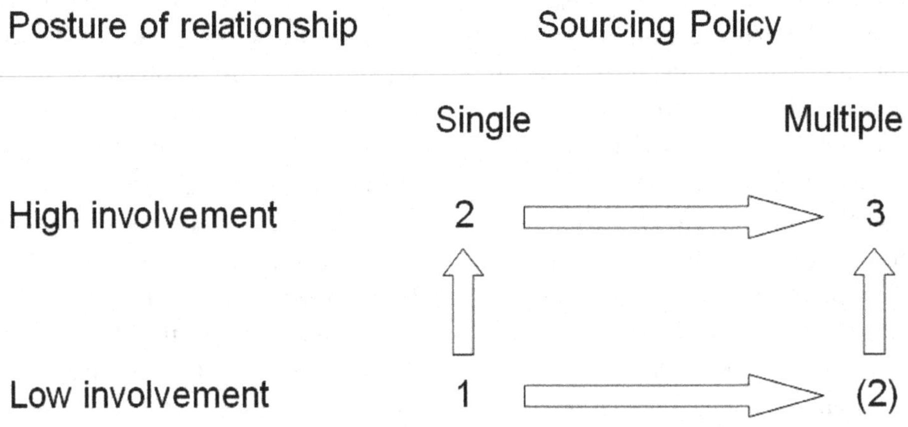

to enter. In contrast, the Swedish armed forces do not choose in which conflict and country to participate. Participation in a peacekeeping operation is a political decision that is taken on grounds where sourcing of the operation is of no relevance. This means that the Swedish armed forces cannot minimize psychic distance in their operations.

Hallén et al. (1982) see psychic distance as a concept to be used on an organizational level, which fits into this case. This can be mirrored in the sourcing decisions on top management level. Analyzing the psychic distance can firstly identify possibilities to reduce or understand differences and secondly support top management decision-making in what to source locally, regionally or from the home country. Swifts' (1999) idea about decreasing psychic distance can be adopted at personal level. In peacekeeping operations where the units are formed voluntary, there is a certain level of psychic closeness due to the fact that they are on voluntary basis. According to Swifts' argumentation, this has a positive effect on the business relation building.

The concepts and models applied turned out to be useful. But, the models need to be adapted, concerning supplier relations to the unique situation in peacekeeping operations.

FUTURE RESEARCH DIRECTIONS

This study is too limited to draw any general conclusions about the Swedish local sourcing in peace support operations. It is necessary to collect empirical material from other Swedish operations as well as from other nations and compare with the results from Liberia. Even if some comparative data have been collected from two other operations, it is not enough to generalize the results about Swedish local sourcing and apply them to other small nations' local sourcing of peacekeeping.

One challenge for future research is to identify criteria for selecting local suppliers, due to the problems with corruption and criminality (military SCR) and requirements on impartiality in the conflict.

Another related area of research of importance for local sourcing is that if local sourcing has an impact on the society, whether the businesses will grow faster or if the sourcing has a marginal positive effect or in worst case only causes shortages of important consumables for the growth of businesses, and thereby hindering the growth of the peace-economy. This study only indicates that it exists possibilities to support the growth of the local economy.

A final important question could be, if the local sourcing has a stabilizing effect on the society. The sourcing activities force the soldiers to get more involved with the local population. This opens up new ways for information, which can support the peacekeeping.

CONCLUSION

Our purpose with this paper is to explore why, what and how the Swedish armed forces develop local sourcing in peacekeeping operations. The empirical base for our study is the UN-mission in Liberia.

Why? On unit level it is possible to identify both explicit and implicit reasons for local sourcing. As for explicit reasons, the obvious risk is being far from home and that deliveries could not only be delayed but also hindered and rendering alternative solutions necessary. Services needed are often only possible to source locally; alternatively it has to be done with own resources, which will affect the operational capability.

Implicitly, a peacekeeping unit shall do activities that de-escalate the conflict level. Local sourcing helps the businesses to grow, which has the intended stabilising effect. The situation differs when compared to regular private business relations. Local sourcing is something that is useful and important to spend time on when other activities are on a low level in a peacekeeping operation.

What? Every operation is unique, and therefore the sourcing needs to be tailored for each operation. Detailed tailoring of local sourcing is possible by using product categorisation. Most important is of course products or services which are difficult to obtain from the home country. It is also possible to identify products available on the local market to safeguard against delayed deliveries. Not of a central role but still important is supply of recreation activities for the troops.

How? The Swedish armed forces did not show any signs of intended strategies on local sourcing in advance of operations. Thereby the activities that did go on were mainly on the unit

level, where the reality stroked and things had to work. The intention when going abroad was to be fully equipped from Sweden. Local sourcing was developed in the country based on existing needs and when opportunities arise. Military units in peacekeeping operations had somewhat a different perspective being customers in a war-torn market compared with regular supplier buyer relations between firms. Theories for private business relations after minor adjustments seem to be useable, to describe local sourcing of the military units in peacekeeping operations.

Theoretically this paper contributes to three areas. First, the theory of psychic distance was developed to understand the internationalisation process, and how firms expanded their business abroad to areas with low psychic distance. In peacekeeping operations, the operational area is not chosen on the basis of psychic distance. The concept as such is useful to understand possible development of supplier relations in local sourcing. The second area concerns the development of supplier relations (Gadde and Snehota, 2000). Over time, in peacekeeping operations most supplier relations are with small firms. The relations develop from single source with low involvement too multiple source with high involvement. This differs from findings in normal supplier relations in business patterns. An important reason to this difference seems to be the size of the firms and the available time military units have for "socialising" with the population. The unit strives to have multiple supplier relations. But this is not always possible, due to the fact that the number of suitable suppliers is limited. Third, Local suppliers are selected ad hoc, but the selection process follows Dwyer et al.'s (1987) first stages in the supplier relation model. Personnel in the unit search the area for possible suppliers and possible candidates are contacted and relations are slowly developed.

The practical implication of this study illustrates that it is important to develop and diversify the Swedish armed forces' sourcing in international operations. The theories concerning post conflict development very clearly indicate the need to support the development of peace-economy and growth of small businesses. Having compared the situation for the NSE, there are important improvements to be made both in relations to the requirements and the local market in the area of operation. If the Swedish armed forces are to follow their logistics doctrine, then they have to develop their sourcing strategy nationally, internationally and locally in the area of operation.

REFERENCES

Agndal, H. (2004). *Internationalisation as a process of strategy and change*. Thesis (PhD). Jönköping International Business School.

Axelsson, B., Rozemeijer, F., & Wynstra, F. (2005). The case for change . In Davis, J. A. (Ed.), *Developing sourcing capabilities* (pp. 3–13). West Sussex, UK: John Wiley & Sons Ltd.

Boutros-Ghali, B. (1992). *An agenda for peace: Preventive diplomacy, peacemaking and peace-keeping (A/47/277-S/24111)*. New York: Secretary-General, United Nations.

Buckley, P. J., & Casson, M. (1976). *The future of the multinational enterprise*. New York, NY: Holmes & Meier.

Dunning, J. H. (1980). Towards an eclectic theory of international production: Some empirical tests. *Journal of International Business Studies, 11*(1), 9–31. doi:10.1057/palgrave.jibs.8490593

Dwyer, F. R., Schurr, P. H., & Oh, S. (1987). Developing buyer-seller relationships. *Journal of Marketing, 51*, 11–27. doi:10.2307/1251126

Ebersole, J. M. (1995). Mohonk criteria for humanitarian assistance in complex emergencies. *Disaster Prevention and Management, 4*(3), 14–24. doi:10.1108/09653569510088032

Ellram, L. M. (1996). The use of case study method in logistics research. *Journal of Business Logistics, 17*(2), 93–138.

Evans, J., Treadgold, A., & Mavondo, F. (2000). Psychic distance and the performance of international retailers- A suggested theoretical framework. *International Marketing Review, 17*(4/5), 373–391. doi:10.1108/02651330010339905

Försvarsmakten. (2005). *Grundsyn Log Fu*. Stockholm, Sweden: Försvarsmakten.

Foxton, P. D. (1994). *Powering war, modern land force logistics*. London, UK: Brassey's Ltd.

Gadde, L.-E., & Snehota, I. (2000). Making the most of supplier relationships. *Industrial Marketing Management, 29*, 305–316. doi:10.1016/S0019-8501(00)00109-7

Ghezán, G., Mateos, M., & Vileri, L. (2002). Impact of supermarkets and fast-food chains on horticulture supply chains in Argentina. *Development Policy Review, 20*(4), 389–408. doi:10.1111/1467-7679.00179

Gullander, S., & Larsson, A. (2000, May). *Outsourcing and location – Comparing industrial parks in the automotive industry and contract manufacturing in the electronics industry*. Paper presented at the Conference on New Tracks on Swedish Economic Research in Europe, Mölle, Sweden.

Johanson, J., & Vahlne, J.-E. (1977). The internationalization process of the firm - A model of knowledge development and increasing foreign market commitments. *Journal of International Business Studies, 8*(1), 23–32. doi:10.1057/palgrave.jibs.8490676

Johanson, J., & Vahlne, J.-E. (1990). The mechanism of internationalization. *International Marketing Review, 7*(4), 11–24. doi:10.1108/02651339010137414

Johanson, J., & Vahlne, J.-E. (2003). Business relationship learning and commitment in the internationalization process. *Journal of International Entrepreneurship*, *1*(1), 83–101. doi:10.1023/A:1023219207042

Johanson, J., & Vahlne, J.-E. (2009). The Uppsala internationalization process model revisited: From liability of foreignness to liability of outsidership. *Journal of International Business Studies*, *40*(9), 1411–1431. doi:10.1057/jibs.2009.24

Kaldor, M. (1999). *New and old wars, organized violence in a global era*. Cambridge, UK: Polity Press.

Kaldor, M. (2003). Civil society and accountability. *Journal of Human Development*, *4*(1), 5–27. doi:10.1080/1464988032000051469

Kamphuis, B. (2005). Economic policy for building peace . In Junne, G., & Verkoren, W. (Eds.), *Postconflict development-Meeting new challenges* (pp. 185–210). Boulder, CO: Lynne Rienner Publishers Inc.

Kane, T. M. (2001). *Military logistics and strategic performance*. London, UK: Frank Cass Publishers.

Korac, M. (2006). Gender, conflict and peace-building: Lessons from the conflict in the former Yugoslavia. *Women's Studies International Forum*, *29*, 510–520. doi:10.1016/j.wsif.2006.07.008

Lambert, D. M., Cooper, M. C., & Pagh, J. D. (1998). Supply chain management: Implementation issues and research opportunities. *The International Journal of Logistics Management*, *9*(2), 1–19. doi:10.1108/09574099810805807

Macrae, J., & Leader, N. (2000). The politics of coherence: The UK government's approach to linking political and humanitarian responses to complex political emergencies. *HPG Research in Focus*, *1*, 1–3.

Markowski, S., Hall, P., & Wylie, R. (Eds.). (2010). *Defence procurement and industry policy: A small country perspective*. Abingdon, UK: Routledge.

Maslow, A. H., & Stephens, D. C. (Eds.). (2000). *The Maslow business reader*. New York, NY: John Wiley & Sons Inc.

NATO. (2007). *NATO logistics handbook*. Retrieved July 15, 2009, from www.nato-otan.org/docu/logi-en/logistics_hndbk_2007-en.pdf

Nordin, F. (2008). Linkages between service sourcing decisions and competitive advantage: A review, propositions, and illustrating cases. *International Journal of Production Economics*, *114*(1), 40–55. doi:10.1016/j.ijpe.2007.09.007

O'Grady, S., & Lane, H. W. (1996). The psychic distance paradox. *Journal of International Business Studies*, *27*(2), 309–333. doi:10.1057/palgrave.jibs.8490137

Pagonis, W. G. (1992). *Moving mountains, lessons in leadership and logistics from the Gulf War*. Boston, MA: Harvard Business School Press.

Penrose, E. (1959). *The theory of the growth of the firm*. London, UK: Basil Blackwell.

Porter, M. E. (1980). *Competitive strategy*. New York, NY: Free Press.

Singer, P. W. (2008). *Corporate warriors* (updated ed.). Ithaca, NY: Cornell University Press.

Skjøtt-Larsen, T. (1999). Supply chain management: A new challenge for researchers and managers in logistics. *The International Journal of Logistics Management*, *10*(2), 41–54. doi:10.1108/09574099910805987

Steinle, C., & Schiele, H. (2008). Limits to global sourcing? Strategic consequences of dependency on international suppliers: Cluster theory, resource-based view and case studies. *Journal of Purchasing and Supply Management*, *14*(1), 3–14. doi:10.1016/j.pursup.2008.01.001

Stöttinger, B., & Schlegelmilch, B. B. (1998). Explaining export development through psychic distance: Enlightening or elusive. *International Marketing Review, 15*(5), 357–372. doi:10.1108/02651339810236353

Swift, J. S. (1999). Cultural closeness as a facet of cultural affinity: A contribution to the theory of psychic distance. *International Marketing Review, 16*(3), 182–201. doi:10.1108/02651339910274684

Trent, R. J., & Monczka, R. M. (2005). Achieving excellence in global sourcing. *MIT Sloan Management Review, 47*(1), 24–32.

Tuttle, W. G. T. Jr. (2005). *Defense logistics for the 21st century*. Annapolis, MD: Naval Institute Press.

Tysseland, B. E. (2008). Life cycle cost based procurement decisions: A case study of Norwegian defence procurement projects. *International Journal of Project Management, 26*(4), 366–375. doi:10.1016/j.ijproman.2007.09.005

UNHCR. (1997). *The state of the world's refugees, a humanitarian agenda*. Oxford, UK: Oxford University Press.

Van Weele, A. J. (2005). *Purchasing & supply chain management* (4th ed.). London, UK: Thomson Learning.

Verkuil, P. R. (2007). *Outsourcing sovereignty*. Cambridge, UK: Cambridge University Press. doi:10.1017/CBO9780511509926

Yin, R. K. (1981). The case study crisis: Some answers. *American Quarterly, 26*(1), 58–65.

ADDITIONAL READING

Atkinson, P. (1997). The war economy in Liberia: a political analysis. *Relief and Rehabilitation Network paper, 22*, 3-31.

Goodhand, J., & Humle, D. (1999). From wars to complex political emergencies: understanding conflict and peace-building in the new world disorder. *Third World Quarterly, 20*(1), 13–26. doi:10.1080/01436599913893

Hallen, L., & Wiedersheim-Paul, F. (1982). *Psychic Distance in international marketing - An interaction approach*. Uppsala: University of Uppsala.

Listou, T. (2008, March). *Current status on military logistics research*. presentation at Nordic Defence Logistic Research Network meeting, Oslo.

Piggee, A. F. (2002). *Transformation-Revolution in Military Logistics*. Carlise Barracks: U.S. Army War College, Retrieved March 12, 2008, from http://www.iwar.org.uk/rma/resources/logistics/Piggee_A_F_02.pdf.

KEY TERMS AND DEFINITIONS

Local Sourcing: is the sourcing in the area of the military peacekeeping operational theatre, delineated by e.g. UN Security Council decision.

Military Logistics: In line with the NATO definition: "The science of planning and carrying out the movement and maintenance of forces. In its most comprehensive sense, the aspects of military operations which deal with: design and development, acquisition, storage, transport, distribution, maintenance, evacuation and disposal of materiel; transport of personnel; acquisition or construction, maintenance, operation and disposition of facilities; acquisition or furnishing of services; and medical and health service support." (NATO, 2007, p. 7). It covers not only internal activities within the defence forces, but also activities within the whole defence supply chain.

Peacekeeping Unit: a national military unit which is engaged in a peace-support operation (following Boutros-Ghali (1992) definition of peace-support operations).

Chapter 8
Military Involvement in Humanitarian Supply Chains

Elizabeth Barber
University of New South Wales, Australian Defence Force Academy, Australia

ABSTRACT

The purpose of this chapter is to demonstrate the multitude of activities that military logisticians can provide throughout the various stages in relief supply chains. Most military joint doctrine identifies humanitarian assistance (HA) as one of the "Military Operations Other Than War" (MOOTW) that military personnel are trained to undertake. Part of this HA involves contributing to humanitarian supply chains and logistics management. The supply chain management processes, physical flows, as well as associated information and financial systems form part of the military contributions that add to other aid in the relief supply chain. The main roles of the military to relief supply chains include security and protection, distribution, and engineering. Examples of these key contributions will be provided in this chapter.

INTRODUCTION

Military supply chains are dedicated to warfare, peace keeping missions and since the 1990s disaster relief and humanitarian aid. Military logistics and their command and control systems are very suitable for operations in disaster areas. The lack of stability, infrastructure, and communications in harsh and/or remote areas are situations that military logisticians are trained and prepared to operate in. Military command and control systems are able to deal with large scale disaster situations or war. They, like other humanitarian agencies, can deploy very quickly. The key contributions to humanitarian supply chains include: security and protection, mass provisions and distribution and engineering reconstruction. Very recently military logisticians have been involved in training host

DOI: 10.4018/978-1-60960-824-8.ch008

nation personnel to take over various tasks of the supply chain activities. These activities can be used across all phases of humanitarian relief from the initial five to ten days of disaster relief through to the longer term to the reconstruction and redevelopment stages.

In the main, military forces enter into humanitarian aid in natural disasters or complex emergencies under a UN specified mandate. The mandates differ markedly between humanitarian agencies and defense organizations. These mandates become morphed within supply chains. Humanitarian agencies have a clear mandate to implement impartial aid programs to all sufferers as an inalienable right where as military involvements especially in peace keeping operations have their inevitably partial and political mandates. The separation of these mission goals must be kept distinct in the humanitarian space but within the supply chains delivering the aid the humanitarian issues seem to have won out in most cases.

This chapter will discuss the various roles military organizations fulfill within humanitarian supply chains. It will discuss some of the pros and cons associated with such involvement and conclude that when military organizations work in close co operation with humanitarian agencies the overall effectiveness and efficiency of the total humanitarian supply chains are improved.

Definitions

The Council of Supply Chain Management Professionals defines supply chain management as: "Supply chain management encompasses the planning and management of all activities involved in sourcing and procurement, conversion, and all logistics management activities. Importantly it also includes coordination and collaboration with channel partners, which can be suppliers, intermediaries, third party service providers, and customers. In essence, supply chain management integrates supply and demand management within and across companies."[1]

Their definition for Logistics Management is: "Logistics management is that part of supply chain management that plans, implements, and controls the efficient, effective forward and reverses flow and storage of goods, services and related information between the point of origin and the point of consumption in order to meet customers' requirements."[2]

A military definition of logistics is given as: "The science of planning and carrying out the movement and maintenance of forces… those aspects of military operations that deal with the design and development, acquisition, storage, movement, distribution, maintenance, evacuation and disposition of material; movement, evacuation, and hospitalization of personnel; acquisition of construction, maintenance, operation and disposition of facilities; and the acquisition of furnishing of services".[3]

The Australian Defence Force Doctrinal Definition of Logistics is: "The science of planning and carrying out the movement and maintenance of forces. In its most comprehensive sense, the aspects of military operations will deal with:

a. Design and development, acquisition, storage, transport, distribution, maintenance, evacuation and disposition of materiel;
b. Transport of personnel;
c. Acquisition, construction, maintenance, operation, and disposition of facilities;
d. Acquisition or furnishing of services; and
e. Medical and health service support."
(Land Warfare Procedures – General 0-1-6 Land Glossary, 2004)

Academics have been defining supply chain management in similar broad terms as the military logistics definition above. One of the most definitive sources on the definitions of supply chain management is given in a comprehensive literature review that provided the following definition of a supply chain. "A supply chain is defined as a set of three or more entities (organizations or individuals) directly involved in the upstream and downstream flows of products, services, finances and/or information from a source to a customer". (Mentzer *et al.*, 2001, p. 4.)

Generally speaking, and for the purposes of this relevance to this chapter, military perceptions of supply chains refer to the supply line from the appropriate warehouse to the 'foxhole' need. Whereas distribution of the final or finished good from the manufacturer or the postponement point to the end customer is seen in business definitions of supply chain management as distribution. Logistics in military terms compares with industry and civilian definitions as the total supply chain. In other words industry perceives logistics as comprising a subset of the total supply chain whereas military perceives supply chains as a subset of logistics. This follows the 'traditionalist' definition of supply chain management being a subset of logistics and as widely used in military doctrine whereas the 'unionist' approach defines logistics as a subset of supply chain management as widely used in industry (cf. Larson & Halldórsson, 2004). There does seem to be a marked difference between business or civilian perceptions and military perceptions of supply chain vs. logistics which in situations of collaboration, co-ordination and co-operation of military in HA supply chain management practices has caused some contention. Nevertheless recently, and possibly due to the extensive outsourcing by military organizations of their supply chain operations to civilian companies, these distinctions are finally becoming blurred and the traditional military definitions are being accepted as differently defined in military doctrinal areas but in practice follows the civilian interpretations. This chapter will follow this trend and use the terms somewhat interchangeably.

BACKGROUND

Traditionally it was natural disasters that requested humanitarian logistic assistance. The world political environment was fairly stable and the separation between conflicts and humanitarian needs was quite distinct. Today the complexities of conflict and resultant humanitarian needs seem often to merge and with these complexities, the link with military involvement in humanitarian aid, including the supply chain operations, has become more central to the United Nations relief support systems. Military activities and their degree of involvement are covering an increasing number of different scenarios and consequently their missions differ markedly between involvement in humanitarian peace keeping missions and support in assisting distribution of supplies to natural disaster areas. This section will distinguish between military involvement in the complex peace keeping missions and the natural disaster response situations.

The Berlin Airlift, just after World War II was a classic early case of military involvement in humanitarian supply chains. The allied militaries and civilians where were contracted to assist provided the whole logistics chain from identifying demand, procurement, warehousing, inventory management, transport and air distribution. Waves of C-47 aircraft provided nightly air drops to the isolated West Berlin at the start of the Cold War. It was one of the largest humanitarian relief operations involving military organizations during 1948-49 when the basic needs of two million people in West Berlin were support by military air lifts after a Soviet land and sea blockade. (Antill, 2001; Hofmann & Hudson, 2009)

Peacekeeping Missions in Conflict Situations

Peacekeeping missions[4] tend to be supported by supply chains which need to be sustained over the long term, whilst disaster relief supply chains experience a huge forward surge into the affected area for a relatively short period before assistance is handed over to other agencies to continue. The complexities of peace support operations include curtailing any combative insurgencies, humanitarian aid and host country reconstruction. Following from the research paper: "Building on Brahimi: A Coalition for Peacekeeping in an Era of Strategic Uncertainty"[5], the Department of Peacekeeping Operations and Department of Field Support, New York published an informal report titled: "A New Partnership Agenda – Charting a New Horizon for UN Peacekeeping", 2009 which expands on these complexities. The UN peacekeeping operations began in 1948 in the Middle East when UN military observers were deployed. Since then and especially after the end of the Cold War the typography of these missions developed from the traditional observations to greater degrees of involvement and the consequent import on supply chain management practices in HA. (see: Franke & Warnecke, 2009, for a tabular list of the 69 UN peace missions since the end of the cold war) The spectrum of peace and security activities in HA ranges from conflict prevention, peacemaking, peace keeping, peace enforcement and peace building. For definitions of these activities see the United Nations Peace Keeping Operation, (2008, Chapter 2, p. 23).[6]

The humanitarian efforts to protect and support the civilians caught up in situations of internal armed conflicts are particularly vulnerable due to their displacement. The UN humanitarian agencies and NGO partners involved require close coordination with the military and other police organizations to ensure protection and support. From the 1990s military forces have had more involvement in humanitarian interventions. Chapter VII of the United Nations Security Council has authorized military intervention in Somalia, Haiti, Rwanda and Kosovo during the 1990s. These peace–keeping efforts in fragile states address threats to peace and security and often to emerging humanitarian tragedies. The trend is now for military forces to include assistance to humanitarian situations in their training doctrine. The UN's Office for the Coordination of Humanitarian Affairs (OCHA) in 2001 published that there is an 'evolution of

military thinking in regard to the provisions of humanitarian aid and services. In NATO and elsewhere there has been an evolution of the doctrine of military-civilian operations, with an increasing tendency for military forces being used to support the delivery of humanitarian aid, and sometimes even to provide this aid directly.' (OCHA, 2001)

There is no doubt that the number and extent of peace keeping missions have surged recently. The first surge was in the early 1990s. The second surge occurred a decade later in 1999-2000 with peace keeping missions in Ethiopia, The Congo, Kosovo and Timor-Leste. The third surge occurred in 2003-4 with five multidimensional operations in Liberia, Burundi, Cote d'Ivoire, Haiti and Sudan. Since then 15000 troops entered Lebanon on 11 August 2006; UNMIT began in Timor-Leste and 17,300 troops entered Darfur on 31 August 2006. At this time it was estimated that over 65,572 military were deployed world-wide in 2006 on UN peacekeeping missions. (Keating, 2006) At present it is estimated that there are over 116,000 deployed personnel across 15 UN peace keeping missions. (Department of Peacekeeping Operations & Department of Field Support, 2009)

When military organizations under the UN Security are involved in peace keeping and other peace related missions it is typically a multinational military involvement. The command and control structures co-ordinate the various national contributions. The supply chain is typically unified to provide all support requirements to all militaries from central distribution hubs inside or peripheral to the area of operations. Once a level of security has been established these lines of supply and distribution channels provide a ready-made structure and network for the physical flows of aid that can be used by other humanitarian agencies. For example, the East Timor campaign saw Australian troops in a peace keeping role as well as a humanitarian role. The recognition of the complexities of conflict and humanitarian needs has seen some military organizations providing written instructions to initiate humanitarian assistance; provide limited governance assistance and restoration of essential services.

Sometimes in conflict situations the initial military input can be quite short term nature. For example, recent Haiti disaster saw civil unrest for a short period of time. In situations of ongoing conflict military involvement could simultaneously be involved in security issues such as curbing aggressive groups as well as assisting in protection of logistical routes, distribution centers and stores. They can be involved in the provision of national assistance in conjunction with host nations and other aid agencies for transport systems and distribution training. Military operations can provide reconstruction and engineering expertise to assist in the reconstruction of the host country which require large construction vehicles and equipment to be shipped into the affected areas. These operations are often long term in nature where the military perform a complex role in providing ongoing peace keeping operations; training of security forces in the hose nation; as well as assisting in the engineering and reconstruction of infrastructure. The classic case in point is in Afghanistan. It has also occurred over the last decade in East Timor.

Natural Disaster Situations

When the type of mission is to provide disaster relief for natural disasters such as earthquakes, floods, hurricanes and tsunamis, the military assistance excels in rapid response. Military organizations have the means and resources to distribute huge volumes of basic food, water and medicine. They have the capability to provide mass shelters and sanitation in a rapid response situation. All of these contributions were demonstrated in the recent Hurricane Katrina devastating New Orleans in 2005. For example, huge Army convoys were mobilized across the United States to deliver provisions to the affected areas. Mission types include: disaster relief, dislocated civilians, technical and support, civic assistance. The combinations of

these different types of missions add to the supply chain complexities.

When the mission type changes to one of dislocated civilians, military assistance can be in the form of camp constructions and the massive distribution of provisions. In the case of Hurricane Katrina the military joined other security forces to protect the displaced civilians, control the looting and violence and other criminal activities. By September 1, over 6,500 National Guard troops had arrived in New Orleans and a subsequent 40,000 assisted in the overall evacuation and security efforts in Louisiana. (Barringer, 2005) Military personnel also undertook search and rescue operations. Military logisticians can provide security missions for the protection and opening up of transport and storage facilities. They can command and control distribution within and into the disaster area in the initial disaster relief stage. This occurred in the initial stages of aid into Haiti.[7]

Military logisticians are most appreciated in the early stages of disaster relief as they can restore communication systems, provide protection for incoming aid and assist with urgent air lifts and drops of aid in inaccessible locations. Military equipment is designed for mass distribution. Consequently the military ships fully loaded with supplies can off load enormous tones of aid relief in devastated areas using made to purpose landing craft. For example, the HMAS Tobruk's assistance to the island of Niuatoputapu suffering from the Samoan hurricane involved ship to shore transport by landing craft. The ship carried relief packages from AusAID, Red Cross as well as a donated fire truck. The ship stayed in the Samoa and Tonga area transporting humanitarian aid and equipment to affected communities for a number of weeks and distributed approximately $13 million of aid.

The logistical support to any natural disaster ranges from rapid deployment of peace and security keeping forces, massive food and water air drops and in some cases, the delivery of specialized pieces of equipment owned by military organizations. As Kovacs and Tatham (2009) stated military organizations are very resource oriented and these resources are often sought after. They can range from surveillance and intelligence gathering equipment to locate displaced and lost people to the provision of transportation assistance in the form of landing craft to remote shore locations. Along the supply chains military personnel can provide protection of stores, distribution and transport routes and use in transit warehousing in ships and distribution centers. Military personnel and equipment are well suited to providing first stage engineering works to re-establish basic infrastructure including road clearance, bridge repairs, sea port and airport and tarmac clearance, site clearance of mines and any other potentially dangerous munitions.

What has been achieved by military involvement, perhaps to varying degrees of success, is to develop reliable initial supply chains into these affected areas. The supply chains will be used in the first instance by military equipment and personnel whether it is inflows to regions of conflict or natural devastation. Over the longer term these supply chains become secure and can continue to be used by humanitarian agencies and possibly strengthened to provide capacity development.

These complexities make it particularly difficult to assess the level of performance and effectiveness of the military involvement as well as the level of effectiveness of the military and humanitarian supply chains. The blurring between the military supply chains being used for humanitarian assistance whilst at the same time as supporting a war effort cannot distinguish performance of one without the other in the case of conflict situations. In natural disaster situations the initial supply chains might be developed by military and police organizations but the overlap of many different organizations using the capacity of these supply chains distorts the efficiencies and effectiveness of the flows. Performance issues of these supply chains are not the only complexity. Military movements into the humanitarian space raise significant issues of principle. Not least of these involve efforts by military organizations in aid distribution and their own military supply chains to achieve the peace keeping objectives.

Characteristics of Military Supply Chains

The main characteristic of military supply chains is the capacity for preparedness. Military dedicate funding to high levels of preparedness. Military organizations generally and their logistics divisions in particular, have centralized decision making in a well organized and rapidly implemented command and control system. They follow a standardized set of procedures typically arising from a joint logistics system and filtered through to the three force levels of air force, army or navy depending on the most efficient means of distribution. The levels of coordination between the various forces and along the hierarchy are tight. Coordination between allied military logistical systems, processes and procedures are also often very compatible. Military personnel are highly trained for deployments in very harsh, often remote locations and within these locations they can be well supported by their logistical systems. Military supply chains have at their disposal rapid movements from specialized aircraft such as their cargo planes and capacity of high volume and mass movements provided by their ships. Armies provide secure warehousing and robust road transportation capable of transporting across harsh terrain without roads. Military logistics systems have highly trained staff, excellent information and visibility throughout their systems and continual upgrades enable these systems to remain at the cutting edge of technology. Kovacs and Tatham (2009) described the differentiation of military and humanitarian models that provide the capabilities needed to respond to large scale disruptions under the headings of physical, human and organizational capital resources.

Military training and subsequent characteristics include:

- Trained to operate in destabilized conditions
- Trained to operate under uncertain demand and limited knowledge conditions
- Provide support and operate in 'green field' situations (meaning can operate in environments without full infrastructure capability, communications and other constraints)
- Trained to operate under harsh and stressful conditions with scheduled rotations
- High ability for rapid sense and respond reactions
- Continually trained from 'lessons learnt' scenarios.
- Trained to operate in rapidly shifting situations which change the demands and required agility of their supply chains.

Disaster relief supply chains are characterized by ambiguous objectives because of the initial lack of knowledge of the disaster area. Thus they are quite dynamic in the crises or disaster relief first stages. For example in the bush fires of Victoria in Australia in 2010 the fires were so wide spread and spread so rapidly it was unknown how many townships were destroyed, how many access roads were impassable and how many Australian were dead or trapped in the disaster area. In this situation the Australian Defence Force was used to air lift in resources and expertise to either produce transitory logistical infrastructure or provide engineering skills to fix or stabilize damaged bridges and road infrastructure for aid to flow into the affected regions.

There are limited resources to enable the supply chains to flow smoothly. Limited resources include staff that may be generally untrained. The military logisticians often find that a teaching role is required to bring some staff up to speed in difficult working environments. One of the problems faced by the military forces is that deployable qualified military staff is needed for other more urgent operations than training.

Civil-Military Contributions (CIMIC) in the Relief Supply Chains

When discussing civil-military cooperation or coordination care must be taken to clarify the specific relationships. Cooperation refers to a maximum state of civil military coordination where there is a range of cooperative relations between the humanitarian community and a military force that is not regarded as a combatant force, typically including joint planning, division of labor and sharing of information. The UN and NATO understanding of cooperation and coordination seem to be reversed, because in the NATO context, cooperation is understood to imply a less binding relationship than coordination, and NATO argues that the humanitarian community will be willing to cooperate, but not coordinate, and therefore they use cooperation. (NATO, 2004; UNOCHA, 2006)

Organizational Involvement

The United Nations Humanitarian Civil-Military Coordination (UN-CMCoord) (UnitedNations, 2005) section is part of the Office for the Coordination of Humanitarian Affairs (OCHA) which emerged as the successor to the UN Department of Humanitarian Affairs in January, 1998. (Weiss, 1998) The DHA was created In April, 1992 in response to the frustrations of major donors to coordinate their activities very effectively in the 1991 Gulf crisis. OCHA is based in Geneva, Switzerland and provides international communities with services ranging from development of guidelines and documentation, facilitating training programs, supporting military exercises and field exercises. UN-CMCoord develops the planning, information sharing and task divisions between military and civilian agencies in humanitarian situations. As logistics link nearly all stakeholders in humanitarian relief operations, the role of UN-CMCoord is vital.

In complex situations OCHA aims to systematically improve the relationships amongst the expanding number of relevant stakeholders. In both natural disasters and conflict situations the military work with entities such as the Department of Political Affairs (DPS), Department of Peacekeeping Operations (DPKO), and the Peace building Support Office (PBSO), the Department of Safety and Security (DSS) as well as other militaries. OCHA aims to improve the coordination and integration of these stakeholders with other humanitarian entities such as IASC organizations and NGOs.[8] Co-ordination roles extend beyond the operational level to the strengthening of integration with wider groups to mutually develop more efficient and effective planning. (Objective 17-1.2) The co-ordination role extends to wider engagements across the before, during and after emergencies and post conflict and recovery and rebuilding periods. (Reference Guide for OCHA's Strategic Framework: 2010-2013)[9]

How this integration occurs in multidimensional peacekeeping missions alongside the humanitarian aid is through three distinct structural models, namely the "Two Feet In"; "Two Feet Out" and "One Foot in, One foot out" model. In the "Two Feet In" model the UN peacekeeping or political mission is deployed with humanitarian agencies in post conflict settings. Current examples are the Timor Leste situation. In the "Two Feet Out" model the situations are highly unstable and the peacekeeping forces are not widely accepted. OCHA remains entirely separate from the mission structure in order to preserve independence and neutrality for their humanitarian work. Current examples include Sudan, MINURCAT in Chad and UNPOS in Somalia. In "One foot in, one foot out" model the peacekeeping forces are deployed in regions emerging from conflicts and crisis. A relatively stable political situation is in place and OCHA holds a clearly identified presence outside the peacekeeping mission structure. Current

example includes UNMIS in Sudan. In all three models OCHA's key role is to facilitate liaisons between UN political and military forces and the humanitarian communities. (OCHA-Integration: Structural Arrangements", 2010).[10]

In summary OCHA coordinates through facilitation. It facilitates civil-military coordination before, during and after humanitarian crisis and further it maintains guidelines on coordination processes.

Civil-Military Operations Center

The focal point for civil-military coordination in the UN lies within the OCHA as a section called the Civil-Military Coordination Section (CMCS).[11] Civil-Military Operations Centers (CMOCs) are usually established within the humanitarian space by the joint force commander to assist in the coordination of military activities and other government agencies, nongovernmental organizations and regional organizations. The structure varies to match the size and composition requirements of any given situation. It is a United States government initiative. These centers are formed during complex humanitarian emergencies such as the post Desert Storm Kurdistan situation in Iraq and the genocide atrocities in Somalia and Rwanda. They were a response to situations where no host national government existed to manage the humanitarian relief activities. They usually had a military officer from the United States Department of Defense in charge and had the overall aim of filling the coordination function not only of logistical support but of all activities. These umbrella type coordination centers eventually were included into U.S. joint doctrine. It is argued that these centers are not suitable where there are host governments with their own military forces to coordinate in conjunction with the UN agencies such as the UN-CMCoord. (UnitedNations, 2005) For example Haiti is a geographically isolated humanitarian disaster. It has a functioning government and associations with appropriate UN

Table 1. Coordination mechanisms and their objectives

Coordination Mechanism	Objective	Responsibility
CERF (Central Emergency Response Fund)	Stand-by disaster response funds	OCHA
CAP (Consolidation Appeal)	Funds mobilization	OCHA
UNDAC (United Nations Disaster Assessment and Coordination)	Disaster needs assessment	OCHA
HIC	Information	OCHA
MCDU	Civil military coordination	OCHA
UNJLC United Nations Joint Logistics Command	Humanitarian logistics	WFP

agencies. Where there is no government within the host nation to take on the coordinating role especially for the distribution of supplies then such centers can be very useful.

Nevertheless the key elements of CMOCs are to share information, divide tasks appropriately and plan the operation. The various guidelines relating to civil-military involvements that OCHA upholds humanitarian principles are:

- Guidelines on the Use of Military and Civil Defence Assets in Disaster Relief 'Oslo Guidelines', adopted in 1999, updated in 2006 and revised 2007.
- Guidelines on the Use of Military and Civil-Defence Assets to support UN Humanitarian Activities in Complex Emergencies 'MCDA Guidelines', 2003
- Discussion Paper and non binding Guidelines on the Use of Military or Armed Escorts for Humanitarian Convoys, 2001
- IASC Reference Paper on Civil-Military Relationship in Complex Emergencies, 2004, and,
- Country specific guidance. (examples include Afghanistan, Iraq, Liberia and Sudan)

The CMCS provides pre deployment training, in-mission training as well as numerous workshops for both military and civilian staff. Training covers both natural disasters and complex emergency

situations. Donors provide assets to CMCS who in turn co-ordinate the use of these assets and calls for personnel with specific expertise and equipment for support.

Supply Chain Flows and Military Involvement

Relief and humanitarian supply chains consist of product, information and financial flows. These flows are linked by systems and processes. Humanitarian sectors dispose of a number of coordination mechanisms through a number of different policy making bodies all of which work together, often with military groups to provide needed funds, information and products. These groups include (see Table 1):

The need for consolidation of governance and processes is heightened with many players using different systems. To prevent overlap and non-interoperability becoming a major constraint to these various logistical flows, preparedness for smoother supply chain operations has increased. The recurrence of disasters underscores the urgency for robust disaster preparedness plans. For example the World Vision has created a phased relief response which typically occurs in three phases: seven day, 30-day and 90-day response. Various UNDAC experienced staff can deploy within hours to carry out assessment of needs to a disaster affected area as he have teams permanently on stand-by to attend disasters anywhere in the world. Often military personnel will join their teams.

The UNJLC ensures the coordination of logistics and information management which in the Sudan for example covers the following key areas:

- Logistics information dissemination
- Logistical planning and common logistics services facilitation
- Geographic information system (GIS) that includes mapping services
- Provide logistics for emergency preparedness and response
- Coordinate the non food items (NFI) and Emergency Shelter Sector. It has managed the NFI common pipeline into the Sudan since 2004.

Military organizations also have well structured systems to absorb all flows and processes of supply chains. When these are linked with UNJLC's systems more cost effective logistical flows are coordinated.

Physical Flows

The military together with various aid agencies are often the first over the disaster area with aerial relief drops, surveillance and intelligence gathering. The military are natural leaders in natural disaster relief as they are consistently in a state of preparedness. Preparedness is the key to a successful response in disasters and military logistics systems develop levels of preparedness with pre-packed kits and high training levels. They thus deploy rapid response along fast and efficient physical flows.

Military forces are well able to provide physical support such as air lifting, tents, medicines, foodstuffs but often rely on contracted medical and engineering teams. Military involvements in humanitarian relief (in peaceful environments) tend to be more likely in cases of large scale sudden events because of their ability for rapid response

and scale of requirements. (Pettit & Beresford, 2005) A characteristic of military supply chains is the level of physical capital resources at its disposal. Although the military acquisition processes are moving towards performance based logistics models where the supplier might still own the equipment and be required to maintain the equipment such as ships and aircraft over the total life cycle of the asset; the military organization has at its disposal very highly sophisticated and large scale physical assets. These assets can be used to literally provide its own physical supply chain to anywhere in the world. The physical items of food, water, medicines, and shelters can flow rapidly in huge volumes along these chains. They can be transported rapidly in huge volumes by aircraft for example, the C17s cargo planes. The seven tone MRH90 helicopters have a maximum weight lift of 10,600 kg.

The physical inflows of donations to a disaster relief area can be arranged in types of solicited and unsolicited goods (see Table 2).

The inflows along military physical supply chains into affected areas would typically fall under the first two types of donations for a natural disaster relief situation. When military are involved in peacekeeping scenarios their supply chains are coordinated in a multilateral or bilateral manner.

Information and Financial Flows

"Logistics and information management are immutable but the humanitarian sector as a whole has not made that connection yet." (Whiting and Ayala-Ostrom, 2009, p. 1087)

The humanitarian supply chain management system SUMA was developed and launched in 1991. Its objectives were to increase the capacity of disaster operators to be able to manage the information of supplies relating to the huge surges of inflows of donations. At this time the donated

Table 2. Physical flows of donations

Multilateral	Solicited by agency and agreed by agency
Coordinated bilateral	Solicited by agency, agreed by recipient society
Other, coordinated	Solicited by other NGOs
Unilateral	Unsolicited goods, without agreement, announced
Non-coordinated bilateral	Unsolicited, but recipient society agrees to receive
Other non-coordinated	Unsolicited, shipped to other NGOs

Sourced: Tuck School of Business at Dartmouth – Glassmeyer/McNamee Center for Digital Strategies

inflows were often unsolicited and sometimes unwanted and the tagging and visibility of these items were unknown. This system attempted to make the information more transparent and accountable to the total system. By 1996 a training establishment known as Fundesuma was created as a non-profit organization. Its goals were to support training in and the maintenance of SUMA. Fundesuma was also responsible for upgrades to the system. This was the beginning of good software availability and a sound database structure for the humanitarian supply chain management processes. (de Ville de Goyet, 2008)

In 2002 the World Food Programme assumed leadership and established the new Logistics Support System (LSS). It was designed as a complement to agency specific commodity tracking systems. This combined system of the old SUMA and the UN Joint Logistics Centre system aims to link with the many tracking systems which are now humanitarian agency specific systems. The LSS basically captures information. They do not manage supplies. These systems are often not in use between disasters. The development of the LSS system established it as part of a permanent national disaster response capacity and as such has had much success. Nevertheless the data needs to be nourished and information sharing on a global scale across disasters and between disasters should be maintained. Systems that are set up during initial disaster responses tend to disappear when the emergency phase has ended without capturing information into the LSS. Nevertheless the techni-

cal capabilities of the LSS often remain unused due to lack of knowledge by users. In Guatemala for example this problem is being overcome by Guatemala government regulations requiring international users to share relevant information. A comparison with military logistical systems shows that military systems are maintained and operated between war and humanitarian use, they do capture previous information and lessons learnt; upgrades are continual together with regular training.

Consequently has been argued that systems that have a dominant stakeholder permanently in control of data such as the national defence department would ensure the ongoing nature of the system. It would also have trained operators. (de Ville de Goyet, 2008)

The Fritz Institute has developed the Humanitarian Logistics Software (HLS) which is provided free of charge. It can assist with mobilization, procurement, transport and tracking decisions. It also connects with financial systems which enables agencies to keep track of budget and donor funds. (Altay, 2008) More relief agencies are adopting this platform and it is rapidly becoming the common platform amongst the aid community.

The information flows of the humanitarian supply chains serves as the information bridge between any preparedness planning and response, between procurement and distribution and between the coordinating headquarters of the logistical flows such as a lead agency, UNJLC or other body and the field operators and recipients. The information flows flow downstream and upstream into the affected areas. It coordinates the sourced donations from donors, purchases from vendors and inventories held in pre-positioned stock; it indicates replenishment needs and tracks the distributions. Figure 1 shows a typical information supply chain for humanitarian relief as identified by the Red Cross.

The military information supply chain systems are well defined, regularly upgraded with advanced technology enabling high levels of tracking and visibility. Military supply chain information systems have a very high level of trained staff operating their logistics systems that are in use all the time. Military supply chain management systems are upgraded regularly. They are regularly used in multinational exercises. Their systems provide good visibility and traceable functions. All inflows and outflows are fully accountable and budgeted. Military systems are all encompassing in that they capture data; manage flows, storage, transportation and warehousing. Not only do these systems provide visibility along supply chains, they are linked to financial systems. Personnel are typically well trained in the logistical systems and their contractual arrangements with their prime contractors. This enables visibility of supply chains well back along the chain which can combine both availability of supplies and records of transactions which leave an audit trail.

The military often provide their own surveillance, search and rescue forces and intelligence gathering. (Bammel & Rodman, 2006/2007) The ease in setting up systems in green field sites that can be linked globally as well as to ships off shore of disaster areas to provide full accounting records is an enviable logistical system to have in chaotic situations.

The military organizational structures suit disaster relief mission type logistics because they work in closely controlled organizational structures where both formal and informal information networks are well established across the organization. The military recognize the worth of the established formal and informal communication flows that can provide the best logistical efficiencies and effectiveness. This is in contrast with the other organizations that often have volunteer members who are unknown to each other. Military personnel are trained to operate efficiently in chaotic conditions. Military personnel operate under a competency-based trust system that is embedded in proven track records.(Cross, 2004)

According to Howden, (Howden, 2009) humanitarian logistics information systems should integrate logistics units into the broader humanitarian supply chain; enhance logistics activities and provide continuous support and create new possibilities for collaboration between organizations. Although the focus seems to still be on the initial response phase of a disaster and prepositioning of supplies in support of the initial response period; the information flows in humanitarian supply chains can extend that focus more broadly. Information systems in military supply chains provide visibility for inventory levels and tracking of all products for all services (Army, Navy and Air Force) not only for their distribution services from warehouse to area of operations, but for their parts and components for maintenance of equipment; purchasing and disposal of assets. The humanitarian logistical information systems need to extend and support the recovery and rebuilding phases in humanitarian disasters to reduce delays, corruption, slippages through damage and theft

Figure 1. Information supply chain in disaster relief

of reconstruction materials etc. They also need to integrate or link the various agencies providing aid.

Humanitarian information systems like the military logistical information systems can assist with the transition from 'action' to 'peace' states of being in that the transitions between the response, recovery, rebuilding and handovers to the host nation can be streamlined by using the same information system throughout the total humanitarian aid cycle. The replacement of short term emergency response teams with longer term recovery and rebuilding teams can see via logistical information systems where the supplies are, what have been used and what is still available for use. It is envisaged that as the humanitarian logistical information systems become more sophisticated and more widely used they will develop along similar lines of military logistical information systems.

MILITARY CONTRIBUTIONS

Rapid Response

One of the main strengths of military supply chains lie in their self supporting autonomy. Military supply chains can provide rapid deployment to remote and harsh green field and damaged sites throughout the world. These rapid deployment facilities are something that humanitarian supply chains can lack. A major contribution that military logistical forces can provide is their specialization of rapid response. It is often called upon by the UN, Red Cross and other NGOs to assist with disaster relief establishment of supply chains and initial distribution services.

Within the new Strategic Framework 2101-2013 of OCHA (OCHA, 2009) there is an aim to improve their suite of surge response solutions. They aim to maintain low vacancy rates for these surge field positions. OCHA aims to improve its Response Division with adjustments to its contractual requirements as well as adding

a roaming surge response capacity that will have the ability to deploy field staff immediately. This is an area were military forces excel. They are trained to deploy quickly and in great numbers. Key contributions include: global reach; reactive and proactive approaches; dedicated systems, personnel and equipment to green field or chaotic situations, protection capabilities; surveillance and intelligence capabilities that enable pre kitting and accurate targeting for logistical air drops. Other key contributions include the long term nature of military logistical teams who are regularly trained and exercised.

Combat support personnel have developed consumption models based on relatively long supply chains that cannot accurately predict support needs. These prediction capabilities compensate the 'mountains of supplies' that are often pushed forward in disaster relief situations. The new sense and respond models rely more on intelligence and surveillance gathering of information in remarkably quick times and aims to ensure that supply chains are more responsive to the required needs of the victims. (Tripp, 2006) For example, in November, 2007 the Oro Province in Papua New Guinea (PNG) was struck by Cyclone Guba killing over 150 people and creating wide spread disaster over the region. Within a day, Joint Task Force 636 from Australia arrived in the province to help the victims. Their level of preparedness permitted them to immediately provide relief food and water provisions. They assessed the health situation on the ground and set up facilities within two days. An engineering team assessed the damage to local infrastructure including the port facilities, roads and availability of water and sanitation and continued to assist the follow-on humanitarian agencies re-establish the basic water and sanitation requirements which prevented the spread of disease.

Air Response

Military aircraft are regularly used to make the first airdrops of humanitarian aid over disaster areas. The C-17 Globemaster aircraft have a long fuel range and can carry three times the load of the older C-130s. For example, the US Air Force under Operation Unified Response Mission responded to the Haiti 7.0 magnitude earthquake on 12th January 2010 by dropping food and water parcels into Port-au-Prince and surrounding areas up immediately after the initial quake. Aircraft are the quickest response to reach the affected areas. They can access extremely hazardous or inhospitable areas that ground transport cannot reach. The US military took control of the Port-au-Prince airport and military support flowed in to maintain security on the island. The military received much criticism as it was seen that the military aircraft were bringing in troops rather than aid provisions.[12] The US military movements choked the airport and prevented other aid aircraft to land due to this congestion. Often they were seen to be turning away donations but some might have been unsolicited supplies that could have clogged the limited airport and warehouse space.[13] Some saw this disaster as an excuse for the United States to politically secure Haiti whilst other saw the military of taking the opportunity to prevent drug smuggling in the area – both at the expense of choking the airport and preventing the distribution of both land and air relief stores. In controlling the airport at Port-au-Prince the US military took the lead of the inflows to Haiti. They coordinated the supply chains so that assets and resources such as aircraft, trucks, food stores and forklifts that facilitated the unloading and distribution of food and medical items and pharmaceuticals were facilitated in a more efficient and timely manner. Cold storage facilities were established for the temperature and moisture sensitive donations such as blood.

Sea Response

Maritime services into disaster areas provide a number of services. Military ships provide huge volumes of aid; they act as aid worker relief stations; they are often secure havens for refugees; they provide hospitalization services. Their auxiliary services of associated landing craft and deck winches and cranes can alleviate the congestion at ports. Their associated helicopter pads and air craft can distribute rapid air aid to isolated coastal areas. The aid supplies that are transported by sea to disaster areas have the advantage of delivering large volumes in one movement. The choke point at ports, for example the Port au Prince during the Haiti disaster, causes the distribution of supplies to be held up to some extent. In the Haiti earthquake disaster the helicopter carrier USS Bataan and three large dock landing ships as well as two US survey/salvage vessels created a sea base for the rescue efforts. The French Navy provided the Navy ship, the Francis Garnier, and an amphibious transporter, the Siroco. The US hospital ship, USNS Comfort, was extensively used for emergency operations of over 600 casualties of the earthquake. The US Navy contribution to the Haiti disaster was 17 ships, 48 helicopters and 12 fixed-wing aircraft together with 10,000 sailors and Marines. The US Navy delivered 32,400 gallons of water and 532,440 bottles of water, over 100,000 meals and massive medical supplies. (Wood, 2010)

When military air forces and navy forces work in co-ordination the initial thrust of aid relief is delivered by air with the back up large volumes for sustainable aid relief being supplied by sea transport. Heavy equipment such as cranes, bulldozers, reconstruction equipment and materials are suitable for navy to transport and land on beaches such as the recent delivery by HMAS Tobruk to the island of Niuatoputapu in Tonga which was one of the hardest hit islands in the Samoan tsunami of October, 2009.(Austin, 2009)

As part of the Australian Government's response to a request for humanitarian assistance from Indonesia in 2009 when earthquakes damaged Padang extensively, HMAS Kanimbla sailed to Padang and anchored off the town of Pariaman. It offloaded heavy engineering equipment and Army personnel for joint TNI-ADF relief operations. The associated Sea King helicopters transported personnel, medical teams and equipment to remote villages. The Joint Task Force set up a temporary field hospital and some medical evacuations occurred from the field hospital to the ship and were then further air lifted into Padang. (McHugh, 2009)

Protection and Safety

UNSC resolutions on the protection of civilians include S/RES1267 of October 15 1999 and continue through to this day. The President of the Security Council has issued statements on protection of civilians in humanitarian disaster areas on February 12 1999 (S/PRST/1999/6) and has consistently stated this need through to May 29 2009 (S/2009/277). Military uniformed personnel involved in the United Nations peacekeeping operations increased from 47,883 in 2001-02 to 72,822 in July 2006. (Keating, 2006)

The need for protection of humanitarian workers risking their lives to assist civilians experiencing disasters has risen recently. In 2009 UN agencies and NGOs in Afghanistan alone were involved in 114 security incidents and eighteen NGO members were killed. Increasing violence, kidnappings, attacks and deaths of humanitarian workers in Pakistan, Sudan, Chad and the Congo have lead to recognition that aid workers will require greater security and protection of themselves, their bases and their supplies. Security measures can be costly and can hamper humanitarian efforts not only in peace keeping missions but in disaster relief situations such as Hurricane Katrina and Haiti earthquakes where looting was common. (OCHA, 2010)

The need for protection of civilians in conflict regions needing humanitarian assistance is also

rising. This has led to a recent study by OCHA and Department of Peacekeeping on protection mandates by peacekeeping missions. It found that "Military forces are sometimes assigned humanitarian assistance missions, often without adequate training, policies or doctrine to integrate into the international response system". (OCHA, 2010, p. 4)(OCHA, 2010)

Generally military involvement in humanitarian space can be in the form of deterrent deployment; peace building; peacemaking; peace keeping; or peace enforcement with the latter implying the maximum force and deterrent deployment implying no military force. Peace keeping and peace enforcement will both involve human right enforcement and humanitarian needs. Both of these require military deployment to the location with or without the consent of the host nation. (Altay, 2008) The military operations over a particular disaster area may cover a range of intensities, namely, peace support, peace-keeping and peace enforcement. Various Chapters from the UN Charter become involved depending on the intensity of coercive operations.

Peacekeeping missions have changed from the implementation of peace agreements. They are now multidimensional and include the monitoring of human rights, protection of civilians and rebuilding the capacity of the host country. The military roles within peacekeeping missions have also changed. Within hostile environments the military forces supporting the humanitarian efforts are very important as they might face a range of hostilities within the one region. There might be low grade civil disorder, terrorist or guerilla activities or even full scale regional wars. For instance in Darfur, relief aid has been persistently hindered by conflicts. In all these cases the military forces need to be extra vigilant and provide at least some of the following services:

- defending the perimeters of the central distribution nodes and initial entry point into the area

- provide escorts for safe passage of relief convoys and medical teams
- assisting in personnel recovery
- protection of refugee camps
- prevention of relief resources transferring to corrupt governmental, political, social personnel for either personal gain or hostile combatant's needs.

Further the involvement of military logistics will change over time. Several dimensions of protection, aid and development might be addressed in parallel with a variety of NGO input and other charities. Composite emergency relief response models involving all parties have been developed. (Pettit & Beresford, 2005) No single model can accommodate all the variables and each disaster relief supply chain will be different with different needs for different forms of protection depending on security issues.

The multidimensional peace building and peace enforcement mission in Burundi, Africa from 2004 and ongoing, not only included protection of civilians from physical violence but also involved: monitoring a ceasefire agreement; collection and disposal of weapons as well as the dismantling of militias. The peace keeping mission in Somalia deployed over four thousand military personnel to expressly protect and keep operational the sea and air ports to enable the incoming humanitarian aid to reach the country. The ports of Mogadishu as well as the distribution centers within Mogadishu were heavily guarded. The military were required to protect humanitarian convoys and distribution centers throughout Somalia. Not only were the ports maintained operational by military peacekeeping forces but these forces were also mandated to secure the lines of communication for humanitarian assistance. Similarly the need for security of movements in Darfur, Sudan from 2007 and ongoing is provided by military peace keeping forces. (Franke, 2009)

Preventing Corruptive Activities

When the 2004 Boxing Day Tsunami hit the South East Asia region the devastation to Indonesia was enormous. Within the region of Aceh there was a history of conflict ranging from sporadic instances of violence to full scale concentrated military operations. The Islamic separatist movement in Aceh, the Free Aceh Movement (GAM) was at the time quite strong in the region. Under Operation Sumatra Assist the devastated areas of Aceh were laid open to GAM corruptive influences. The Indonesian and Australian Defence forces recognized the potential for the threats posed by GAM and protected the many NGOs and aid agencies including their supplies and provisions from looting. The continual threat of conflicts flaring up was a major concern which was effectively dampened by military involvement. It was the sovereign military influences that adapted to local threats which allowed the humanitarian mission to continue. Protection of the agencies involved in Operation Sumatra Assist by military included discouraging and/or preventing illegal arms flows within and across region borders during the chaos. It dispersed militias without open conflict and disposed of illegal weapons. It ensured that the distribution systems were protected and assisted in the fair and equitable distribution of aid.(Greet, 2009)

Security of relief ships to Gaza, prevention of looting in the Hurricane Katrina aftermath and in the aftermath of Haiti's recent earthquake as well as the corruption in supply lines into Somalia and some African states all require increased funding to prevent these corruptive practices.

Protection and Relief Supply Routes in Failed States

Fragile states are characterized by poor governance, weak economies, poor infrastructure, displaced groups of inhabitants and generally fractured societies. Iraq, Afghanistan, Somalia and some African states and Pacific regions are examples of fragile states and these tend towards corruptive activities. In 2003 the Solomon Islands in the Pacific exhibited a number of conditions of a fragile state. Under Operation Anode the Australian Defence Force was deployed to protect the various security, police and NGOs. The military provided the security umbrella on the ground and also provided the ships for the transport of large volumes of provisions to support the humanitarian workers. The military involvement in this region was to achieve military goals which in turn created the environment for humanitarian outcomes. It was a military led reconstruction task that aimed for the creation of an environment conducive to long term capacity building.(Greet, 2009)

The UN operation in Somalia (UNOSOM 1, 1992-93) initially failed. It was established under the UN Security Council to monitor the cease fire in Mogadishu. The mandate was for military to provide protection and security for UN personnel, supplies, equipment, distribution ports (both sea and air) and to ensure safe deliveries of aid supplies to distribution aid relief hubs. The mission initially under the UN Chapter 6 was insufficiently robust. It did not permit troops to provide adequate protection to the aid relief convoys entering into intense conflict regions. UNOSOM 1 was subsequently replaced by a US-led coalition (UNITAF) which was more robust and provided the necessary protection. UNITAL was succeeded by UNOSOM 2 (1993-95) under a UN Chapter 7 mission giving authorization to use enforcement measures to establish and ensure secure environments throughout Somalia for HA operations.

The other area where military can assist is to prevent human trafficking occurring along the logistical aid routes (and elsewhere) of displaced children, orphans, women and young men which could emerge within failed states.

Engineering Support

Military engineering support to humanitarian supply chains covers some distinct segments but essentially its overall role is to reinstate essential engineering services. Within humanitarian supply chains this included repairing or rebuilding infrastructure such as warehousing, sanitation and water provision and transportation facilities. The following mini case studies demonstrate the importance of engineering support to humanitarian supply chains as well as the diversity of support.

Lashkar Gah, Afgaanistan. In July 2006 the 28th Engineering Regiment of the UK Army was deployed under the International Security Assist Force (ISAF) to the Helmand province in Afghanistan to begin the reconstruction of humanitarian efforts. The province was considered not stable to support a civilian driven redevelopment. This regiment started with security of the transport routes, repairing a couple of bridges and improving road pavement to enable supplies to flow more easily and constructed, with the able assistance of local contractors, warehousing for storage. Once the initial foundations for support for redevelopment were achieved the local contractors and other aid agencies continued the redevelopment program. Nevertheless the 28th Engineering Regiment was retained to supervise and oversee the contractual arrangement to ensure that the build facilities were not only robust from an engineering perspective but the contracts were financially auditable and commercially sound. The engineers took on the role of project engineer managers. (Allen, 2007)

Bagram Airfield, Kandahar, Afghanistan. The Bagram Airport is the main military airport into Afghanistan Area of Operations. This example demonstrates the fine line between military involvement in conflicts and humanitarian air relief missions. During 2009-2010 more than $220 million is currently being spent on airfield construction projects at the Bagram Airfield. This is due to the long term U.S. military commitment to ensure security and stability in the area and to enable more

International Security Assistance Forces (ISAF) to assist the war efforts in Afghanistan. Nevertheless many military planes are bringing into the region massive loads of humanitarian relief. These planes use the Bagram airport as their point of entry to the humanitarian supply chain into the region. The USACE project engineers are executing this huge project. In early 2009 the helicopter ramp pads were causing some obstructions to the war fighter plane needs. To ensure uninterrupted capability to these war fighter planes the helicopter ramp constructions were staged. Once finished it is expected that many rapid deployments for humanitarian relief operations to the local people in nearby regions will be staged from this airport. (Jackson, 2009)

Samoa. Engineers from the U.S. Army Corps of Engineers (USACE) were rapidly deployed to Samoa in response to the earthquake and subsequent tsunami disaster on 29th September, 2009. Personnel expert in handling debris, emergency power support and establishing water systems were deployed from a mounting base in Honolulu. They assisted with the installation of FEMA (US Federal Emergency Management Agency) generators at critical life saving and life sustaining facilities, sewer and water treatment plants, police, fire and medical facilities. USACE provides essential support for Emergency Support Function (ESF-3) under the U.S. Department of Homeland Security's National Response Framework. The diversity of the engineering support to ESF is shown in the seven divisions, namely, ice, water, emergency power, debris removal, temporary housing and facilities, temporary roofing and structural safety assessment. The FEMA generators also provided power for communication systems used to coordinate the inflows of aid. (Justinger, 2009)

As a follow on from this field work a team of the USACE Reachback Operations Center (UROC) trained key personnel at a Hawaii-based interagency exercise. Included in this exercise were members of the 565th Forward Engineering Support Team (FEST) in the use of multiple

technologies that can serve assets globally in the aftermath of any storm or natural disaster. These included communications and technologies such as the Automated Route Reconnaissance Kit and the Ike/Geospatial assessment tools for engineers. The UROC team covered the coordination and logistics for area reconnaissance and conducted site assessments of supply chain critical infrastructures. The overall result was training for overall hurricane preparedness for engineers.

Gulf of Mexico, USA: Hurricane Katrina. Although the U.S. military was very heavily involved in numerous ways supporting the disaster relief efforts associated with Hurricane Katrina which hit the Florida coastline in 2005, one interesting logistical support effort was undertaken by the Naval engineers. The Expeditionary Unit Water Purification Program (EUWP) has been used in a number of humanitarian missions. The U.S. Navy's response to drinkable water for affected residents led to the Navy engineers using their on board desalinization plants to provide 100,000 gallons of purified water per day from the turbid Gulf of Mexico. This engineering feat replaced the daily convoy of tankers shipping water into the area and thus cleared some of the choke points and congestion on the main road highways accessing the region.(Jackson, 2009)

Australia: Floods at one end and Bush Fires at the Other. In February 2009 the worst bush fires ever were experienced in Victoria whilst at the same time extensive cyclones and flooding occurred in northern Queensland. There was massive devastation over huge areas at both ends of the country. The Australian Defence Force and in particular the No 1 Airfield Operational Support Squadron (1AOSS) and the Army's 3rd Combat Engineer Regiment (3CER) provided heavy plant and equipment to clear debris and repair washed roads and bridges. Engineers and work crews battled elements to establish sanitation, ablutions and laundries for the thousands made homeless in northern Australia. The town of Normanton was isolated for six weeks and engineers were unable to clear the airfield for landed supplies so all provisions were air dropped in by helicopter. With approximately sixty percent of Queensland under water during the month it was difficult for the engineering crews to maintain the critical air fields for the establishment of aerial distribution hubs. All roads and rail access were flooded. The water pipelines were damaged and drinking water was a critical provision that had to be transported over large distances.

At the other end of the country in Victoria the fire ravaged area of Kinglake and Marysville also required the 1AOSS and 3CER teams. Once roads were cleared and checked by the engineering teams they were accessible and the Army sent convoys of trucks with provisions, tents, and engineering equipment. 1AOSS established communication networks and shelter centers which had make shift laundry facilities and ablutions and 3CER set up clean water facilities. (Edwards, 2009)

Challenges

The role of logistics in humanitarian relief has received more attention in recent times. (Whiting & Ayala-Ostrom, 2009) Most military organizations have doctrinal documentation now covering military involvement in humanitarian activities. The key challenge of impartiality in most instances, even in failed states scenarios, appear not to impact so adversely on the humanitarian supply chains and the military support of those chains. The reactions of other neutral agencies and the host nations and its people in response to military involvement in humanitarian assistance still pose challenges and this is an area for more research.

Non Impartiality of Military Involvement

The host nation typically does not have total control over their uncertain environments, purely by the nature of the disaster making the environment unstable. In some instances the host nation can be

opposed rather than receptive to the humanitarian efforts of assistance and sometimes military assistance by some countries can be refused or be a very sensitive issues, such as the case as the 2004 train explosion at Ryongchon in North Korea. (Pettit & Beresford, 2005)

Afghanistan. A strong argument against using military logistics in disaster relief for humanitarian situations is the perception of lack of neutrality. The military involvement in relief in Afghanistan during 2004 caused the complete withdrawal of Medecins Sans Frontieres (MSF) citing the strategic interests of the US military compromised their ability to work in the region (Pettit & Beresford, 2005).

Other areas where military logistical support may be challenged to meet requirements include:

• Medical supplies are directed to war wounds not refugee diseases
• Command and control practices and processes might not fit well with NGOs and international humanitarian agency's management systems (Pettit & Beresford, 2005)

Somalia. During 2009 the number of people in need of assistance in Somalia rose from 3.2 million to 3.8 million. This is over half of the total population. Over 1.2 million experienced acute livelihood insecurity and starvation.[14] In January, 2010 the UN World Food Programme (WFP) suspended some of its food distributions in southern Somalia due to escalating attacks and demands by armed groups making it unsafe to continue. Incidents of extortion, looting, assaults and kidnapping against aid workers led to many workers leaving and the reduced staff could not continue. The protection of NGOs and their supply chains would facilitate the provision of aid substantially. Once consistent and reliable supply chains are established enabling aid to reach displaced people the stability of the region would increase. However the concern is that if a military escort of armed peacekeepers protected the UN aid workers Al-Shabab and affiliated organizations would be more inclined to attack. Aid workers have been killed in Somalia but some aid has been accepted as it is desperately needed and it is viewed in some instances that aid workers pose no threat. However if they were to travel with a protective military convoy would imply a serious risk. Thus attempts to make aid workers work safer with military protection has severe risks in some humanitarian situations and would well backfire. (Anonymous, 8 January 2010)

Papua New Guinea. In 2001 the U.S. Peace Corps pulled out of Papua New Guinea (PNG) due to security concerns. Today it provides ongoing International Military Education and Training (IMET) program as well as an NIV/AIDS training program. In May 2005 over two hundred line police officers from Australia left PNG following the Papua New Guinea's Supreme Court's decision to remove Australian police office of immunity. Since then Australia has remained the largest bilateral aid donor to PNG and is continuing ongoing negotiations for a scaled down version of the previous Enhanced Cooperation Program providing the police officers. Security is still a major issue through the small nation.

The issue of sharing classified information across national boundaries is often a problem for coalition military members. Robust connectivity is largely dependent on goodwill and on fostering personal relationships and informal networks. The issue of sharing information, especially informal information, with aid agencies is not such a problem, although incompatible technology often is.

Personnel and Emphasis on Training

The highly trained military personnel are at a state of readiness for deployment. (Kovacs and Tatham, 2009) They are continually trained for their particular areas of expertise whereas humanitarian supply chains often rely on volunteers who are untrained or inconsistently trained. Military

logisticians are well trained in their logistical systems that they use for deployment into disaster relief areas. These military systems are not compatible with other military logistical systems or the various aid agencies' systems. Compatibility of logistical systems is a challenge not only facing all participants of the humanitarian supply chains but the security of these systems is also a challenge. At present there is no one common training centre that assists in training in logistical systems most suited to providing humanitarian aid. In the United States for Hurricane Katrina the Battle Command Sustainment Support System (BCS3) was used with an RFID system was used to identify the distribution routes from the Federal Emergency Management Agency (FEMA) to the State of Louisiana. It gave visibility of supply movements (in-transit tracking) of emergency supplies such as water, medical materials and food. This system had fully trained personnel following all emergency relief aid flowing into the area. It even enabled the management of relief convoys to efficiently manage the timeliness of distribution and deliveries. The combination of highly trained staff and in-transit visibility permits disaster relief distribution networks to not only integrate supplies and improve service levels but to synchronize logistical support so that facilities and their capacities can be utilized with appropriate priorities. The use of BCS3 could also calculate anticipated logistical needs required into the disaster region. (Henderson, 2007)

Interoperability of Military Organizations with other Humanitarian Agencies

The lack of common doctrine, procedures and processes is a further challenge. Military forces are very strict on doctrinal procedures and the command and control hierarchy is well known and obeyed by each military commander. The problem of interoperability between NGO, military and UN doctrine, procedures and controls is a challenge faced by all stakeholders. Training encompassing appropriate umbrella logistical systems and organizational cultures is a major challenge. Humanitarian and military organizations differ markedly in their structures, capabilities, levels and areas of expertise which leads to difficulties in encompassing training in a generalized sense compared with training a highly specialized expert in a particular core service or geographical location. One way that has been attempted to meet this challenge is the UN cluster approach that groups specialized humanitarian organizations into one cluster and then develops a level of interoperability. The Inter Agency Standing Committee (IASC) endorsed the cluster approach in 2005 and since then a logistical cluster has been used. The WFP is the logistics cluster lead at a global level but when aid missions vary in scale so too does the logistics cluster operations. Logistics clusters range from global reach down to field level.(IASC, 2006) They can vary from sharing common storage facilities and transportation capabilities to just sharing information.

The establishment of the UN CMCoord assists the interaction and coordination between military assistance and other humanitarian agencies. This capability includes personnel, equipment and support.

FUTURE RESEARCH DIRECTIONS

A few of the major roles of military organizations in humanitarian supply chains have been discussed. Further research is needed in many of these areas but in particular, protection and security enforcement along supply chains require further research. There is a fine line between protecting the contractors and preventing corruptive practices in distribution and inventory. There is often uncertainty of donor contributions, i.e. uncertain supply into humanitarian regions. At the same time the demand for assistance is also unknown. Thus there develops a mismatch between supply and

demand which military organizations can assist to alleviate. The areas of uncertainty in humanitarian supply chains need further research. The use of military involvement in formal and informal communication networks in chaotic situations need further research. The areas of capacity building and engineering support that military organizations provide in disaster areas and fragile states also require further depth of research.

CONCLUSION

Humanitarian operations in general and humanitarian supply chains in particular are not just an extension of military participation. Military organizations of the future need to be equipped and trained to under both peacekeeping and warfighting. This duality of requirements will affect future policies, training needs and organizational aspects of most military commands. The nexus of civil/military interfaces will remain strong in the key area of supply chain management in humanitarian issues. The 'spirit' of the terms 'comprehensive approach'; 'clusters'; 'whole of government'; 'inter-agency approach' all support a harmonized or unified approach to resolving problems of providing efficient supply chain and logistical support to humanitarian aid. The duality of most military organization to train for war fighting and peacekeeping and humanitarian efforts will affect future policies, organizational and training needs of military forces. It will also challenge some aspects of an evolving culture in military forces. Adaptive civil/military interfaces remain a key area for improvement but military organizations are changing to encompass this need.

This chapter has demonstrated the increasingly contributing military assets in humanitarian emergencies. The resultant attention to civil-military relations has culminated in a series of guidelines and research activity. It does seem that military involvement in providing and assisting with the provision of humanitarian aid is an 'inevitable trend' that is likely to continue. (Hofmann & Hudson, 2009)

REFERENCES

Allen, G. W. (2007). Reconstruction and development on Operation Herrick 5 utilising local contractors. *The Royal Engineers Journal, 121*(2), 84–89.

Altay, N. (2008). Issues in disaster relief logistics. In Gal-el-Hak, M. (Ed.), *Large-scale disasters: Prediction, control, and mitigation* (pp. 121–146). New York, NY: Cambridge University Press.

Anonymous. (8 January 2010). UN aid agencies will not abandon Somalia despite insecurity, says official. *UN News Centre.* Retrieved 12 February, 2010, from http://www.un.org/apps/news/story.asp?NewsID=33433&Cr=somali&Cr1

Antill, P. (2001). *Military involvement in humanitarian aid operations.* HistoryofWar.org

Austin, T. (2009). Tobruk's proud mission. *Navy News.* Retrieved 12 February, 2010, from www.defence.gov.au/news/navynews

Bammel, J. L., & Rodman, W. K. (2006/2007). Humanitarian logistics: A guide to operational and tactical logistics in humanitarian emergencies. *Air Force Journal of Logistics, 30/31*(4/1), 1-42.

Barringer, F., & Longman, J. (2005). Police and owners begin to challenge looters. *New York Times.*

Cross, P. (2004). *The hidden power of social networks.* Boston, MA: Harvard Business School.

de Ville de Goyet, C. (2008). *The use of a logistics support system in Guatemala and Haiti.* Washington, DC: World Bank.

Department of Peacekeeping Operations & Department of Field Support. (2009). *A new partnership agenda: Charting a new horizon for UN peacekeeping.* New York.

Edwards, K., & Mathews, L. (2009). Fires and floods: What a month for 1AOSS. *Combat Support Spring,* 12-14.

Franke, V. C., & Warnecke, A. (2009). Building peace: An inventory of UN peace missions since the end of the Cold War. *International Peacekeeping, 16*(3), 407–436. doi:10.1080/13533310903036467

Greet, N. (2009). ADF experience on humanitarian operations: A new idea? *Security Challenges, 4*(2), 45–63.

Henderson, J. H. (2007). *Logistics in support of disaster relief.* Bloomington, IN: AuthorHouse.

Hofmann, C. A., & Hudson, L. (2009). *Military responses to natural disasters: last resort or inevitable trend?* Humanitarian Exchange Magazine.

Howden, M. (2009). *How humanitarian logistics Information Systems can improve humanitarian supply chains: A view from the field.* Paper presented at the Proceedings of the 6th International ISCRAM Conference, Gothenburg, Sweden.

IASC. (2006). *Logistics cluster - About the logistics cluster.* Retrieved 12 February, 2010, from http://www.logcluster.org/about/logistics-cluster/

Jackson, J. M. E. (2009). Desalination technology increases naval capabilities, meets humanitarian needs. *The Military Engineer, 101*(662), 36.

Justinger, L. (2009). USACE deploys response teams following Pacific tsunami. *The Military Engineer, 101*(662), 22–23.

Keating, C., Weschler, J., Ward, C. A., Claude, A., & Fernandes, F. R. (2006). *Twenty days in August: The security council sets massive new challenges for UN peacekeeping.* New York, NY: United Nations Security Council.

Kovacs, G., & Tatham, P. (2009). Responding to disruptions in the supply network - From dormant to action. *Journal of Business Logistics, 30*(2), 215–229. doi:10.1002/j.2158-1592.2009.tb00121.x

Larson, P. D., & Halldorsson, A. (2004). Logistics versus supply chain management: An international survey. *International Journal of Logistics: Research and Applications, 7*(1), 17–31. doi:10.1080/13675560310001619240

McHugh, G. (2009, 29 October). Ready to help Padang assist. *Navy News*. Retrieved from www.defence.gov.au/news/navynews

Mentzer, J. T., DeWitt, W., Keebler, J. S., Min, S., Nix, N. W., Smith, C. D., & Zacharia, Z. G. (2001). Defining supply chain management. *Journal of Business Logistics, 22*(2), 1–25. doi:10.1002/j.2158-1592.2001.tb00001.x

NATO. (2004). *AJP-9 NATO civil-military cooperation (CIMIC) doctrine.*

OCHA. (2009). *Reference guide of OCHA's strategic framework 2010-2013.* New York, NY: United Nations.

OCHA. (2010). *Annual plan and budget.* Geneva, Switzerland: United Nations.

Pettit, S. J., & Beresford, A. K. C. (2005). Emergency relief logistics: An evaluation of military, non-military and composite response models. *International Journal of Logistics: Research and Applications, 8*(4), 313–331.

Tripp, R. S., Amouzegar, M. A., McGarvey, R. G., Bereit, R., & George, D. (2006). *Sense and respond logistics: Integrating prediction, responsiveness, and control capabilities.* RAND Corp.

United Nations. (2005). *United Nations humanitarian CMCoord concept endorsed by the IASC.* Rome, Italy: Inter-Agency Standing Committee (IASC).

UNOCHA. (2006). *Guidelines on the use of military and civil defence assets to support United Nations humanitarian activities in complex emergencies.* Brussels United Nations Office for the Coordination of Humanitarian Affairs.

Weiss, T. G. (1998). Civilian-military interactions and ongoing UN reforms: DHA's past and OCHA's remaining challenges. *International Peacekeeping, 5*(4), 49–70.

Whiting, M. C., & Ayala-Ostrom, B. E. (2009). Advocacy to promote logistics in humanitarian aid. *Management Research News, 32*(11), 1081–1089. doi:10.1108/01409170910998309

Wood, R. A. (2010, 13 January). Vinson deploys to respond to Haiti earthquake. *Navy Military News.*

KEY TERMS AND DEFINITIONS

Air Response: Airdrops of humanitarian aid over disaster areas.

Military Involvement: The participation of armed forces in disaster relief through peacekeeping operations.

Military Supply Chain: The supply line from the appropriate warehouse to the need. The military perceives the supply chain as a subset of logistics.

Peacekeeping : A spectrum of peace and security activities ranging from conflict prevention: peacemaking, peace keeping, peace enforcement to peace building.

Protection: Ensuring the security of people (civilians and humanitarian workers), logistics facilities (warehouses, bases) and transportation routes and modes, and of items.

Rapid Response: A surge deployment situation of the military in a remote and harsh green field and damaged site.

Security Management: Curbing aggressive groups as well as assisting in the protection of logistical routes, distribution centers and stores.

ENDNOTES

1   http://cscmp.org/aboutcscmp/definitions.asp

2   http://cscmp.org/aboutcscmp/definitions.asp

3   http://www.logisticsworld.com/logistics.htm

4   There are various levels of peace missions, namely, conflict prevention, peacemaking, peace enforcement, peace building and peace keeping. All have different requirements. Peace keeping is the term used broadly to distinguish these missions from natural disaster missions requiring humanitarian aid.

5   www.un.org/Depts/dpko/dpko/newhorizon.shtml)

6   http://pbpu.unlb.org/pbps/Library/Capstone_Doctrine_ENG.pdf

7   www.logcluster.org/ops/hti10a/situation-report-concolidated-3-february-2010

8   http://ochaonline.un.org/OCHAHome/tabid/5837/language/en-US/Default.aspx

9   http://ochaonline.un.org/OCHAHome/tabid/5837/language/en-US/Default.aspx

10 http://ochaonline.un.org/OchaLinkClick.aspx?link=ocha&docId=1165031

11 http://ochaonline.un.org/cmcs

12 http://news.bbc.co.uk/2/hi/americas/8469800.stm

13 http://www.doctorswithoutborders.org/press/release.cfm?id=4176

14 www.un.org

Chapter 9
Challenges of Civil Military Cooperation / Coordination in Humanitarian Relief

Graham Heaslip
National University of Ireland - Maynooth, Ireland

ABSTRACT

The term civil military coordination (CIMIC) suggests the seamless division of labor between aid workers and international military forces. The images of humanitarian organizations distributing food and medicines under the protection of military forces, or aid workers and military working together to construct refugee camps, set up field hospitals, provide emergency water and sanitation, et cetera, has become more frequent. The media coverage from crises such as New Orleans, Kosovo, the tsunami in Asia, Pakistan, Liberia, Sierra Leone, Chad, and more recently Haiti and Japan, has heightened the expectation of a smooth interaction between humanitarian organizations and military forces. Due to fundamental differences between international military forces, humanitarian and development organizations (in terms of the principles and doctrines guiding their work, their agendas, operating styles, and roles), the area of civil military coordination in disaster relief has proven to be more difficult than other interagency relationships. This chapter will identify the many factors that render integration and collaboration problematic between diverse organizations, and especially so between civilian and military agencies. The chapter will conclude with proposals to improve CIMIC within disaster relief.

INTRODUCTION

The provision of aid by military forces is not a new phenomenon. From the Napoleonic Wars, the First and Second World Wars (particularly the Marshall Plan after World War II) the Berlin Airlift

(1948-9), and up until the present day, including the Congo, Bangladesh, Ethiopia, Sudan, Iraq, the former Yugoslavia, Rwanda and Mozambique (Doel, 1995). It seems that when disasters, either natural or man-made occur, governments often turn to the military for help as the military have

DOI: 10.4018/978-1-60960-824-8.ch009

certain resources immediately to hand, such as food, medicine and fuel as well as logistical resources of transport, communications and human assets with which to distribute them (Weiss & Campbell, 1991).

The proliferation of conflict since the end of the Second World War and particularly the end of the Cold War in many parts of the developing world (Croft & Treacher, 1995) has led to new thinking regarding the provision of assistance by the military in humanitarian aid operations. The emergencies that now develop are usually characterized by a large media presence (the 'CNN' factor, e.g., Haiti), which increases the pressure on national decision makers to respond with a large international aid effort and in many cases, a large military presence to keep the peace (James, 1997). This sort of effort is usually targeted at the immediate saving of lives, requiring massive logistical and material support and the employment of considerable logistical assets.

However, the involvement of the military in such operations is not without challenges (e.g., different organizational cultures) and a balance must be sought in allowing the humanitarian aid agencies a free hand in utilizing the available military resources (Heaslip, 2010). The 'actors' (see section 'Crowded Stage') in humanitarian aid have differing management styles and administrative structures and whilst the supply chain appears straightforward, the complexities in the relationships that occur as well as the impact of having different structures and procedures may conspire to frustrate the establishment of collaborative supply chain strategies.

Cooperation, coordination and collaboration between the military and relief partners, particularly Non Governmental Organizations (NGOs), is often uneven and uncertain (Heaslip *et al.*, 2007a). NGOs can be difficult partners, especially for the military. Although knowledge has grown in the last decade, military officers and NGO officials often have little understanding of each other's institutions and operating procedures. Many

military officials lack an understanding of the distinct charters and doctrines of NGOs, failing to recognize that what works with the International Rescue Committee (IRC) will not work with the International Committee of Red Cross's (ICRC) (Heaslip *et al.*, 2008a). In turn, humanitarian aid organizations criticize the military for not understanding their hierarchies. The military may not be familiar with important NGOs in the Area of Operations (AOR) and the reason for this lack of knowledge could be perceived as being institutional (Heaslip *et al.*, 2008a).

Practical realities on the ground have gradually necessitated various forms of civil military cooperation, coordination and collaboration for humanitarian operations (Whiting, 2009). These developments demand increased communication cooperation, coordination and collaboration between humanitarian actors and require improved knowledge of each others mandates, capacities and limitations (Whiting, 2009).

Civil military coordination essentially deals with two aspects of military support to civilians, namely the provision of security, e.g. a military escort for a humanitarian convoy, and secondly the provision of military assets, including skills, knowledge and manpower, e.g. equipment such as trucks or helicopters, and skills and knowledge such as medical and engineering expertise (Heaslip, 2010). In the field, the area of civil military coordination is even more difficult than other interagency relationships given fundamental differences between international military forces and humanitarian and development agencies in terms of their agendas, operating styles, roles, and the principles and doctrines guiding their work.

The trend in peace-building since Kosovo in 1999 is towards greater integration of international efforts and the necessity for collaboration between relief, development and security organizations (Olson & Gregorian, 2007). By the late 1990s, from key donor countries, to United Nations (UN) agencies, to Non Governmental Organization (NGO) networks, a common understanding emerged that efforts for peace must become more strategic and coordinated if they are to have the ambitious impacts they intend. This consensus has led to efforts in recent years to promote explicit communication, coordination and even formal integration among interveners to achieve greater impact (Olson & Gregorian, 2007). This push for more unified efforts has led to innovations such as UN Integrated Missions, which combined the political, peacekeeping, and humanitarian arms of the UN system under a unified command.

Many donor countries have now synchronized the foreign assistance arms of government in what has been variously called the "joined up approach," the "whole of government approach," or, the "3-Ds" approach, referring to defense, development and diplomacy (OECD, 2005, p. 13-14). The goal has been to use military, political and humanitarian/development instruments in a more synchronized and presumably more effective manner to achieve security, development and peace in conflict affected countries. This chapter identifies the many factors that render integration and collaboration problematic between diverse assistance agencies, especially between civilian and military agencies and it will conclude with proposals to improve CIMIC within humanitarian relief.

BACKGROUND

The delegates who attended the Hague Peace Conference in 1899 and 1907 attempted to impose international controls on the conduct of war and within this framework considered two strands – peacekeeping and humanitarian assistance – which have threaded through the century to modern times (Laurence, 1999). For the greater part of the twentieth century both of these mechanisms developed largely in isolation from each other, but the crises of the nineties have thrown them together in an uneasy alliance (Laurence, 1999).

In 1991 the almost unthinkable happened, the Soviet Union collapsed, the Warsaw pact dis-

solved, taking the menace of global war with it. The end of the Cold War brought a dramatic increase in conflicts within states and a similar increase in international peace operations, as complex multi-actor interventions to end civil wars and build peace have come to be known (Olson & Gregorian, 2007). While in 1998 the UN deployed 14,000 peacekeepers worldwide, today over 90,000 military and civilians are in the field in 16 UN missions (CIC, 2006). New international 'coalitions of the willing' led by regional organizations like North Atlantic Treaty Organization (NATO) in Kosovo and European Force (EUFOR) in Chad, emerged in place of UN involvement in some contexts. More fundamentally, the nature of UN involvement expanded dramatically from classical peacekeeping characterized by the monitoring of ceasefires. New expanded UN missions involved solidifying fragile truces, building the capacity and legitimacy of states emerging from conflict, holding elections, demobilizing and reintegrating combatants, and sometimes direct administration of a territory for a period (De Conning, 2007). These changes encompassed a broadening of the goals and sectors that were involved, a deepening of the engagement with the internal workings of societies, and a lengthening in terms of the stages of conflict when such missions would be deployed, with preventive and post-conflict state-building missions a major new focus (Lund, 2003).

Whether operating in a war zone or responding to a humanitarian disaster both circumstances have many similarities. In both situations the organizations involved operate in a destabilized environment with a reduced communications and transport infrastructure (Long & Wood 1995), limited initial knowledge about the situation (Tomasini & van Wassenhove 2004a) and, in particular, limited knowledge about the location and numbers of affected people and their needs (Özdamar *et al.*, 2004). Furthermore, the operations themselves are also likely to be conducted in the spotlight of media, public and political attention, and in the presence of significant numbers of casualties and displaced persons. Recently, some military organizations (for example the Nordic countries and Ireland) have started to see disaster relief as being in their normal sphere of activities, at least when it comes to domestic affairs.

There is much debate over the role of the military within the humanitarian context. Within this environment, the military do not seem to participate in any systematic, overarching and coordinated humanitarian plan and are forced to dabble in the provision of humanitarian assistance in an *ad hoc* manner. There are guidelines, such as the *Guidelines on the Use of Military and Civil Defence Assets in Disaster Relief – Oslo Guidelines* (United Nations, 1994), which set out occasions when the military should engage in humanitarian activities. However, these guidelines "are in fact rarely observed" (Gill *et al.,* 2006, p. 41). The provision of humanitarian assistance by the military continues to generate passionate debate between the military and non military actors (Barry & Jefferys, 2002), in which there is seldom any meeting of minds.

NEW BEGINNING

For groups that do try to engage in civil military cooperation, there is often a strong sense of frustration on both sides particularly with the energy invested in trying to establish good communication and clear understanding of each other's positions. Another key issue is that field-level coordination is vulnerable to directives coming from the policy level (organizational headquarter level), and a reported lack of two-way information flow between the field and policy levels (Olson & Gregorian, 2007). Furthermore, field cooperation often depends on developing good personal relationships, but with frequent turnover of personnel (especially in the humanitarian community) such relationship building must be repeated between replacement personnel (Barry & Jefferys, 2002).

In situations of humanitarian relief, the interlinked nature of security, aid and development is inescapable, with security necessary to enable progress on development, and immediate relief and longer-term development gains necessary to solidify the peace by giving people a stake in the new stability (Olson & Gregorian, 2007). Roles and mandates often overlap as military forces engage in aid provision and governance support, major donor representatives work directly with provincial and local governments, and development actors participate in the security sector reform spectrum (of disarmament, demobilization and reintegration (DDR), justice reform, police reform) (De Conning, 2007). Whether international military forces, UN, humanitarian and relief agencies choose to explicitly work together or not, the outcomes of their efforts in such settings are deeply intertwined.

Given this interdependence, improving how military, UN, humanitarian and development actors interact in such settings is critical to increasing the probability that their independent efforts will lead to positive outcomes, and to increasing the chances that some level of constructive coordination may be developed.

The Crowded Stage

A number of actors interact in humanitarian relief (Seiple, 1996; Mackinlay, 1996). In their work Kovács & Spens (2008) identified the various actors involved in the humanitarian aid supply network, see Figure 1.

For the military the first step toward better coordination is to gain a better understanding of the relief community. The actors in the relief community vary tremendously in their capabilities, size, and attitudes, with considerable implications for cooperation with the military and for the success of the overall relief effort. The number of disparate actors involved in providing humanitarian assistance complicates efforts to improve coordination. At times, everyone and no

Figure 1. The Humanitarian Aid Supply Network [Source: Kovács & Spens (2008, p. 223)]

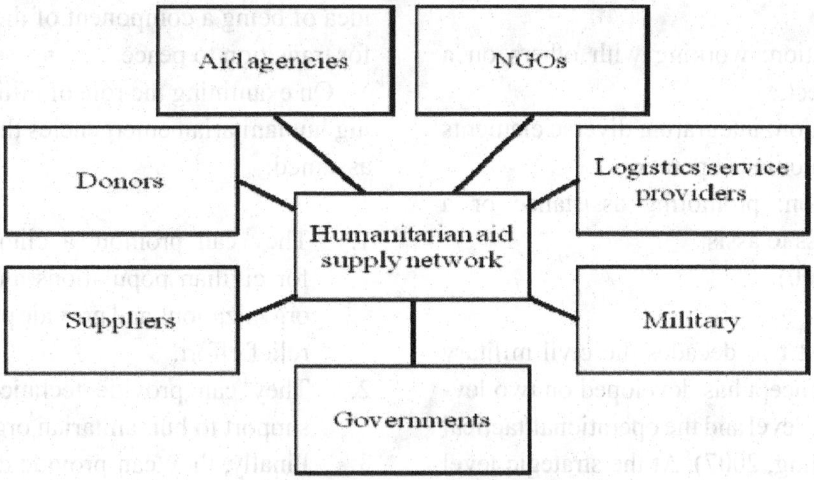

one may seem to be in charge. Understanding these various players is a precondition to coordinating activities.

Although the humanitarian sector disposes of a number of coordination mechanisms and has a policy making body (Table 1), no agency has the authority to take the lead and coordinate the actions of others. Moreover, as Stephenson (2004) contends, a strong competition among agencies for resources, competition for media attention, high staff turnover, and the different organizational backgrounds, cultures and incen-

tives do not create a conducive environment for inter-agency coordination.

Conceptual Confusion

Civil-Military Cooperation (CIMIC) has grown in importance within the military as a necessary tool to assist the commander in the successful completion of the mission. The second "C" in CIMIC can stand for Collaboration, Co-ordination, Co-operation or cynically, Confusion, but because of the connotations for each, delicacy and political

Table 1. Humanitarian coordination mechanisms

Coordination Mechanism	Objective	Responsibly
Central Emergency Response Fund (CERF)	Stand-by disaster response funds	United Nations Office for Coordination of Humanitarian Affairs (OCHA)
Consolidated Appeals Process (CAP)	Funds mobilization	OCHA
United Nations Disaster Assessment Coordination (UNDAC)	Disaster needs assessment	OCHA
Humanitarian Information Centre (HIC)	Information	OCHA
Military and Civil Defence Unit (MCDU)	Civil military coordination	OCHA
United Nations Joint Logistics Centre (UNJLC)	Humanitarian logistics	World Food Programme (WFP)

Source: Adapted from OCHA (2003) and Tomasini, R. M. & Van Wassenhove, L.N. (2004b)

correctness must be displayed. Therefore a few definitions may be useful:

- Collaboration: working with others on a joint project.
- Coordination: integrating diverse elements in an harmonious operation.
- Cooperation: promoting assistance or a willingness to assist (OED, 2000)

Over the last two decades the civil military coordination concept has developed on two levels, the strategic level and the operational/tactical level (De Conning, 2007). At the strategic level it refers to a type of mission construct and when it is used at the operational/tactical level it refers to a specific function within the military force. These multiple identities and meanings have caused considerable confusion. At the strategic level civil military coordination is used to suggest a multidimensional, whole of government or comprehensive approach, where various civilian and at least one military entity are engaged in a joint initiative or mission (De Conning, 2007).

At the operational/tactical levels, 'civil military coordination' is used to refer to the specific policies, modalities, structures and tactics that are used to manage the relationship between the military and other components of an operation (De Conning, 2007). At this level the focus has predominantly been on the humanitarian military relationship, and two distinct sets of policies have developed over the years: one policy set that deals with the relationship from a military perspective, e.g. NATO Civil Military Cooperation (CIMIC) doctrine, and another, dealing with it from a humanitarian perspective, i.e. UN Humanitarian Civil Military Coordination (UN CMCoord).

CIMIC has two identities, each with different motives (Gordon, 2001; Studer, 2001; Zandee, 1999). The *political identity* of CIMIC can manifest itself at the lowest level as building 'consent' to the presence of a military force and therefore providing a means of protection (Gordon, 2001). The *humanitarian identity* of CIMIC contains the idea of being a component of the broad strategy for transition to peace.

On examining the role of military forces during humanitarian emergencies three roles can be assumed:

1. They can promote a climate of security for civilian populations and humanitarian organizations and provide protection for the relief effort.
2. They can provide technical or logistical support to humanitarian organizations.
3. Finally, they can provide direct assistance to populations in need. (Rietjens *et al.*, 2007)

A further examination of point two demonstrates that similar to governments, military forces can have a monopolistic position or control over certain resources. The cooperation of national military staff and the use of military infrastructure and assets such as airports, warehousing facilities, helicopters, vehicles and classified information facilitate response (Moore & Antill, 2002). Similarly, access to internationally available military resources helps augment response capacity and ensure a speedier delivery of relief items to the disaster area and their eventual distribution to the afflicted populations. During conflicts, an operational dialogue with combatant forces ensures access to logistics resources and guarantees safe and secure land, sea and air operations (Tatham & Kovács, 2007).

Access to such resources increases the level of operational flexibility, can improve operating costs and level of accuracy with respect to the storage and timely transportation of relief items. The military are not monolithic however, and the use of the word 'military' must be treated with caution (Connaughton, 1996). This is because a military force may take many different forms. Force size, structure, capability, and posture may vary enormously and there are also disparities in military competence and professionalism.

When analyzing specific civil military coordination definitions, policies and doctrines it should first be noted that these have all been developed for the operational/tactical level. There seem to be two main conceptual streams, i.e. 'cooperation' and 'coordination', and an acronym soup of specific functions and titles: NATO CIMIC, EU CIMIC, CIMCO, CMO, CMCoord, etc (see key terms and definitions) (De Conning, 2007).

From a UN perspective, coordination refers to a spectrum of relations that range from coexistence to cooperation. This UN coordination concept has been developed in the context of humanitarian civil military coordination, where coexistence refers to a situation where the minimum necessary information is being shared between the humanitarian community and a military combatant force (De Conning, 2007). This would typically include sharing of information about security, movement of humanitarian convoys and the management of shared resources, e.g. a port or airport. Cooperation refers to a maximum state of civil military coordination where there is a range of cooperative relations between the humanitarian community and a military force that is not regarded as a combatant force, typically including joint planning, division of labor and sharing of information (OCHA, 2003). The UN and NATO understanding of cooperation and coordination seem to be reversed, because in the NATO context, cooperation is understood to imply a less binding relationship than coordination, and NATO argues that the humanitarian community will be willing to cooperate, but not coordinate, and therefore they use cooperation (NATO, 2004).

A solution to the alphabet soup and different views would be to adopt a pragmatic approach based on the recognition that 'CIMIC', as an acronym, is now so entrenched in the military culture that it would take a disproportional effort to dislodge it. A practical approach would be to capitalize on the fact that most military officers should be familiar with the 'CIMIC' acronym, and probably understand it broadly to refer to civil military relations.

CHALLENGES FOR CIVIL MILITARY COOPERATION/COORDINATION IN HUMANITARIAN RELIEF

Operational humanitarian organizations regardless of their size or area of specialization (refugees, children, food, etc.) face a number of challenges. Each of the following subsections presents issues that negatively impact CIMIC in logistics operations during humanitarian relief missions. Those in management/command appointments have to consider not only the effect the barriers listed below have on their agency, partners and beneficiaries, but as leaders they are challenged to direct and control logistical operations within their remit.

Coordination

To stage its response, the military and humanitarian organizations need to coordinate their activities with other humanitarian organizations and key stakeholders with a view to manage fit, flow and sharing resource dependencies (Kovács & Tatham, 2009). As time is crucial during an emergency response, the military and humanitarian community need to avoid parallel efforts and duplications along the supply chain by combining their efforts and activities with other stakeholders. Joint contingency planning helps reduce lead times and avoid reliance on expensive transport options (e.g. airlifting) or last-minute sourcing (Heaslip *et al.*, 2007b). This planning further helps in the development of alternative plans and ensures the mobilization of the right range of relief items. During joint contingency planning sessions, implementation strategies such as prepositioning - strategic placement of food or Non Food Items (NFIs) throughout a country/

area accessible for distribution to recipients - act as key coordination mechanisms.

Humanitarian logisticians often have little or no advance notice of when and from/to where they have to move and store what type of material and in what quantities. To come to terms with unpredictable disasters, the evolving logistics challenges and the specificity of each operation and country, humanitarian organizations (e.g., World Food Programme (WFP)) formulate flexible plans that can accommodate frequent and numerous last minute changes. Where possible, they build redundant capacity into their operations (Tomasini & Van Wassenhove, 2004b). In terms of transport modes, to reach remote areas the humanitarian community call upon a diverse and combined transportation portfolio. This includes trucks, motorcycles, boats, and helicopters as well as unconventional modes such as donkeys, elephants and airdrops. In WFP operations, trucking often replaces airlifts after the initial emergency phase (Tomasini & Van Wassenhove, 2004b). Ocean transportation is used to move significant quantities of cargo (e.g. food) at low cost.

With such an influx of aid workers, military and others entering the fray, and often having differing objectives, cooperation, coordination and collaboration among the organizations taking part in the response is paramount. In a disaster the scale of that in Haiti, the UN activates the cluster group system (which was developed after Hurricane Katrina and the Asian Tsunami in order to improve coordination). These cluster groups (UNJLC, 2008) include:

- Water, sanitation and hygiene (Wash) cluster: chaired by UNICEF (United Nations Children's Fund)
- Camp co-ordination and management cluster: chaired by IOM (International Organization for Migration) for natural disasters

- Emergency shelter cluster: chaired by IFRC (International Federation of Red Cross and Red Crescent's) for natural disasters
- Logistics cluster: chaired by WFP (World Food Programme)
- Emergency telecoms and IT (Information Technology) cluster: chaired by UNICEF / WFP
- Health cluster: chaired by WHO (World Health Organization)
- Nutrition cluster: chaired by UNICEF
- Early recovery cluster: chaired by UNDP (United Nations Development Programme)
- Protection cluster: chaired by UNHCR (United Nations High Commissioner for Refugees) / UNICEF

Relief items differ in terms of their urgency (water), life saving contribution (medicine), volumes required (food), complementary (kitchen kits) or simple (clothing) function, continuous (food) or one-of-a-time (vaccines) usage, degree of substitutability, and need for specialized input (medical personnel for medicines and medical care). The different value and characteristics of relief items together with the mismatch between available transport and handling capacity and the volume of cargo moving into a region call for the prioritization of relief item movements into the disaster theatre (Heaslip, 2010). Prioritization – movement of food, NFIs and people in accordance with humanitarian priorities – helps overcome a sharing dependency. By participating in joint contingency planning (e.g., Kosovo, Haiti) the military and humanitarian organizations are able to identify potential problems associated with the movement of food and NFIs and the potential security threat to the movement of people (refugees, returnees, humanitarian and media operators, etc.) as well as providing the transportation and human resources to solve the problems.

Organizations do not necessarily share a common understanding of the requirements or objectives of coordination. This problem is compounded by the fact that many organizations do not have established and well-defined working relationships with one another (Olson & Gregorian, 2007). Difficulties also arise when organizations are stretched beyond their traditional areas of expertise, as when development agencies take on security sector reform and the military becomes involved in state building operations. Few models are available on best practice in combining and sequencing assistance, development, state building, security and stability in so-called "failed state" (De Conning, 2007).

Despite its many advantages, military coordination with potential partners in a humanitarian crisis is often difficult because there is no official structure to oversee/manage/coordination of activities. Particularly at the operational level, coordination among NGOs, International Organizations (IOs), donor governments, and military forces lacks structure. Because the structures often vary considerably from crisis to crisis, establishing relationships and procedures is difficult.

Response to humanitarian crises is fractioned and organized organizationally along mandates and functional lines (Kent, 1987; Borton, 1993). Therefore, when a large scale emergency strikes, a large number of actors converge to the site. Once in the operating theatre, the actors tend to compete over the same range of resources at exactly the same time. Given the plurality of humanitarian organizations, the sector has long identified the need for effective ways and means to ensure inter-agency coordination. However, some aid organizations consider their independence a higher priority than coordination with other organizations. They are not prepared to follow the lead of another organization, particularly if it is the military (Heaslip *et al.*, 2008a). There is also a fear among NGOs of cooption and marginalization in some crisis regions where military forces have an overwhelming presence.

Some agencies argue that coordination (or integration, as in the UN system) is, by definition, a threat to humanitarian action because it undermines impartiality and represents a fundamental threat to the operational flexibility and physical safety of aid workers (Olson & Gregorian, 2007). The counter argument posits that humanitarian space can be better protected through integrated structures as opposed to a fragmented approach and that the humanitarian perspective will have a more effective voice when at the same table with other elements of an integrated mission. There are also questions about the capability of humanitarian and aid organizations to provide for their own security in highly dangerous settings, the ethics of leaving some area without assistance because they are too dangerous for aid workers, and the ability of soldiers to provide quality aid (Olson & Gregorian, 2007, De Conning, 2007).

Culture

It is easy to conceptualize civil military coordination as an effort by humanitarian and military organizations to synchronize their operations, and yet it has been described as a "contested concept with many different, competing definitions and doctrines that describe essentially the same activity…" (De Coning, 2007, p. 6). Military culture and civilian cultures do not generally mesh seamlessly in relief settings. There are inherent stressors between them owing to differences in mandates, objectives, methods of operation and vocabulary. When the cultural differences confront each other on the ground, the inability to communicate effectively, caused by a lack of mutual understanding, creates tension. The tension manifests itself in five distinct areas – expectations, perceptions, resources, missions and values (Slim, 2006).

Schein (1996) describes three groups within organizational culture namely, strategic, operational and tactical and how a lack of lack of alignment among the three groups can hinder learning in an organization. Within the organization culture manifests itself at three levels: the level of deep tacit assumptions that are the essence of the culture, the level of espoused values that often reflect what a group wishes ideally to be and the way it wants to present itself publicly, and the day-to-day behavior that represents a complex compromise among the espoused values, the deeper assumptions, and the immediate requirements of the situation (Schein, 1996). Achieving consistency and coherence between the field level (tactical level) and the Headquarters (HQs) or policy level (strategic level) within an organization is another key element of internal coordination. Organizational culture at the different levels and inadequate awareness of the factors affecting decisions at each level commonly results in a disconnect between HQs and the field. In many organizations, the trend has been towards greater delegation of decision-making authority from HQ to the field in order to increase flexibility and responsiveness to rapidly changing circumstances.

As personnel from humanitarian organizations interact more regularly with military personnel, culture clashes become apparent. Former- Head of United Nations Protection Force (UNPROFOR) Civil Affairs, Cedric Thornberry, explains that the lack of agency cooperation can be largely blamed on a 'two-way lack of familiarity for the attitudinal abyss which separates aid workers from the military' (Siegel, 2003; Duffey, 2000). Aid workers often distrust the military, and the military similarly are suspicious of aid workers. Such unfamiliarity between organizations inevitably encourages the dissemination of ill-informed stereotypes. The military is frequently characterized as an insensitive, ill-informed, controlling, and inflexible war–machine, while personnel of some humanitarian organizations are seen as sandal-wearing, undisciplined, and uncoordinated liberals (Duffey, 2000). Operationally, aid agencies tend to be flexible whereas the military functions in a top-down manner, the durations of stay of aid agencies can be for many years, the military, on the other hand, prefers well defined end states and exit strategies, aid agencies have a culture of independence while the military is hierarchical, and soldiers are armed when deal-

ing with local actors while aid and development workers are not (Ferks & Klem, 2006). Recently researchers and practitioners, in addition to aid organizations, militaries and governments, have instigated different forms of cooperation (Kovács & Spens 2007), the Humanitarian and Emergency Logistics Programme (HELP) in the UK, and Humanitarian Logistics (HUMLOG) in Scandinavia are some of the forums encouraging dialogue.

With continuous and multiple points of interface, military personnel and humanitarians interpret the world through the lens of their own culture. Lack of familiarity with the differences embedded in the organization cultures is a breeding ground for misunderstanding and poor coordination and cooperation (Siegel, 2003; Duffey, 2000). In many circumstances the use of a different language and terminology further obscures understanding, compounded by different interpretations of the same terms of reference (Heaslip *et al.*, 2008a).

Organizations will not learn effectively until they recognize and confront the implications of the three occupational cultures (strategic, operational and tactical). Until military and humanitarian personnel discover that they use different languages and make different assumptions about what is important and until they learn to treat each others culture as valid and normal, organizational cooperative efforts will continue to fail. Some solutions include taking the concept of culture more seriously. Instead of superficially manipulating a few priorities and calling that "culture change" (Schein, 1996) both the military and humanitarian community must recognize and accept how deeply embedded the shared, tacit assumptions of military and humanitarian personnel are. Avenues to communicate across the cultural boundaries have to be found. This is achievable by establishing some communication that stimulates mutual understanding rather than mutual blame. The creation of communication by learning how to conduct cross-cultural "dialogues" needs to be encouraged. If the military and humanitarian organizations (who come from different cultures) sit in a room together, which is hard enough, they must reflectively listen to themselves and to each other, which is even harder.

Resource and Capability Gap

Immediately after a humanitarian emergency is declared humanitarian organizations regardless of their size or area of specialization (refugees, children, food, etc.) face a number of challenges. The first challenge facing a humanitarian organization immediately after a humanitarian emergency is declared is how to bridge the relief resource and capability gap, in other words uncertainty, which is often significant. To stage a response and overcome this gap, military and humanitarian organizations depend on their supply network composed of a number of loose partnerships with a range of actors (Kovács & Spens, 2007). Once in the operating theatre, humanitarian supply chains tend to compete over the same range of resources at exactly the same arch of time.

In humanitarian crises uncertainty can stem from many elements, such as the organization itself, or the nature of demand. For example, uncertainty may arise from inherent characteristics such as what and how much material is demanded, product traits, process fluctuations, and supply problems (Van der Vorst & Beulens, 2002). Van der Vorst & Beulens also recognize how decision complexity, supply chain configuration and control structures, long forecast horizons, poor information reliability, and agency culture may create uncertainty (2002). Regarding uncertainty, Sowinski (2003) quotes Lynn Fritz, founder of the Fritz Institute: "...disasters are the embodiment of randomness. You don't know when they're going to happen, where it's going to happen, and who's going to be affected" (Sowinski, 2003, p. 19).

One of the most significant organizational capability resources of the humanitarian model is the local presence of a humanitarian organization. Not only does this determine the focus of the organization on particular activities (e.g. a focus on region-specific disasters), but also, impacts on the potential minimum speed of the organization to respond to a particular disaster (Kovács & Tatham, 2009). Humanitarian organizations with a local presence or "chapter" have access to local knowledge about beneficiaries and their needs and customs, something the military lack, and the humanitarian organizations are not dependent on the declaration of a state of emergency before they can deliver aid to beneficiaries, unlike the military. On the opposite end of the humanitarian spectrum, international aid agencies and even the United Nations Joint Logistics Centre (UNJLC), which has a mandate to provide logistics information services during the immediate response to large-scale disruptions, can only start operating in a disaster-struck region upon the fulfillment of certain political criteria such as a declaration of emergency.

This problem (of uncertainty) is amplified by distance. Long & Wood (1995) observed that often the office coordinating the aid mission is far away from the actual disaster site and must make assumptions about the types and quantities of aid that should be supplied. The same is true for the military where national headquarters make assumptions as to the requirements for troops being deployed. Once response teams and advance military parties are in place at the disaster site, the supply pipeline can transition from a "push" system to a "pull" system based on more accurate needs assessments and communications back to headquarters and donors (Long & Wood, 1995). These assessments should also include anticipated needs (PAHO, 2000). If supplies are "pushed" through a system, quantities are dictated by an upstream authority with little or no input from the customer. In a "pull" system, quantities are determined at the point of consumption (Rodman, 2004).

The Asian tsunami highlighted another element of uncertainty. Well intentioned donors generate supplies and manpower support for the relief effort that are of the wrong type or condition.

This places an added burden on the relief effort as variability in quantity, quality, and suitability of products as well as overloading the process of sorting, storing, and distribution become an unnecessary headache for the relief community. The United Nations Disaster Management Training Programme (DMTP) observes that:

"... consistently, many of the internationally supplied relief goods flown into countries ... prove to be inappropriate and unnecessary ... [and] may even be a barrier to more important deliveries" (DMTP, 1993, p. 10).

Financial Resources

Unfortunately for many disaster-struck areas, such as Haiti, funding is often focused on short-term disaster relief (the immediate response phase). Yet the phase that frequently gets little public and media attention and is probably the most important, is the preparation phase. There is recognition that disaster preparedness enhances disaster response efficiency and effectiveness (Kovács & Spens, 2007). However, donors prefer to fund emergency activities and are often reluctant to cover core costs necessary to strengthen organizational capacity and capability. On the other hand, military organizations receive funding in order to be prepared for a disruption (i.e. the development of a capability), whilst in humanitarian organizations such funding is only received in response to a disruption (Kovács & Tatham, 2009). Funding for organizational support and infrastructure is often neglected under donor demands that as much aid as possible is visibly pushed to victims. Thus, distribution channels may suffer as warehouses, equipment, communications infrastructure, and training remain unimproved or deteriorating. As a result, there may be aid available, but the humanitarian organization may be incapable of effective delivery in a timely manner due to limiting factors in the distribution process.

Investing in disaster preparedness allows the military to focus on physical and manpower resources, i.e. training, but this is inherently lacking for humanitarian organizations (Kovács & Tatham, 2009). Humanitarian organizations may, for example, be reluctant to spend money on a sophisticated information system that would actually improve their efficiency in the long run. Earmarking funds specifically for the affected population can also lead to a lack of parts and service support for the truck and planes required to move material aid and lack of funding for unallocatable costs such as headquarters expenses (Pettit & Beresford, 2005).

Donors tend to evaluate humanitarian organizations on the percentage of funds used on direct relief activities (Binder & Witte, 2007). As such, organizations with higher overhead costs, for example, higher investments in support activities such as IT, are often considered less efficient. In contrast, those with lower overhead costs – the most-valued indicator of efficiency – are often rewarded with additional voluntary donations. To improve its operations, a humanitarian organization has to overcome the prevailing donor mind-set and ensure investment in the area of disaster preparedness. Thus it is unsurprising that historically humanitarian organizations have been criticized for under-funding the preparation phase (Kovács & Tatham, 2009).

In effect, the earmarking of funds violates the humanitarian ethos of impartiality by placing stipulations on how relief is administered. This is a political and administrative problem that has underlying implications for NGOs since earmarking aid focuses the relief agency on delivering the majority of aid to the most affected populations in the most visible crises, not on promoting efficiency (Binder & Witte, 2007). In effect, in spite of their non-profit nature, NGOs need to compete, the quality of their field programs affects their capacity to gather government grant funding and their public visibility affects their private contributions. Donor earmarking of funds and stipulation of how materiel aid is distributed also inhibits rapid progress during relief operations and is a method to influence the global policy of humanitarian organizations because of the fear that funding will be "turned off" if donor stipulations are not met (Binder & Witte, 2007).

Human Resources

In addition to financial resource mobilization and management, another critical element of an effective humanitarian response is the mobilization and deployment of material and human resources (Forman & Parhad, 1997). Poor or nonexistent training ultimately affects the quality of any operation, particularly a relief operation. The unpredictable nature of emergencies makes it difficult to retain well trained employees, and those who have been trained are often volunteers who can only work for short periods before they must return to their "real world" jobs. Organizations may experience as high as 80% annual turnover in field logistics personnel (Thomas, 2003), further compounding personnel issues. This results in a constant influx of untrained personnel, inexperienced in the particulars of logistics within the organization and relief as a whole. Natsios makes a dramatic point by stating that:

"... the rolling tide of complex emergencies has caused organisations to be drawn into each new major crisis before completing work on the last... this has meant that NGOs and UN organisations are increasingly sending inexperienced staff to the field to run massive operations that even seasoned managers would find intimidating". (Nastios, 1995b, p. 417-418)

The military has several virtually unique resources. They can both protect and defend themselves and break down the resistance of others with violence, they have rapid access to strategic and tactical transport resources, they can be self-sufficient for a longer period, and they have

specialized aircraft capacities, maritime resources, reconnaissance, intelligence capacities, and an effective communications network (Eriksson, 2000). The military logistic network and machinery are extensive and speedy (claiming delivery in about five days from project approval, compared to five months for the EU and UN) (Pugh, 2001).

Humanitarian organizations, especially NGOs and development organizations, who maintain a presence within a country, will often be the first on the scene when a disaster strikes. Natsios (1995a) posits that most recruits (in humanitarian organizations are trained on the job, with few standardized instructional resources, "...and where NGO doctrine does exist, it comes out of generally shared experiences and responses, is seldom written down, and is not always followed uniformly" (1995a, p. 70).

Thomas points out that there may be problems with employee reliability (2003) stemming from lack of training. There is a notable lack of employees who are knowledgeable in supply chain or logistics management. In the logistics area, the challenges facing humanitarian organizations are the formal qualification of logistics staff (Oloruntoba & Gray, 2006), optimization of their logistics activities and the integration of activities across business functions. Long notes that "most people from development agencies... have backgrounds in public policy or third world development, and professional logisticians are rare" (1997, p. 27). Recently however, the setting of standards and framework agreements, to the training and education of staff has improved (Kovács & Tatham, 2009).

The "lack of universally accepted performance indicators" (Macrae, 2002, p. 5) makes gauging mission success and learning lessons from the operation difficult. The Sphere Project handbook, though accepted as a reference for performance measures in the humanitarian sector, lacks thorough detail, particularly on cooperation and coordination. The book advocates the integration of "many different players" (Sphere Project, 2004) in humanitarian relief but doesn't state specific measures for evaluating performance of cooperation. Even in the United Nations humanitarian system, "evaluation and lessons learning is something that the IASC [Inter-Agency Standing Committee] has yet to tackle in a serious way" (Jones & Stoddard, 2003, p. 17). Without performance standards, employees have no means to gauge their success and no reference for making their operations better.

Lack of funding for back-office infrastructure and processes and the need to upgrade the logistics function including its information and knowledge management aspect (Van der Laan et al., 2007) have attracted the first wave of structured business-humanitarian partnerships. Binder and Witte (2007) conclude that the role of business in humanitarian relief is becoming more prominent even if it remains a limited phenomenon.

Alliances between humanitarian actors and commercial corporations are also increasing (Beausang, 2003). Both NGOs and Corporations are pursuing these partnerships suggesting potential benefits are perceived by organizations in both sectors (Yamamoto, 1999). Simultaneously, as nation-state support for humanitarian issues declines, the corporate domain is increasingly highlighted as a source of potential donors and partners while stakeholders increasingly drive corporations to demonstrate participative support for societal requirements in line with the triple bottom line, of profits, people and the environment (Goddard, 2005). Business contributions range from ad-hoc donations to donations provided through a partnership structure (Binder & Witte, 2007). To overcome the underinvestment in preparedness capability and to improve upon the response function, humanitarian organizations have to go beyond the generic and tied resources offered to them by their traditional donors and tap into specialized and additional resources. In this regard, select resources within business are relevant. Business can contribute to a humanitarian organization's relief operation with their specialized resources and expertise in a number of ways: sharing of physical logistics resources (e.g., airplanes, trucks, warehouses, etc.), donation of company products (e.g. food and NFIs), secondment or allocation of personnel, access to organizational capability and resources (e.g., tracking and routing systems), (Thomas & Kopczak, 2005).

In terms of HR resources and development, a business can contribute to the professionalization of the humanitarian sector in general (e.g. TNT's partnership with WFP) (Tomasini, & Van Wassenhove, 2004b) and the logistics function in particular by supporting research on disaster management, delivering formal training, establishing networking initiatives, etc. Business can facilitate the transfer of sound and relevant supply chain practices from the commercial sector to the humanitarian community. By using for example, IT solutions that generate, store, manage, and transfer information along the supply chain, humanitarian organizations can improve the management of their dynamic supply chains and their preparedness capabilities (Tomasini, & Van Wassenhove, 2004b).

Infrastructure Degradation

Even before a crisis situation has arisen, the quality of the infrastructure of a potential host country, its topography and its political situation are all factors that often conspire against efficient logistical operations. Inadequate transportation, housing, shelter and communications (see section below for further analysis) are further barriers to effective delivery of aid.

In the DMTP logistics handbook, it states "the overall effectiveness of relief logistics often depends on the level of prior investment in both the transport and communications infrastructure and how far relief requirements have been considered in the planning" (DMTP, 1993, p. 12). Rapid onset of a disaster may degrade the country's existing infrastructure to the point where delivery of aid is severely hampered, as in the recent case of Haiti. As noted by Gooley (1999, p. 82), "Often...

transportation infrastructure is in poor condition and cannot handle the huge numbers of refugees, military vehicles, and relief shipments that pour into these areas in times of disaster".

System-wide, the military commanders and humanitarian managers could encounter delivery options ranging through ships, aircraft, rail, and trucks. At the same time, those routes may closed or clogged (Moody, 2001) limiting distribution:

"Accurate assessment of the road infrastructure is critical...a road may be a five-foot wide strip of mud only inches above the water line that can accommodate only scooters and livestock, or it can be an eight-lane highway pocketed with bomb craters" (Long & Wood, 1995, p. 225).

Previously, the shelter sector was understood mainly in terms of distributing tents or plastic sheeting: temporary shelter before the reconstruction of permanent housing. In Haiti many of the public buildings, residential housing and utilities were destroyed by the earthquake. There was severe damage to hospitals, schools and the Port-au-Prince's two seaports. This infrastructure damage resulted in large scale displacement of people from their homes, into makeshift and overcrowded shelters. In some areas, up to 50% of buildings are completely destroyed and most others damaged, but repairable. Shelter for the hundreds of thousands of survivors remains a desperate race against the clock as the rainy and hurricane seasons looms (Heaslip, 2010).

Aid agencies, such as World Vision whose mandate is to respond in some way to any disaster around the world, has created a phased relief response which typically occurs in three phases: seven-day, 30-day, and 90-day. For example, during the first phase of the emergency, flyaway kits are provided. These can sustain up to 2,000 people for seven days. The second phase involves sending family survival kits, which can support up to 5,000 people for 30 days. The third phase is related to reconstruction and it involves long-term

rehabilitation. For example, in the aftermath of the earthquake in El Salvador, reconstruction assistance was provided by fixing damaged homes and also by constructing new homes for displaced families (Sowinski, 2003).

In humanitarian crises not alone does the destruction affect private houses and social institutions: schools, hospitals, health centers, but in some cases (e.g., Afghanistan) water systems are heavily damaged and, as a consequence villages are without a water supply. Here the military and aid agencies (e.g., Oxfam) have an opportunity to collaborate to alleviate the suffering of the population.

The obstacles mentioned above must be dealt with on a case-by-case basis due to the unpredictable effects of disasters and the vulnerability of the infrastructure.

Communications

A major barrier to delivery of aid is poor communication. Not only are there obvious difficulties associated with speaking to someone using a different language, but as in Haiti the communications infrastructure may be crippled. Relief agencies may not be able to communicate upstream with headquarters or donors during a disaster. Military forces however, can supply specialized capabilities such as communications equipment and information technology and information sharing capabilities (Moore & Antill, 2002). Moreover the military possess the capability to establish a communication network from a green field site. With the priority being the completion of completing an accurate assessment in the immediate aftermath of a humanitarian crisis a fully functional communications and information systems network, plays an important role in delivering the right information regarding the right amount of aid to be delivered to the right people (Kovács & Spens, 2007). In the humanitarian community OCHA is tasked with the role of coordinating the assessments, the dissemination of information regarding the affected areas and the appeals process, however sometimes it is not in a position to provide the communications infrastructure to deliver the necessary information (e.g., Haiti) and in these situations the military has a key role in filling the communications gap.

Technologies are indispensable tools for many essential mission tasks. Modern technologies can extend the range of observation and communication, improve the safety of personnel, and enhance the efficiency and effectiveness of the mission (Wheatley & Welch, 1999; Pettit & Beresford, 2005). Military organizations can have specialized divisions and units (e.g., the army, navy, logistics, and communications) which need to be interoperable, but they still belong to the same defense force. Humanitarian actors, on the other hand, can be quite specialized even within the UN family (with WFP delivering food, UNHCR shelter, WHO medication, etc.). For civil-military cooperation, technologies of both organizations should be compatible. As noted by Rietjens (2006) the use of incompatible communications equipment (field phones, satellite phones, short wave radios) was a widespread problem in the former Yugoslavia. Some UN military contingents possessed more technically advanced equipment than that of NGOs or even other military contingents, making communications in the field difficult and often impossible (Beauregard, 1998; Kovács & Tatham, 2009).

Long & Wood explain that organizational language and terminology may hamper the aid process. For example, some organizations estimate need on a family basis whereas others use a per person basis (Long & Wood, 1995). Organizations may use different names and definitions for transportation modes, supplies, the composition of worker teams, etc. Long & Wood observe that:

"Ironically, inter-organizational relations are usually a challenge to the relief effort instead of a source of support. Each organization has its own operating methods and goals, and it is only with great effort that they coordinate their plans and share resources". (Long & Wood, 1995, p. 216)

This is an indication that organizational and cultural language may lead to procedural difficulties (Long, 1997). This inability to coordinate effectively is common during emergency response and is only made worse by disputes between organizations, and reluctance to share information which will ultimately lead to duplicated efforts and wasted resources (PAHO, 2000). To overcome this humanitarian organizations and military organizations should settle on a common language regarding relief missions (Barry & Jeffrys, 2002).

Communications problems exist long after the effects of a disaster are mitigated. Sowinski states that a lack of funds at the end of a humanitarian action often limits recording of best practices and tracking of information on complex supply chain conditions. It thereby hampers learning opportunities and institutional memory regarding successes and failures (Sowinski, 2003). As the money runs low and the relief mission and its workers fade into the background, it is understandable that events could slip by unrecorded. The military for their part record all aspects of their mission and debrief national headquarters on their return (Petit and Beresford, 2005).

Personality

Olson & Gregorian (2007) suggest that success in cross-sectoral collaboration, particularly in the realm of civil-military relations, often depends on the personalities of the field level personnel and the liaison structures that are established. Given the high rates of staff turnover in the field, particularly amongst the relief and development community, reliance on individuals becomes a risky business. Uncooperative attitudes are not uncommon within and across organizations. This may result from competition for resources, for power, and for notoriety, but it may also arise from personal likes and dislikes or stereotyping (Heaslip *et al.,* 2007a).

Coordination between military forces and humanitarian organizations is found to be driven primarily by personalities rather than well-developed standard operating procedures (Brocades-Zaalberg, 2005). Since efforts are person-dependent, they vary within and between different military contingents. Beauregard (1998) in his study of civil-military activities during a number of disasters identifies six principal factors that hamper coordination and cooperation. These include differences in cultures and ideologies, differences in organizational structures and chain of command, communication breakdowns due to incompatible equipment or absence of communication procedures, refusal by humanitarian organizations of military assistance to protect independence and impartiality, and the threat or use of force by the military. He concludes by suggesting a range of solutions (training, better communication and consultation processes through events that improve mutual understanding, liaison teams) to improve the civil-military relationship.

SOLUTIONS AND RECOMMENDATIONS

During humanitarian relief operations, strongly motivated people in both camps (i.e. civil and military) usually find ways to surmount barriers that they encounter, but valuable time is lost inventing and reinventing these solutions. At a philosophical level, it is noted that NGOs are uncomfortable with the military, but in the field there is often effective cooperation. As personnel security in relief operations becomes a growing concern for the relief community it follows that interaction with the military is set to grow. Enhanced CIMIC requires a greater effort of all actors involved in humanitarian relief. The points discussed here are not in any way intended as the

ultimate solution, but rather as a starting point for further discussions.

Boundaries

Humanitarian organizations adhere to the principles of neutrality, impartiality, and humanity. A key challenge for the humanitarian community is to clarify for themselves and the military how humanitarian principles apply to activities in settings such as humanitarian relief. The military do not apply these principles in quite the same way and consequently close cooperation between humanitarians and other actors can lead to the perception that the humanitarians have become 'tainted', reducing their ability to gain access to those in need.

The principle of neutrality and the division of tasks from the military is guarded by agencies engaged in humanitarian work. The military have the logistical capacity to deliver aid in certain circumstances where humanitarian actors cannot gain access (due to degraded infrastructure, security threats) so, should the military not intervene? To ensure clear delineation the military should focus its efforts on establishing security and resist the temptation to promote its mission in humanitarian or development terms. Local populations appreciate security and stability in humanitarian crises.

Military and humanitarian organizations need to establish 'clear blue water' between one another otherwise boundaries become blurred, which results in lives becoming endangered, particularly the beneficiaries and aid workers. The solution is likely to involve closer contact and joint training exercises before deployment to foster understanding and trust between the various organizations.

Planning

The military have a very clear idea of what is meant by levels of strategic, operational and tactical planning, this is not necessarily true of the aid community. There are a number of reasons for this, the use of relatively flat organizational structures in the aid community do not promote hierarchical management as in the military. What is needed is a way of working side by side that advances the portion of their respective work that is mutually supportive, but that is different from integration.

Finance

Funding structures can impede the humanitarian community in delivering aid in a timely and efficient manner. The military often have access to substantial discretionary funds, whilst the humanitarian relief community must go through much longer processes to secure funding. Mechanisms need to be developed to facilitate the swift transfer of funds from donors to NGOs that would allow NGOs to respond more quickly to urgent needs in areas where there is an international military presence and preclude the need or temptation for militaries to fill these roles themselves.

A more positive development has been the donors' insistence on receiving detailed plans from agencies they fund; this encourages aid agencies to plan more strategically, by outlining their objectives, outputs, impact, activities and outcomes. This approach encourages the organizations receiving funds to manage projects effectively and efficiently and work towards a strategic end that benefits the broader stakeholder community and complements, rather than duplicates or erodes, the work of other intervening agencies.

Human Resources

The issue of selection and recruitment remains the cornerstone for effective logistical management in CIMIC. As in business logistics the quality of people is critical in the delivery of humanitarian aid, particularly those that have the skill set to operate complex logistical systems. The implications for training and the costs of solutions are important in this regard. Innovation in problem solving is needed, along with the importance of understanding the sources of uncertainty in the humanitarian supply chain.

Calls persist for more cross-fertilization to take place within military, UN agency and NGO training programs and contingency planning. The skills and approaches used and the personalities of the people involved matter to coordination outcomes. As well as strong logistical knowledge and skills, recruitment should emphasize skills that include negotiation, conflict management, leadership capabilities and/or abilities, as well as interpersonal and communication skills, if civil and military personnel are to bridge major organizational divides and promote coordination.

Efforts are underway to encourage military and civilian personnel to participate in joint workshops and exercises (e.g. Exercise Viking – a European interagency exercise) to study subjects of mutual interest, and to learn about each other's perspectives. There could be merit in conducting joint contingency planning, particularly logistical planning.

Communications

In many cases, the use of different 'language' and terminology seem to obscure any understanding of common objectives between the actors. Indeed it is questionable whether the various actors even have common objectives; perhaps it is more correct to say that each organization will have its own objectives, but that they should all work towards a common purpose or vision.

There needs to be mechanisms for collecting information (e.g. inventory requirements) but equally importantly, for producing useful knowledge from that information. There must be an expedient means of conveying knowledge to those who need it. To assist in the distribution of information, transparency needs to permeate across all organizations military and humanitarian which will provide answers rather than obscuring questions due to institutional resistance.

Furthermore, as language and communication has proven to be a barrier it may prove beneficial to use neutral language when describing coordination meetings. This gesture would avoid misperceptions and mistrust.

FUTURE RESEARCH DIRECTIONS

Sowinski posits that a lack of funds at the end of a humanitarian action often limits recording of best practices and tracking of information in humanitarian relief. It thereby hampers learning opportunities and institutional memory regarding successes and failures (Sowinski, 2003). As the money runs low and the relief mission and its workers fade into the background, it is understandable that events could slip by unrecorded. The military for their part record all aspects of their mission and debrief national headquarters on their return (Pettit & Beresford, 2005). This transfer of knowledge and information is crucial in building and developing efficient and effective logistical relief for the next humanitarian crisis. Research should be conducted in analyzing mechanisms to encourage humanitarian agencies to capture information pertaining to the success and failure of a relief effort. This research should include cross referencing with the military to determine where improvements and greater efficiencies can be found.

An added dimension of many humanitarian operations with "wide variation in the quality of field programs and the technical competence of staff" is that beneficiaries and donors often have no way to gauge the effectiveness and accountability of humanitarian agencies at the field level (Natsios, 1995b, p. 409). The "lack of universally accepted performance indicators" (Macrae, 2002, p. 5) makes measuring mission success and learning lessons from humanitarian relief difficult. The Sphere Project handbook, though accepted as a reference for performance measures in the humanitarian sector, lacks thorough detail, particu-

larly on cooperation and coordination. The book advocates the integration of "many different players" (Sphere Project, 2004) in humanitarian relief but doesn't state specific measures for evaluating performance of cooperation. Even in the United Nations humanitarian system, "evaluation and lessons learning is something that the Inter-Agency Standing Committee (IASC) has yet to tackle in a serious way" (Jones & Stoddard, 2003, p. 17). Further research of performance management and performance indicators is necessary. Without performance standards, the humanitarian community have no means to gauge their success and no reference for making their operations better.

One avenue that can be pursued is to adopt models for cooperation between the military and the civilian actors (Currey, 2003; Gourlay, 2000). These models can synthesize what has been accomplished in previous operations and can foster a theatre specific *modus vivendi* between military formations and the variety of civilian actors. For those involved in humanitarian relief, models can contribute to the development of checklists, an increased understanding of (potential) conflicts in the process of cooperation, and elements for procedures to increase the performance of the cooperation. Models can provide guidance about how partners may foster and manage relationships that will achieve favorable outcomes (Tuten & Urban, 2001). By conducting further research in this area it offers researchers a framework for future empirical studies to confirm or refute the legitimacy of the model.

CONCLUSION

A number of common threads emerge from the themes explored in this chapter, namely boundaries, planning, finance, human resources and communication.

Coordination between the military and relief partners, particularly NGOs, is often uneven and uncertain. NGOs can be difficult partners, especially for the military. There is a wide gap in organizational culture, and NGOs are inhibited by their concern for neutrality and impartiality. Attempting to bridge what are perceived to be cultural and operational disconnects, may very well dilute the consolidated product/service that an inter-agency effort delivers.

A further theme explored is an incremental approach to planning, based on transparency between agencies and a continuous cross-fertilization process of examining best practice and extending existing networks. NGOs by virtue of public expectation (re: allocation/use of donations) do not have the same emphasis on pre-crisis preparation/planning, making cooperation before a crisis difficult. There is an evident lack of mutual familiarity, and NGOs are often reluctant to share information with the military.

The mobilization of finance has become a contentious issue. Donors prefer to fund emergency activities and are often reluctant to cover core costs necessary to strengthen organizational capacity and capability. As a result, the humanitarian organization may be incapable of effective delivery of aid in a timely manner due to limiting factors in the distribution process. Earmarking funds specifically for the affected population can also lead to a lack of parts and service support for the trucks and planes required to move material aid and lack of funding for un-allocatable costs such as headquarters expenses. By receiving funding to prepare and train for disasters the military can focus on physical and manpower resources, i.e. training, but this is inherently lacking for humanitarian organizations.

Whether in the field or at headquarter level people remain the most critical asset. Humanitarian agencies need to priorities when it comes to logistical recruitment. Well thought out strategies for recruitment, retention and career planning must be developed. The identification, selection and development of qualified logistical personnel needs support from not only senior logistical managers but also a champion from senior management in the respective humanitarian agencies.

Lastly, communication and information sharing in this age of impressive technological sophistication is absolutely vital for real-time activity to be responded to by real-time solutions. The challenges to better civil military coordination are numerous but not insurmountable. In reality, during major humanitarian operations, strongly motivated people in both camps usually find ways to surmount these challenges, but valuable time is lost inventing and reinventing these solutions. In recent years relationships have improved, but considerable progress is necessary before both sides can realize the advantages of improved cooperation.

REFERENCES

Barry, J., & Jefferys, A. (2002). *A bridge too far: Aid agencies and the military humanitarian response*. Humanitarian Practice Network (HPN) Paper, No. 37.

Beauregard, A. (1998). *Civil-military cooperation in joint humanitarian operations: A case analysis of Somalia, the former Yugoslavia and Rwanda*. Waterloo, Canada: Ploughshares Monitor.

Beausang, F. (2003). *Is there a development case for United Nations-business partnerships*. LSE Working Paper Series. ISSN: 1470-2320

Binder, A., & Witte, J. M. (2007). *Business engagement in humanitarian relief: Key trends and policy implications*. London, UK: Humanitarian Policy Group, Overseas Development Institute.

Borton, J. (1993). Recent trends in international relief system . *Disasters*, *17*(3), 187–201. doi:10.1111/j.1467-7717.1993.tb00493.x

Brocades-Zaalberg, T. (2005). *Soldiers and civil power: Supporting or substituting civil authorities in peace support operations during the 1990s. Amsterdam*. Amsterdam: University.

CIC (Centre on International Cooperation). (2006). *Annual review of global peace operations, 2006*. Boulder, CO: Lynne Rienner Publishers.

Connaughton, R. (1996). *Military support and protection for humanitarian assistance: Rwanda, April-December 1994*. Occasional Paper No.18, Camberly, UK: Strategic and Combat Institute.

Croft, S., & Treacher, T. (1995). Aspects of intervention in the South . In Dorman, A. M., & Otte, T. G. (Eds.), *Military intervention: From gunboat diplomacy to humanitarian intervention*. Dartmouth, NH: Dartmouth Publishing.

Currey, C. J. (2003). *A new model for military/ nongovernmental relations in post-conflict operations. Carlisle*. PA: U.S. Army War College.

De Conning, C. (2007). Civil-military coordination practices and approaches within United Nations peace operations. *Journal of Military and Strategic Studies, 10*(1).

Disaster Management Training Programme. (1993). *Logistics* (1st ed.). New York, NY: United Nations Development Programme/Department of Humanitarian Affairs.

Doel, M. T. (1995). Military assistance in humanitarian aid operations: Impossible paradox or inevitable development? *Royal United Services Institute Journal, 140*(5), 26–32.

Duffey, T. (2000). Cultural issues in contemporary peacekeeping. *International Peacekeeping, 7*(1), 142–168.

Eriksson, P. (2000). Civil-military co-ordination in peace support operations – An impossible necessity? *The Journal of Humanitarian Assistance.*

Ferks, G., & Klem, B. (2006). *Conditioning peace among protagonists: A study into the use of peace conditionalities in the Sri Lankan peace process*. Netherlands Institute of International Relations, Clingendael Institute, Conflict Research Unit.

Forman, S., & Parhad, R. (1997). *Paying for essentials: Resources for humanitarian assistance*. Paper prepared for meeting at Pocantico Conference Centre of the Rockefeller Brothers Fund. New York.

Gill, T., Leveillee, J., & Fleck, D. (2006). *The rule of law in peace operations*. General Report of the seventeenth Congress of the International Society for Military Law and the Law of War, 16-21 May, Scheveningen, Holland.

Goddard, T. (2005). Corporate citizenship and community relations. Contributing to the challenges of aid discourse. *Business and Society Review, 110*(3), 269–296. doi:10.1111/j.0045-3609.2005.00016.x

Gooley, T. B. (1999). In time of crisis, logistics is on the job. *Logistics Management and Distribution Report, 38*, 82–86.

Gordon, S. (2001). Understanding the priorities for civil-military co-operation (CIMIC). *The Journal of Humanitarian Assistance.*

Gourlay, C. (2000). Partners apart: Managing civil-military co-operation in humanitarian interventions. *Disarmament Forum, 3*, 33–44.

Heaslip, G. (2010, 19 January). Civil military coordination. *Irish Times*, p. 13.

Heaslip, G., Mangan, J., & Lalwani, C. (2007a). *Humanitarian supply chains, the Irish defence forces and NGOs – A cultural collision or a meeting of minds*. CCHLI International Humanitarian Logistic Symposium, Cranfield, United Kingdom, November 2007.

Heaslip, G., Mangan, J., & Lalwani, C. (2007b). *Integrating military and non governmental organisation (NGO) objectives in the humanitarian supply chain: A proposed framework.* Logistics Research Network, Hull, United Kingdom, September 2007.

Heaslip, G., Mangan, J., & Lalwani, C. (2008a). *Strengthening partnerships in humanitarian supply chain.* Nordic Logistics Research Network (NOFOMA), Helsinki, Finland, June 2008.

James, A. (1997). Humanitarian aid operations and peacekeeping. In Belgrad, E. A., & Nachmias, N. (Eds.), *The politics of international humanitarian aid operations.* Westport, CT: Praeger.

Jones, B., & Stoddard, A. (2003). *External review of the inter-agency standing committee.* New York, NY: Centre on International Cooperation, December 2003.

Kent, R. C. (1987). *Anatomy of disaster relief: The international network in action.* London, UK: Pinter.

Kovács, G., & Spens, K. (2007). Humanitarian logistics in disaster relief operations. *International Journal of Physical Distribution and Logistics Management, 37*(2), 99–114. doi:10.1108/09600030710734820

Kovács, G., & Spens, K. (2008). Humanitarian logistics revisited. In Arlbjørn, J. S., Halldórsson, Á., Jahre, M., & Spens, K. (Eds.), *Northern lights in logistics and supply chain management* (pp. 217–232). Copenhagen, Denmark: CBS Press.

Kovács, G., & Tatham, P. (2009). Responding to disruption in the supply network – From dormant to action. *Journal of Business Logistics, 30*(2), 215–228. doi:10.1002/j.2158-1592.2009.tb00121.x

Laurence, T. (1999). *Humanitarian assistance and peacekeeping: An uneasy alliance.* London, UK: The Royal United Services Institute for Defence Studies.

Long, D. (1997). Logistics for disaster relief: Engineering on the run. *IIE Solutions, 29*(6), 26–29.

Long, D. C., & Wood, D. F. (1995). The logistics of famine relief. *Journal of Business Logistics, 16*(1), 213–229.

Lund, M. (2003). *What kind of peace is being built? Assessing the record of post-conflict peacebuilding, charting future directions.* Ottawa, Canada: International Development Research Centre.

Mackinlay, J. (Ed.). (1996). *A guide to peace support operations.* Providence, RI: The Thomas J. Watson Jr. Institute, Brown University.

Macrae, J. (2002). Analysis and synthesis. In Macrae, J. (Ed.), *The new humanitarianisms: A review of trends in global humanitarian action. Report to the Humanitarian Policy Group.* London, UK: Overseas Development Institute.

Moody, F. (2001). Emergency relief logistics: A faster way across the global divide. *Logistics Quarterly, 7*(2). Retrieved on March 9, 2007, from http://www.lq.ca/issues/summer2001/articles/article07.html.

Moore, D. M., & Antill, R. D. (2002). Opportunities and challenges in logistics for humanitarian aid operations: A role for UK Armed Forces? *Proceedings of the Logistics Research Network Conference,* ILT, Plymouth, September.

NATO. (2004). *AJP-9 NATO civil-military cooperation (CIMIC) doctrine.* Retrieved 10 April, 2007, from http://www.nato.int/ims/docu/AJP-9.pdf

Natsios, A. S. (1995a). The international humanitarian response system. *Parameters, 25,* 68–81.

Natsios, A. S. (1995b). NGOs and the UN System in complex humanitarian emergencies: Conflict or cooperation? *Third World Quarterly, 16*(3), 405–419. doi:10.1080/01436599550035979

OCHA. (2003). *Guidelines on the use of military and civil defence assets to support United Nations humanitarian activities in complex emergencies*. Geneva, Switzerland: OCHA.

OECD. (2005). *Paris declaration on aid effectiveness*. Retrieved on September 12, 2007, from www.oecd.org

OED. (2000). *Concise Oxford English dictionary*. Oxford, UK: Oxford University Press.

Oloruntoba, R., & Gray, R. (2006). Humanitarian aid: An agile supply chain? *Supply Chain Management – . International Journal (Toronto, Ont.), 11*(2), 115–120.

Olson, L., & Gregorian, H. (2007). Interagency and civil-military coordination: Lessons from a survey of Afghanistan and Liberia. *Journal of Military and Strategic Studies, 10*(1). Retrieved on June 26, 2008, from http://www.jmss.org/2007/2007fall/articles/olson-gregorian.pdf

Ozdamar, L., Ekinci, E., & Kucukyazici, B. (2004). Emergency logistics planning in natural disasters. *Annals of Operations Research, 129*, 217–245. doi:10.1023/B:ANOR.0000030690.27939.39

Pan American Health Organization. (2000). *Manual logistical management of humanitarian supply*. Washington, DC: PAHO.

Pettit, S. J., & Beresford, A. K. C. (2005). Emergency relief logistics: An evaluation of military, non military and composite response models. *International Journal of Logistics: Research and Applications, 8*(4), 313–332.

Pugh, M. (2001). *Civil-military relations in peace support operations: Hegemony or emancipation?* Plymouth, UK: University of Plymouth.

Rietjens, S. J. H. (2006). *Civil-military cooperation in response to a complex emergency: Just another drill?* Doctoral Dissertation, University of Twente, Enschede, the Netherlands.

Rietjens, S. J. H., Voordijk, H., & De Boer, S. J. (2007). Co-ordinating humanitarian operations in peace support missions. *Disaster Prevention and Management, 16*(1), 56–69. doi:10.1108/09653560710729811

Rodman, W. K. (2004). *Supply chain management in humanitarian relief logistics*. MSc Thesis, Air Force Institute of Technology, Wright-Patterson Air Force Base, Ohio, US.

Schein, E. (1996). Three cultures of management: The key to organizational learning. *Sloan Management Review*, (Fall): 9–20.

Seiple, C. (1996). *The US military/NGO relationship in humanitarian interventions*. Carlisle Barracks, PA: Peacekeeping Institute Centre for Strategic Leadership, U.S. Army War College.

Siegel, A. (2003). Why the military think that aid workers are over-paid and under-stretched. *Humanitarian Affairs Review, 1*, 52–55.

Slim, H. (2006). Humanitarianism with borders? NGOs, belligerent military forces and humanitarian action. *The Journal of Humanitarian Assistance*.

Sowinski, L. L. (2003). The lean, mean supply chain and its human counterpart. *World Trade, 16*(6), 18.

Sphere Project. (2004). *The humanitarian charter and minimum standards in disaster response*. Oxford, UK: Oxfam Publishing.

Stephenson, M., Jr. (2004). *Making humanitarian relief networks more effective: Exploring the relationships among coordination, trust and sense making*. Paper prepared for Delivery at the National Conference of the Association for Research on Non-Profit Organizations and Voluntary Action (ARNOVA). Los Angeles, California.

Studer, M. (2001). The ICRC and civil-military relations in armed conflict. *International Review of the Red Cross, 83*(842), 367–391.

Tatham, P., & Kovács, G. (2007). The humanitarian supply network in rapid onset disasters. In A. Halldorsson & G. Stefansson (Eds.), *Proceedings of the 19th Annual Conference for Nordic Researchers in Logistics*, (pp. 1059-1074). NOFOMA 2007, Reykjavik, Iceland.

Thomas, A. (2003). Fritz Institute: Leveraging private expertise for humanitarian supply chains. *Forced Migration Review, 21*, 64–65.

Thomas, A., & Kopczak, L. R. (2005). *From logistics to supply chain management – The path forward in the humanitarian sector.* Fritz Institute.

Tomasini, R. M., & Van Wassenhove, L. N. (2004a). Pan-American health organisation's humanitarian supply management system: Depoliticization of the humanitarian supply chain by creating accountability. *Journal of Public Procurement, 4*(3), 437–449.

Tomasini, R. M., & Van Wassenhove, L. N. (2004b). The TPG-WFP partnership: Looking for a partner. (INSEAD case study 06/2004-5187).

United Nations. (1994). *Guidelines on the use of military and civil defence assets in disaster relief.* Geneva, Switzerland: United Nations.

UNJLC. (2008). *UNJLC training material.* Copenhagen.

Van der Laan, E., de Brito, M. P., & Vermaesen, S. (2007). Logistics information and knowledge management issues in humanitarian aid organizations. *Proceedings of the SIMPOI/POMS conference*, Brazil, August 8-10.

Van der Vorst, J. G. A. J., & Beulens, A. J. M. (2002). Identifying sources of uncertainty to generate supply chain redesign strategies. *International Journal of Physical Distribution & Logistics Management, 32*(6), 409–430. doi:10.1108/09600030210437951

Weiss, T. G., & Campbell, K. M. (1991). Military humanitarianism. *Survival, 33*(5), 451–465. doi:10.1080/00396339108442612

Wheatley, G., & Welsch, S. D. (1999). The use and limitations of technology in civil-military interactions. In The Cornwallis Group (Eds.), *Volume IV: Analysis of civil-military interactions.* Nova Scotia: The Lester B. Person Canadian International Peacekeeping Training Centre.

Whiting, M. (2009). Enhanced civil military cooperation in humanitarian supply chains. In Gattorna, J. (Ed.), *Dynamic supply chain management* (pp. 107–122). Surrey, UK: Gower Publishing.

Yamamoto, T. (1999). Corporate – NGO partnership: Learning from case studies. In Yamamoto, T., & Gould, K. (Eds.), *From corporate–NGO partnership in Asia-Pacific.* Japan Centre for International Exchange.

Zandee, D. (1999). Civil-military interaction in peace operations. *NATO Review, 47*(1), 11–15.

ADDITIONAL READING

Dorman, A. M. (2007). Western Europe and military intervention. In *Military Intervention: From Gunboat Diplomacy to Humanitarian Intervention*, Dorman, A M & Otte, T G., Dartmouth Publishing, Dartmouth.

Galbraith, J. R. (1973). *Designing Complex Organizations.* Reading, MA: Addison-Wesley Publishing Company.

Heaslip, G., Mangan, J., & Lalwani, C. (2008b). Modelling a humanitarian supply chain using structured analysis and design techniques (SADT)', *Logistics Research Network (LRN)*, Liverpool, United Kingdom, September 2008.

Thomas, A. (2004). *Humanitarian Logistics: Enabling Disaster Response.* Fritz Institute.

KEY TERMS AND DEFINITIONS

CIMCO: Civil Military Coordination. A new concept being developed by the European Union (EU) to replace the original EU definition of Civil Military Cooperation (CIMIC).

Civil-Military Coordination: Activities that pertain to the harmonization of activities across armed forces and humanitarian organizations in particular in disaster relief. The term suggests the seamless division of labor between aid workers and international military forces.

Cluster: A group of organizations concerned with similar relief items and services globally and/or in a disaster area. A cluster system was set up by the humanitarian reform in 2005.

CMO: Civil Military Operations. The activities of a commander that establish, maintain, influence, or exploit relations between military forces, governmental and non governmental civilian organizations and authorities, and the civilian populace in a friendly, neutral, or hostile operational area in order to facilitate military operations, to consolidate and achieve operational United States objectives. Civil military operations may include performance by military forces of activities and functions normally the responsibility of the local, regional, or national government. These activities may occur prior to, during, or subsequent to other military actions. They may also occur, if directed, in the absence of other military operations. Civil military operations may be performed by designated civil affairs, by other military forces, or by a combination of civil affairs and other forces.

EU CIMIC: European Union Civil Military Cooperation (CIMIC). The EU's definition of CIMIC is the coordination and cooperation, in support of the mission, between military components of EU led Crisis Management Operations and civil role players (external to the EU), including national population and local authorities, as well as international, national and nongovernmental organizations and agencies.

IFRC: International Federation of Red Cross and Red Crescent's. The IFRC's mission is to improve the lives of vulnerable people by mobilizing the power of humanity. It focuses on four core areas: promoting humanitarian values, disaster response, disaster preparedness, and health and community care.

IOM: International Organization for Migration. IOM is the leading international organization working with migrants and governments to provide humane responses to migration challenges.

NATO: North Atlantic Treaty Organization. NATO was signed in 1949 created an alliance of 12 countries committed to each other's defense. By 1999 it had 19 member countries. Together with non-member countries and other international organizations it is involved in peacekeeping and crisis management tasks.

NATO CIMIC: North Atlantic Treaty Organization Civil Military Cooperation. The NATO definition of Civil Military Cooperation (CIMIC) is the coordination and cooperation, in support of the mission, between the NATO Commander and civil populations, including national and local authorities, as well as international, national and nongovernmental organizations and agencies.

OCHA: United Nations Office for Coordination of Humanitarian Affairs. OCHA mobilizes and coordinates the efforts of the international community to meet the needs of those exposed to human suffering in disasters and emergencies. OCHA is a Geneva-based, non-operational UN agency that reports to the UN Secretary General. It coordinates the launch of a consolidated emergency appeal among the donor community, liaises between local governments and humanitarian agencies and provides a number of common services to the humanitarian community. The major ones include setting policies for Civil Military Coordination (CMCoord) services, the United Nations Disaster Assessment Coordination team (UNDAC) and the Humanitarian Information Centre (HIC). Sphere Project Handbook: Accepted as a reference for performance measures in the humanitarian sector.

Organizational Culture: A combination of tacit assumptions (the essence of culture), espoused values (reflecting what a group wishes ideally to be and the way it wants to present itself publicly) and day-to-day behavior.

CMCoord: United Nations Humanitarian Civil Military Coordination (UN CMCoord). The essential dialogue and interaction between civilian and military actors in humanitarian emergencies that is necessary to protect and promote humanitarian principles, avoid competition, minimize inconsistency, and when appropriate pursue common goals. Basic strategies range from coexistence to cooperation. Coordination is a shared responsibility facilitated by liaison and common training".

UNDAC: United Nations Disaster Assessment Coordination. The UN Office for Coordination of Humanitarian Assistance's (OCHA) Disaster Assessment Coordination team (UNDAC) is a rapid response tool (members arrive quickly and stay no longer than two weeks) deployed for "sudden onset of natural disaster". Its mandate consists of the preparation of a consolidated appeal for the emergency on behalf of the government and UN humanitarian organizations (Inter-Agency) for donor funding.

UNDP: United Nations Development Programme. UNDP is the UN's principal provider of development advice, advocacy and grant support. It has six priority practice areas: Democratic governance, poverty reduction, crisis prevention and recovery, energy and environment, IT and communications, and HIV/AIDS.

UNHCR: United Nations High Commissioner for Refugees. UNHCR was established in 1950 and is dedicated to leading and coordinating international action to safeguard the rights and well being of refugees worldwide.

UNICEF: United Nations Children's' Fund. UNICEF was established in 1946 and advocates the protection of children's rights to help meet children's basic needs and expands their opportunities to reach their full potential.

WFP: World Food Programme. Set up in 1963, World Food Programme is the United Nations' frontline agency in the fight against global hunger.

WHO: World Health Organization. WHO was established in 1948 and is the UN Agency dedicated to attaining the highest levels of health for all people. World Vision: A Christian relief and development organization working for the well being of all people, especially children.

Chapter 10
Developing and Maintaining Trust in Hastily Formed Relief Networks

Peter Tatham
Griffith University, Australia

Gyöngyi Kovács
HUMLOG Institute, Hanken School of Economics, Finland

ABSTRACT

Although there is a vast body of academic and practitioner literature championing the importance of trust in long-term business relationships, relatively little has been written discussing the development and maintenance of trust in networks that are formed at short notice and that often operate for a limited period of time. However, some models of trust and trusting behavior in such "hastily formed relief networks" (HFRN) do exist, and the aim of this chapter is to consider the theoretical application of one of the most prominent examples – that known as "swift trust" – to a post-disaster humanitarian logistics scenario. Presented from the perspective of a HFRN, this chapter presents a discussion of the practical application of the swift trust model.

INTRODUCTION

Pick up almost any supply chain management textbook, and it is a near certainty that one or more sections will be devoted to the challenges of achieving trust between the parties that form the supply chain or network. For example, Christopher's (2011)

DOI: 10.4018/978-1-60960-824-8.ch010

definition of supply chain management (SCM) focuses firmly on this area suggesting that SCM is "the management of upstream and downstream relationships with suppliers and customers to deliver superior customer value at less cost to the supply chain as a whole" (p. 3). Indeed, the area of trust can be seen as a core concept in supply chain management (Barratt, 2004; Mentzer *et al.*, 2001), and particularly in the literature relating to supply chain collaboration (e.g. Skjøtt-Larsen *et al.*, 2003). Whilst, in the opposite sense, Fawcett *et al.* (2008) list a lack of trust as one of the most significant barriers to effective management of supply chains and networks.

Given the importance of trust in many walks of life, the last three decades have seen significant research into inter-organizational, intra-organizational and inter-personal trust from a range of perspectives that includes economic, psychological and sociological (Rousseau *et al.*, 1998). Yet, in general, such research has focused on the development and maintenance of trust in long-term relationships with much of the relationship management literature focusing on trust in this context where it often seen as interrelated with risk (Das & Teng, 2001), being the obverse of control (e.g. Grey & Garsten, 2001; Knights *et al.*, 2001). Studies on trust in different types of temporary networks, on the other hand, are relatively scarce – even though there is a recognition that different types of such networks exist including planned ones in project industries, virtual teams and even agile virtual networks (Bal & Teo, 2001; Sarkis *et al.*, 2007), "minimal organizations" (e.g. fire fighting teams, Weick, 1993), "emergent multi-organizational networks" (NRCNA, 2006), "emergent response groups" (Majchrzak *et al.*,2007), and hastily formed networks (HFN Research Group, 2006).

Studies that consider trust in any type of temporary network stress the importance of trust to develop at the very beginning of a project (Bal & Teo, 2001). This is the more important in the case of disaster relief in the absence of prior rules, a common training or a common history. At the same time the link between trust and higher (team) performance can be established in similar veins to supply chain collaboration literature. In specific, Uhr and Ekman (2008) argue that the challenge of developing and maintaining trust in disaster relief networks is significant and can have a major bearing on the success of such relief – or, to put it more starkly, a failure in this regard has

the potential to lead to unnecessary loss of life and/or distress to those affected by the disaster. The initial development of trust is at the heart of Meyerson *et al.* (1996) article that outlines the psychological processes at work in the formation of inter-personal trust and coined the phrase "swift trust" to describe them. However, to date, there has been only limited consideration of how the swift trust model might be applied in a hastily formed network in a post-disaster context, that is, in a hastily formed relief network (HFRN).

With the above introduction in mind, the aim of this chapter is to further the understanding of how, from a theoretical perspective, the concept of swift trust might be used to develop and maintain inter-personal trust in hastily formed networks. To achieve this, the chapter will begin by discussing hastily formed networks in the context of disaster relief in greater detail before a model of swift trust, developed from the work of Meyerson *et al.* (1996), will be used to consider the implications from the perspective of the leader of a HFRN. In the final section, a suggested route for further research to assist the application of the concept is developed.

BACKGROUND

Supply chain management literature describes trust as a basis of collaboration which, in turn, is propagated and leads to high firm and supply chain performance, and as such is an alternative path to the focus on control mechanisms for key assets and resources in traditional management literature. In embracing trust and collaboration in this way, supply chain management offers the possibility for flexible ways of organizing at the same time as observing the traditional aims of effectiveness and efficiency.

Disaster relief is an activity that epitomizes high flexibility requirements. Not surprisingly, relief supply chains have been described as "most agile" (Oloruntoba & Gray, 2006) or "fully flexible"

(Gattorna, 2006). In addition to the sheer number of companies involved, relief supply chains also interact with each other in back-office planning operations, in sharing suppliers and forming purchasing consortia to, most importantly, organizing relief activities in the field where members of different humanitarian organizations come together in the aftermath of a disaster. Although such organizations may specialize in different activities, or clusters of activities (such as providing shelter or health care, providing water and organizing sanitation etc.), all observe the overall aim of humanitarian logistics which is that of meeting the end beneficiaries' requirements (cf. Thomas and Mizushima, 2005). To achieve this, the members of these organizations quickly form a network that needs to meet regularly, share information and work together. In other words, they constitute a hastily formed network.

Hastily Formed Networks

The concept of a hastily formed network (HFN), as described by the HFN Research Group (2006) has five elements. It is a network of people

- established rapidly,
- from different communities,
- working together in a shared conversation space,
- in which they plan, commit to, and execute actions,
- to fulfill a large, urgent mission.

Importantly, hastily formed networks should be distinguished from Weick's (1993) "minimal organizations" (such as fire fighters) whose members are likely to share a common aim, background, approaches and working practices. A further stream of literature that is concerned with temporary networks is that of "virtual networks", and virtual teams. These are teams of individuals coming together from different companies, working on a particular common goal (Bal & Teo, 2001). By

contrast, the individual in a HFN, whilst sharing the same high level goals, may have not worked with other members previously nor have undergone the same training – nor in fact share particular goals set for the team deployed in a disaster area. At the same time, the conversation space for an HFRN is determined by the co-location of its key members, unlike the distributed teams of virtual networks.

Other concepts similar to HFNs are those of "emergent (or emerging) multi-organizational networks" (NRCNA, 2006) or "emergent response groups" (Majchrzak *et al.*, 2007). Head (2000) describes these as (1) crisis driven, (2) task-orientated, (3) self-evolving, (4) time-sensitive, (5) composite and (6) temporary. Majchrzak *et al.* (2007) emphasize their self-evolving nature suggesting that their membership has no pre-existing structure, roles, tasks or expertise. In other words, an emergent response group develops, migrates, reorganizes, gains and looses membership in an unstructured way. Thus, there is a clear difference between such groups and HFRNs as the aims, policies, doctrine and role of the latter organizations are unlikely to change. In summary, the disaster response situation brings together both organizations with their pre-fixed aims and policies, as well as individuals who do not know each other, do not belong to the same organization, and have not undergone the same training. Therefore the concept of HFNs is preferred to other, similar concepts in the humanitarian context – albeit the swift trust model is applicable to all of the above (cf. Majchrzak *et al.,* 2007).

In a disaster relief context, therefore, HFRNs can be described as "co-located teams in short term local projects" (Fitzgerald, 2004, p. 162) with inter-sectoral partnerships that link humanitarian organizations (i.e. aid agencies and NGOs) to governments, local communities, businesses (suppliers and logistics service providers) and the military, in order to form the humanitarian aid supply network (Kovács & Spens, 2008). But while project-based temporary networks are characterized by clear starting and ending dates (though the same companies and, indeed, their competitors can be involved in several projects in a row), the key characteristic of an HFRN is its quick formation which reduces the opportunity for set procedures and even trust-building tools that Bal & Teo (2001) discuss for virtual teams. Furthermore, in most cases, neither the members nor the network will have a shared history or a shared future beyond the relatively short term (Järvenpää & Leidner, 1999). Thus, trust building in HFRNs needs to follow a different pattern from trust in long-term relationships, i.e. that described by Meyerson *et al.* (1996) as "swift trust" (and McKnight *et al.*, 1998, as "initial trust").

Hastily Formed Relief Networks

Having introduced, the concept of a hastily formed network in a general sense, the aim of this section is to situate it within the context of disaster relief. In doing so, it will be appreciated that there has been a significant increase in the focus on improving the response to rapid onset disasters following such high profile events as the 2004 Indian Ocean tsunami, the 2005 Pakistan earthquake, Hurricane Katrina (2005), Cyclone Nargis (2008), the Wenchuan earthquake (2008) and the Haiti earthquake (2010). Tsunamis and earthquakes are indeed rapid onset disasters in the sense of occurring with little or no prior warning – while, for example, hurricanes are more predictable and often cyclical in nature. But rapid onset disasters do not need to have a natural causality as terrorist attacks or infrastructure failure (such as the explosion of BP's oil drilling rig that has resulted in the enormous oil slick in the Gulf of Mexico in 2010) can also be seen as falling into this category. The key point is that a common feature of such disasters is their sudden occurrence (compared to, say, the evolution of a drought), and equal need for quick response.

With this in mind, it is clear that the effectiveness of the relief depends, in part, on the speed of logisticians' ability to "procure, transport and receive supplies at the site of a humanitarian relief effort" (Thomas, 2003, p. 4). Furthermore, it is suggested that humanitarian logistics represents a major contribution to the disaster relief effort, with estimates suggesting that this represents at much as 80% of the cost of an operation (van Wassenhove, 2006). It is argued, therefore, that humanitarian organizations are in effect logistics organizations, albeit with, typically, a particular mandate and a target set of beneficiaries.

Smith and Dowell (2000) therefore characterize relief supply chains in rapid onset disasters as "incident organizations" (i.e. similar to Fitzgerald's, 2004, "collaborative entities"). In other words they represent teams of previously independent individuals, groups and organizations that come together temporarily on the basis of a particular event. Whilst it is suggested that such a newly formed team works towards a common goal and needs to combine the resources of otherwise independent organizations (Smith & Dowell, 2000), Fitzgerald (2004) argues the importance of a "catalyst" in the form of an individual or organization that coordinates (and ideally, leads and monitors) the entire team. In high-impact international disasters, this would be the role of Logistics Cluster (or the United Nations Joint Logistics Centre, UNJLC, as part of the Logistics Cluster), and this is exemplified by the deployment of a UNJLC team to coordinate the overall relief to Cyclone Nargis in May 2008. This included team members (secondees) from nine different humanitarian organizations and was based in two distinct locations (Bangkok and Yangon), as well as the head office in Rome (UNJLC, 2008). This was in effect a HFRN consisting of individuals of at least ten organizations which acted as the coordinating agency for the relief efforts of many more.

Interestingly for HFRNs it is not only the complexity of the relief supply chain, but also the calls for coordination (Oloruntoba and Gray, 2006; van Wassenhove, 2006) as humanitarian organizations compete for financial and material resources (i.e. from donors) as well as media attention (Oloruntoba & Gray, 2009; Kovács & Spens, 2010). Were the aftermath of a disaster to reflect normal market economics, such competition might well be viewed as healthy, but in the humanitarian context, it results in a reduction in the propensity for organizations to cooperate and this, in turn, can lead to inefficiencies through duplication and/or overlap. This is a significant challenge not least because of the sheer numbers of humanitarian organizations (and the individuals working within them) involved in disaster relief – Roberts (2001) putting the number at 30,000 international humanitarian organizations worldwide. As a result, a rapid onset disaster can by typified by the descent of many such organizations into the disaster area leading to huge coordination challenges – 72 inter-agency coordination meetings were held weekly in Banda Aceh alone (Völz, 2005).

Trust in Disaster Relief

In discussing various models of trust, it is appropriate to begin with a definition of the concept – although, as observed by Rousseau *et al.* (1998, p. 72) in a cross-disciplinary review, it is a "…a 'meso' concept integrating micro level psychological processes and group dynamics with macro level institutional arrangements." Indeed, McKnight *et al.* (1998, p. 474) go further by suggesting that "… the word "trust" is so confusing and broad that it defies careful definition", whilst Kramer (1999, p. 571) notes that "a concise and universally accepted definition has remained elusive." Thus the following definition is adopted in this chapter:

Trust is present when the one party has a fundamental belief that the other can be relied upon to fulfill their obligations with integrity, and will act in the best interests of the other.

Importantly, this definition of trust focuses on inter-personal, rather than intra- or inter-organi-

zational relationships. To support this viewpoint, it is argued that decisions within organizations are made by individuals and, therefore, the level of intra/inter-organizational trust can be seen as reflecting those individual relationships. Thus, the extent of the intra/inter-organizational trust will be developed and shaped by the inter-personal components and, for this reason, this chapter will concentrate on this latter unit of account, notwithstanding the problem of distinguishing trust invested in people from trust placed in institutional mechanisms as Knights *et al.* (2001) point out.

That said, it is the nature of HFRNs that such institutional mechanisms may at best be tenuous given the ad hoc nature of the organization/ network and, especially in a post-disaster situation, the speed with which the crisis can unfold. Nevertheless, it is considered to be a reasonable assumption that all members of the HFRN, be they located in the field, in a local subordinate headquarters or in the main (remote) headquarters, will be working to common goal or goal(s) set by the organization, and in line with the philosophy and ideals of that organization. Thus, instances in which members of the HFRN are actively working against each other are perceived to be relatively limited. This latter point is important because the above definition does not imply bi-lateral trust; rather, it is suggested that trust exists when A trusts B, and the fact that this is not reciprocated (or is reciprocated at a lesser level) does not obviate the existence of the A->B relationship.

But how does such inter-personal trust develop in the very short time frame of the creation of an HFRN? The work of Meyerson *et al.* (1996) introduced earlier has been expanded into a more general framework by Hung *et al.* (2004) who suggest that there are three different routes to the achievement of trust, namely the peripheral, the central, and the habitual. If seen in sequence, the peripheral route refers to the early establishment of trust, the central route to its further development in relationships with a long-term perspective, and the habitual route to a next level where trust is based on patterns that have developed in long-term relationships.

Thus, the peripheral route reflects the early stages of a relationship in which individuals meet either physically or virtually to form a team or organization. Trust at this stage is based on (peripheral) cues such as those provided by third parties. Hung *et al.* (2004) argue that the underlying psychological mechanism at work here is that the peripheral route to trust involves less cognitive effort than making one's own judgments and is, therefore, the preferred route in the initial stages of a relationship. Once teams or organizations have formed, individuals are able to cognitively engage in consideration of the other party's perceived ability, integrity and benevolence, and this may lead to the development of trust through the central route. The final route, that of habitual trust, reflects the historical build up of successful trust transactions and often leads to strong emotional bonds (Hung *et al.*, 2004).

The peripheral route of developing trust is key to the HFRN as it reflects the post-disaster scenario in which many individuals descend on the affected location from a variety of organizations (and associated organizational cultures) to form a multiplicity of networks. The composition of these is likely to be further complicated by the presence of individuals in the disaster area as well as those in headquarters who are connected to the network by virtual means, and who manage parts of it remotely, e.g. when headquarters manage not only their own but also national staff of implementing partners (ALNAP, 2008). Such virtual vs. face-to-face members of networks are, of course, also prevalent in other industries (see the review of agile virtual enterprises in Sarkis *et al.*, 2007). However, whilst other industries can often provide the time and space to develop long-term relationships in such mixed virtual and face-to-face networks, the nature of HFRNs calls for a closer investigation of the peripheral route of trust development. Important to HFRNs is the assertion from agile virtual enterprises that

trust is essential from the very beginning of the project, if it cannot be already built before (Bal & Teo, 2001).

Developing Swift Trust

The suggestion that trust can be formed by the peripheral route stems from the work of Meyerson *et al.* (1996) who coined the term "swift trust" to describe the need to manage the issues of vulnerability, uncertainty, risk and expectations that surface with the formation of a HFRN. Such networks "exhibit behavior that presupposes trust, yet traditional forms of trust – familiarity, shared experience, reciprocal disclosure, threats and deterrents, fulfilled promises and demonstrations of non-exploitation of vulnerability – are not obvious in such systems" (Meyerson *et al.*, 1996, p. 167). In developing this concept, Hung *et al.* (2004) suggest that their equivalent (the peripheral route) has five elements that influence trust formation (see Fig.1):

- third party information,
- dispositional trust,
- rule,
- category, and
- role.

Whilst Kramer (1999) also see historical trust as another condition, Hung *et al.* (2004) purposefully exclude this element in the peripheral route, as they argue that the peripheral route to trust formation is based on limited prior interaction among the members of the network.

The other routes to trust discussed by Hung *et al.* (2004), namely the central, and habitual routes, result from the maturation of the trusting relationship. However, HFRNs are by definition not drawing on previous trusting relationships, thus the focus in this chapter is on the peripheral route to trust, i.e. swift trust (see Figure1). Subsequent developments in the trusting relationship reflect the level of trust itself and the trusting behaviors

Figure 1. Developing swift trust (based on Hung et al., 2004, p.4)

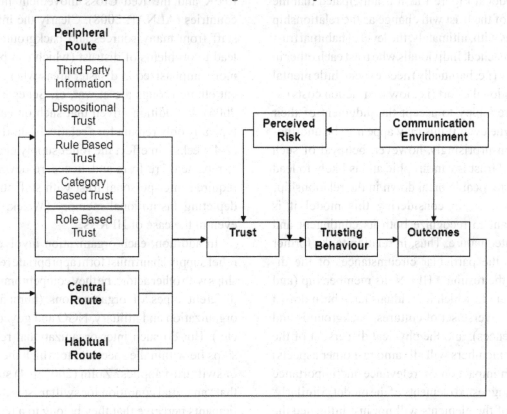

and subsequent outcomes that are generated. Importantly, it is argued that, in addition to the simple feedback loop generated by improved knowledge of the others in the relationship, the process is mediated by the perceived level of risk which, in turn, reflects the communications environment within the network.

In addition, trusting behavior is mediated by the perceived risk of the possible gains and losses of any interaction in the network, so that a high perceived risk may even lead to the deliberate withholding of relevant information. However, Hung *et al.* (2004), whose research focused on virtual teams, also see the communication environment as means to exercise social control. Importantly, in the HFRNs of humanitarian logistics, not only is the network as a whole, and the associated communication environment, composed of both face-to-face elements (of logisticians of different organizations on the ground) and virtual ones (with

remote headquarters), but also the communications links may well be of poor quality as a result of the impact of the disaster on this element of the infrastructure. This incorporation of a virtual element adds a further layer of complexity because, as Järvenpää and Leidner (1999) observe, such networks are composed of "individuals with differing competencies who are located across time, space and culture" (p. 791), which is why Handy (1995) questions whether virtual teams can ever function effectively in the absence of frequent face-to-face contact.

The literature also indicates that trusting behavior leads to specific outcomes, e.g. better network performance, and it is on this basis of the expectation of an improved outcome that investment into the development of trust in relationships can be justified. Furthermore, Laaksonen *et al.* (2009) show that trust can decrease the transaction costs of a relationship. This thought is captured in the model at Figure 1 as it is anticipated that the nature of the trust will change as the relationship matures with, ultimately, the level of habitual trust being reached. Individuals who trust each other in this way (i.e. habitually) need expend little mental or emotional effort (i.e. low transaction costs) as they are happy to accept the judgment of their colleague even if this may appear to be unorthodox. Unsurprisingly, however, betrayal of such habitual trust is catastrophic and is likely to lead to an irrevocable breakdown in the relationship.

However, in considering this model it is important to recognize both its contingent and integrated nature. Thus, in relation to the former aspect, the particular circumstances of the disaster, the resultant HFRN, its membership (and the extent to which individuals have been drawn from a diverse set of cultures, backgrounds and experiences), and the physical dispersion of the HFRN members will all (amongst other aspects) have an impact on the relevance and importance of the suggested elements of the model. Similarly, many of the elements will not just influence the peripheral route to the development of trust, but also will, in all probability, apply in the later stages.

SWIFT TRUST IN HASTILY FORMED RELIEF NETWORKS

Given the plethora of actors in relief supply chain management in the aftermath of a disaster, it will be appreciated that issues of the development and maintenance of inter-personal trust will apply both within a humanitarian organization and between organizations. Furthermore, the strategic approach to the response will differ from country to country for, as Drabek (1985) noted, the United States has a much more decentralized system than other countries. Nevertheless, whilst some disaster response staff within a humanitarian organization are permanent employees, many are drawn from a wider network of logisticians who form an "on call" roster such as those held by Oxfam,

RedR and the Red Cross movement in many countries (ALNAP, 2008). Clearly the influx of staff from many sources and backgrounds will lead to problems of distrust (which can be even more emphasized in disaster areas where different ethnic groups are at war, cf. Scheper *et al.*, 2006). In addition, given that such on call staff typically only remain for a relatively short period (2-4 weeks), in effect the relief supply chain has to repeatedly re-form and, hence, re-develop the required inter-personal trust with staff turnover depleting institutional memory (Weick, 1988), even in the case of HFRNs.

In addition, each organization involved in a relief supply chain must form appropriate relationships with other actors, be they competitors or even different types of organizations (humanitarian organization and military, NGO and government etc.). But do such inter-organizational relationships lie within the spectrum to which the model of swift trust applies? Zolin (2002, p.4) suggests that "an initial condition for swift trust is that participants perceive that they belong to a team, i.e. that they perceive a shared goal." Humanitarian organizations do, indeed, share the overall goal of alleviating the suffering of beneficiaries, yet at the same time they also compete for funding and media attention (Kovács & Spens, 2010). The competition aspect is an unfortunate, but inevitable, outcome of the funding regime in which donors provide the majority of support after a disaster has taken place (Oloruntoba & Gray, 2009). Apart from the fact that this is unquestionably inefficient in the longer term, it also leads to a desire on the part of humanitarian organizations to be seen to be delivering aid. The subliminal message being that success breeds success and that a given humanitarian organization should be favored above others in terms of donor funding (The Lancet, 2010).

On the other hand, and notwithstanding elements of competition, humanitarian logisticians engaged in operations relating to the same disaster can develop a sense of belonging to the same team even though relationships between their organizations may not be formalized. Thus, at the operational level there seems to be broad agreement over the need to support the beneficiaries, to operate within the humanitarian charter, and follow minimum standards of the Sphere Project which lays down a set of values and behaviors that are designed to guide humanitarian response (Sphere, 2004). In short, whilst humanitarian organizations do see themselves as part of a broad community responding to a disaster, it is argued that a greater measure of inter-personal and inter-organizational trust will improve the efficiency and effectiveness of that response. In this regard, Denning (2006, p.18) notes: "The more overwhelming the event, the more likely turf-asserting tendencies will occur and interfere with the effectiveness of the network.", although it is unclear whether any research has been conducted that can substantiate this proposition. On the organizational and inter-organizational level, a lack of risk-sharing, credit- and cost-sharing essentially inhibits collaboration; yet inter-personal trust can still develop in the absence of these mechanisms.

A second key point is, again, the interdependence of the trusting relationships. In essence, there is no one trustor or trustee, but each member of the network engages in a relationship with the other members. What is more, the outcomes of any interchange between the parties will affect each party but, potentially, in different ways. Thus, to the extent that the concept of swift trust incorporates certain elements (see Figure 1) and that these are capable of promotion and/or maintenance of trust, the actions proposed in this paper should apply to all actors in the relief supply chain, and particularly, those in HFRNs.

Thirdly, the level of trust within a relationship is by no means static. Indeed, as Hung *et al.* (2004) argue, the peripheral route to trust can give rise to the central route in a next stage. Over time, the relatively fragile swift trust can thus develop to the robust habitual form. On the other hand, trust can also decline, and the different routes to trust are not necessarily forming a direct sequence.

Furthermore, there is no absolute level of trust at any given time in a relationship, rather, parties may trust each other in relation to one issue, but not another.

Finally, there is good evidence to suggest that, subject to any negative impacts of the perceived success of prior alliances (Gulati, 1995), individual members of a network often act as if trust was in place and this leads to self-fulfillment (Jones & George, 1998). The very act of forming a network may of itself trigger an initial level of trust, where a positive assumption about the trusting behavior of others becomes the baseline position (Meyerson *et al.*, 1996). However, Coppola *et al.* (2004) and Ben-Shalom *et al.* (2005) suggest that this baseline is also affected by the expectations of trust that members import from other settings with which they are familiar. In the HFRN, similar familiar settings include previous interactions with other humanitarian organizations and their logisticians in other disasters.

With the above discussion in mind, in the following sections will discuss the elements of swift trust (see Figure 1) from the perspective of an HFRN.

Third Party Information

Third party information enables the formation of trust based not on the, as yet, unidentified capabilities of an individual, but on their prior reputation and/or the reputation of their employing organization. Clearly such information about reputation is important as it helps mitigate the risk of unreliability or incompetency of the other party, whilst the role of third parties is important because of their ability to diffuse relevant trust information (Kramer, 1999). Uzzi (1997, p.48) suggests that a third party contributes to the formation of embedded ties or networks (or, as in our case, HFRNs) as a "go-between" that "transfers the expectations and opportunities of an existing embedded social structure to a newly formed one, furnishing a basis for trust and subsequent commitment to be offered and discharged".

Importantly, and as discussed earlier, in the disaster relief context, the HFRN itself does not (by definition) have a shared history, but the individuals within it may have carried out similar roles under different circumstances (such as in different disasters). This provides reputational evidence of how individuals behaved and are, thus, expected to behave in the new HFRN. In such a case, third parties play a crucial role in substantiating the effectiveness of such individuals and organizations. In essence, whilst the leader of the HFRN may not know member "A", he or she is known to member "B" who, in turn, is known to the leader. Thus "B" can provide third party testimony of the competence/ability (and, hence, trustworthiness) of "A" on which the leader can draw.

Within the existing community of humanitarian logisticians (as with all such communities of shared interest), it is inevitable that third party information will be exchanged and, depending on its content, it may have a positive or negative impact on the development of swift trust. Moreover, databases on humanitarian logisticians who are available to provide support in the event of an emergency can also include such third party information about each individual. There is, thus, potential for an organization to provide a repository of individuals' names, qualifications and experience. Were such a central database to be developed and maintained, it could provide useful and neutral third party information to inform the development of HFRNs.

More broadly, it is suggested that humanitarian organizations have a responsibility to "advertise" the skills of their employees (or teams of employees) both within the organization itself and between organizations. The aim here is, obviously, not to develop an elitist mentality, but rather to support the formation of trust by emphasizing that individuals are likely to have the appropriate skills in advance of their demonstration of these.

Such a suggestion raises the issue of the competence of a particular humanitarian organization to achieve its mandate. As discussed earlier, there are a vast number of humanitarian organizations world-wide (Roberts, 2001). Notwithstanding the assertion by most (if not all) that they adhere to the Sphere standards, informal discussion with those active in the field would indicate that there is a considerable degree of variability in the levels of competence displayed. That such concerns have not been formally documented is unsurprising, but it does raise the question of whether some form of certification of humanitarian organizations should be introduced – and, indeed, this subject is beginning to appear on the public agenda (Stocking, 2010). From the perspective of this chapter, such an approach would inform the development of swift trust on the basis that a particular organization has been judged competent and, by implication, so too are its staff. On this basis, an *a priori* assumption of trust can be made.

Dispositional Trust

Dispositional trust is another element that forms part of the peripheral route to trust. This refers to the general disposition of an individual to trust other people, in other words, that some people are more trusting than others. There are ample differences between individuals' general predispositions to trust documented in the literature (e.g. Kramer, 1999; Hung *et al.*, 2004) and, as trust in HFRNs is developed between individuals, each individual member's predisposition to trust impacts on the formation of inter-personal trust in the HFRN in the round.

From the perspective of a potential leader of an HFRN, it would, for example, be totally impractical to attempt to select individuals on the basis of their trusting disposition or even their cultural background. At the same time, the very nature of an HFRN prevents a "leader" from selecting individuals. Even the Logistics Cluster cannot offer such leadership, nor, as a matter of fact, a single such leader, as cluster leads in the field may be exchanged on the basis of their rotation which is not always in tune with the length of disaster relief activities. Thus, the simplest and most obvious prescription is for the leader of the HFRN to be constantly aware of the need to ensure that individuals recognize the existence of such important differences in the comparative approach of their colleagues. In this way, differential dispositions to trust can be taken into consideration when organizations are forming up and trust is being developed.

Rule

The presence of rules, under which heading one can include processes and procedures, is deemed by Kramer (1999) to be of considerable significance in supporting the development of swift trust. Put simply, the suggestion here is that, by following such rules, individuals are deemed by their peers to be trustworthy (Greenberg *et al.*, 2007). More explicitly, Kramer (1999, p.579) suggests that "explicit and tacit understandings regarding transactional norms, interactional routines and exchange practices provide an important basis for inferring that others in the organization are likely to behave in a trustworthy fashion". In short, the present of rules, and the adherence to them, is a guard against maverick behavior which has the potential to destabilize an organization and reduce the level of inter-personal and inter-organizational trust. Indeed, this perspective has considerable resonance with the work of other researchers such as Grey and Garsten (2001) who conceptualize trust as enabling individuals to behave in a predictable way.

However, when it comes to initial the development of swift trust, rule-based behavior refers to issues such as the normality of the situation and, potentially, the assurance of organizational structures (cf. Hung *et al.*, 2004). But for humanitarian logisticians, the normality of the situation may well be the situation of disaster relief which is, almost by definition, a highly fluid and uncertain. Thus, there is unlikely to be one organizational structure

that will optimize the output of the HFRN. That said, the development of common approaches, sets of rules (Bal & Teo, 2001) and the general concept of "structuration" (Butcher *et al.*, 2008) has clear relevance to HFRNs as it would help to ensure that individuals who join the network from different humanitarian organizations can make the transition with the minimum of effort. In this respect, coordinating initiatives such as the logistics operational guide (the LOG) of the Logistics Cluster, and the work of the Chartered Institute of Logistics and Transport (CILT) in the development of a common "Need Assessment" template are clearly important. Such initiatives point towards the long term possibility of developing organizational structures that can underpin the rule-based development of swift trust in the HFRN.

Rules in the business context can refer to both pricing mechanisms and contracts – leading to the rise of "contractual trust" as a type of trust in the commercial context (cf. Fynes *et al.*, 2005). Humanitarian organizations do, indeed, employ contracts with their global suppliers and logistics service providers, but the fluid nature of the evolving post-disaster scenario would, unquestionably, make the prior-development of contracts for their on-call workforce a massive challenge. The alternative approach of attempting to write a contract in the immediate aftermath of a disaster is perceived to be equally challenging as it would doubtless (and, arguably correctly) be viewed by the members of the HFRN as a bureaucratic sideshow that detracted individuals from the time-sensitive business of saving lives.

Therefore, in the absence of such an approach, HFRNs have to resort to other types of rules. Here, Greenberg *et al.* (2007) note that rules, processes and procedures need to relate not just to the management of a particular office (i.e. the underpinning bureaucracy of the organization), but also to inter-personal communication. Thus, it is the areas of communication rules that the leader of an HFRN may be able to target in nurturing the development of inter-personal trust within his or her team.

In this regard, the focus of the Logistics Cluster on the development of forms and standards of communication among humanitarian organizations (in addition to its role of operational coordination) can be clearly seen as supporting the development of rule-based trust. However, in the disaster relief context, the development of well documented processes and procedures (i.e. "rules") is counter-cultural. Those working within humanitarian organizations are, understandably, output and outcome focused; their *raison d'être* is the relief of hardship and suffering of those affected by a disaster and adherence to "bureaucracy" is seen as a diversion from this real objective. On the other hand, when responding to a major disaster, humanitarian organizations almost universally are forced to use staff who are not part of their core teams, i.e. those from "on call" rosters and other augmenters. In all probability these additional resources will have had limited experience of working within the particular humanitarian organization and, therefore, will have even more limited exposure to that organization's rules. This results in the potential for inadvertent maverick behavior with its concomitant negative effect on the development of inter-personal trust. Obviously there is a balance to be struck here, as it could be argued that such behavior in the guise of strong leadership could be valuable in cementing relationships within a team. However, from the perspective of the "swift trust" model there is clear benefit in the advance development and exposition of clear simple and easy to follow rules that will help ensure new comers can fit into the organization and become effective both speedily and with the minimum of effort.

Category

According to the swift trust model, individual members of a HFRN are also likely to unconsciously categories other team members as be-

longing to the same or a different social group or category. It is stressed that such membership (and the resultant perception by their fellow team members) is often a fact over which the individual has no choice, and may include simple differences like their gender or race, as well as more complex ones such as their parent organization. Within the context of disaster relief, this is potentially a highly divisive area – indeed, evidence of the negative effects of such categorization has been noted by Zolin (2002, p.7) who observed: "difficulties in establishing interpersonal working relationships between [US Military] and [NGOs] due to perceived differences in organizational goals, strongly held negative organizational stereotypes and perceived ideological differences".

Such trust judgments may be based on stereotypes of gender, ethnicity, religion, race or age. Given that such stereotypes undoubtedly exist, the implication of the swift trust model is that when the trustor and trustee belong (or perceive that they belong) to different categories, this will have a negative impact on the development of trust. To the extent that both are, say, logisticians or both belong to the same humanitarian organization, this negative impact is more likely to occur in inter- rather than intra-organizational trust situations. The challenge is to develop mechanisms to overcome this issue through advanced dialogue and understanding. Excellent examples of this can be found in the ongoing exchanges between the Irish Defence Forces and Irish NGOs in which the latter give presentations to the former on a regular basis, and the former conduct training and education courses for the latter. It is not just the content of the discussions that is important, but the associated knowledge and understanding of each others' perspectives and concerns that will help to break down potential "category" barriers.

From the perspective of a HFRN, in the same way as for the area of dispositional trust, the challenge for the leader is to recognize that the sorts of stereotype that have been mentioned above will exist. In many cases, however, the categorization will take place within an individual's mind in an unconscious way, making its management even more complex. However it is strongly advocated that such challenges are best met by recognizing the existence of the problem and its implications for restraining the development of the desired inter-personal trust.

Role

In the context of the formation of swift trust, using roles as the basis for making initial assumptions has the benefit of being de-personalized. In other words, the trustor can make an assessment of an individual's ability based on the fact that they are fulfilling a particular role rather than through specific knowledge about their competence, motives etc. (Kramer, 1999). A typical example of role-based trust is the positive predisposition of individuals to trust a medical doctor for her/his medical expertise even in their first consultation is based simply on the fact that the doctor holds the relevant professional qualifications. Role-based trust can therefore, be seen as "competence trust", as it is based on the confidence that the other partner carries the competence to perform her/his task (cf. Fynes *et al.*, 2005).

In the context of an HFRN, the fact that a particular humanitarian organization is employing an individual in the role of, say, a logistician leads others to assume that the individual has been judged to have the relevant competencies and capabilities, and can therefore be trusted. However, for this means of developing trust to be effective, there is a very clear onus on humanitarian organizations to fulfill their side of this notional bargain – in other words only to employ staff who does, indeed, possess the relevant competencies etc. As Bal and Teo (2001) suggest for virtual teams, the performance and competence of individuals impacts on their in- or exclusion from informal communication and meetings. Thus the competence of an individual plays an important role for trusting the individual, and ultimately, for

the performance of the HFRN. That said, it is far from established which competences are required or expected from humanitarian logisticians (Tatham *et al.*, 2010). Nevertheless, to the extent that humanitarian organizations are clearly embarking on a series of programs designed to improve the competence levels of their staff (Walker and Russ, 2010), there is clear potential for swift trust to be based, in part, on the possession by an individual of the relevant qualification.

It is therefore suggested that humanitarian organizations should continue to press ahead with their training and certification schemes and that, whenever possible, individuals with the appropriate qualifications should be employed as permanent or on call team members. It is, of course, recognized that there are significant challenges associated with such international certification schemes including ensuring the achievement of a common standards and, indeed, that any examination accurately tests for the existence of the right skills. However, it is argued that such hurdles are not insuperable, and that the balance of benefit lies with pursuing such an approach. Such an approach might, for example, be based on existing schemes such as those provided by the UK Chartered Institute for Logistics and Transport (CILT) which are delivered in concert with the NGO RedR. In any event, successful anchoring of skills and experience on an internationally recognized framework would provide a valuable underpinning for the element of role within the swift trust model and, hence, support the development of the desired inter-personal trust.

The Impact of the Communication Environment

In considering the swift trust model (Figure 1), it is not only important to note the five elements of peripheral trust, but also to distinguish between trust and trusting behavior. Hung *et al.* (2004) depict the latter as being mediated by the perceived risk of potential gains (or losses) of acting on the basis of inter-personal trust (see also Meyerson *et al.*, 1996; Kramer, 1999; Hung *et al.*, 2004; Ben-Shalom *et al.*, 2005). In the disaster relief context, perceived risks can encompass physical danger as well as the loss of reputation as a result of depending on the behavior of other members of the HFRN. In essence, the act of trusting is one in which the trustor is prepared to increase their vulnerability to the actions of others. It follows, therefore, that if level of perceived risk is greater than the level of trust, the individual is less likely to engage in trusting behavior (Hung *et al.*, 2004).

This aspect of the swift trust model is related to various streams of literature on the psycho-logical, physiological and organizational aspects of perception. For example, Laaksonen *et al.* (2009) argue that rules such as contracts or pric-ing mechanisms help to codify the level of risk and ensure a mutual perception in a business context. In the absence of contracts and pricing mechanisms (such as in the swift trust model), the communication environment takes their place (cf. Hung *et al.*, 2004). The scenario surrounding HFRNs precludes lengthy contractual discussions and associated understanding of financial and reputational risk. Rather, an assessment is made by an individual of the impact of trusting his or her colleague but, critically, it is suggested that this is heavily impacted by the effectiveness of the communications environment (Järvenpää & Leidner, 1999).

The two extremes of such a communication environment might be characterized as a face-to-face office conversation, and a telephone call on a poor line between an operator in the field and his or her headquarters located in another country, away from the various mental stimuli of the operational situation. Put simply, through the ease of communication and the presence of additional non-verbal clues in the former scenario, it will be easier for an individual to determine whether or not to trust their informant than in the latter. Furthermore, the effect of the communica-tions medium in virtual environments operates

in both directions and so, from the headquarters perspective, the perceived risks are increased due to a reduction in the degree of control individuals from within the headquarters can exert (Järvenpää *et al.*, 1998). Other examples for increasing the perceived risks in virtual environments include role ambiguity and role overload as a result of a lack of face-to-face communication (Järvenpää *et al.*, 1998).

The importance of communication is also emphasized by Weick (1993) in his analysis of the Mann Gulch disaster in which 13 US fire fighters lost their lives. One of key organizational failings was the near absence of communication between the team members and consequential reduction in the level of intra-team coordination. In short, the lack of communication in the early stages of the development of this temporary group heightened its vulnerability to disruption. When stressed by the advancing wild fire, the inter-team ties (which, in part, reflect the level of inter-personal trust) were insufficient to prevention fragmenta-tion of the group and a reversion to self-interest (or, perhaps more accurately, self-preservation). This point is equally emphasized by Drabek (1985) whose analysis of emergency response organizations in the United States indicates that cross-agency communication was perceived to be the greatest weakness and the source of most difficulties. In summary, there would appear to be broad support for the proposition that the clarity of the communications environment has an effect on the formation of trust and, by extension, the view of Hung *et al.* (2004) that computer medi-ated communications environments increase the perceived risk and, hence, reduce the propensity to convert trust into trusting behavior.

Once again, in terms of mitigating these problems and difficulties, the key would appear to lie in an understanding of the problem (i.e. the effect of the perception of risk) on the actions of individuals, and the role that the effectiveness of inter-personal communication has to play. Clearly a number of technological solutions (such as the use of video-conferencing) may help overcome the inherent defects of simple computer-based interaction (e-mails etc.), but there would also appear to be support for attempting to achieve face-to-face communication (e.g. visits to the field by headquarter staff) wherever possible. Once again, however, the role of the HFRN leader would appear critical in ensuring that team members are aware of this facet of the problem through appropriate guidance, training and education.

Implications of the Swift Trust Model

Whilst swift trust is a model that focuses on trust in inter-personal relationships, the model has important implications for individual members of the HFRN, the leaders of the HFRN (e.g. cluster leads), the HFRN as an entity, as well as the organizations from which its members are drawn. Table 1 summarizes these implications for the different levels.

Much of the discussion of the different elements of swift trust has focused on the perspective of the leader of an HFRN. In doing so, it needs to be acknowledged that such leadership may run counter to the values of individual HFRN members and, indeed, member organizations themselves. Various conflicts based on mandates and the very elements of swift trust can constitute barriers to not only for trust to emerge, but also for leadership to be effective. Categories not only refer to stereotypes but also hierarchies in organizations. Command and control structures, while applicable to some actors of the humanitarian aid supply network, constitute rather the exception than the rule. Hence, coordination in the HFRN is based on many-to-many communication and not established rules of command. Nevertheless, there are some established mechanisms for e.g. cluster leads globally and in disaster areas. Leadership here is thus attributed foremost to current and actual cluster leads in the field.

Turning to the logistic implications of the swift trust model, it is suggested that the establishment of meeting procedures and a common set of performance indicators that are to be reported to the HFRN will facilitate the adherence to common rules. Also, process standardization across humanitarian organizations will facilitate the emergence of trust, as individuals will be able to trust a common course of action whilst agreeing to distribute different activities in the HFRN. Last but not least, establishing pipeline (and indicator) visibility across organizations will facilitate not only the adherence to common rules, but also supports a focus on the common aim rather than individual differences in stereotypes as well as in the understanding of the job of a humanitarian logistician.

FURTHER RESEARCH DIRECTIONS

It is hoped that this chapter has successfully demonstrated that, in a disaster relief operation which typically includes a number of HFRNs, the ability of the members to work together has far-reaching consequences for the success or failure of the disaster response. But, given the nature of a HFRN, this leads to the question of how such inter-personal trust can be created and developed. With this in mind, the chapter has focused on the model of swift trust that originated in the work of Meyerson *et al.* (1996) with the aim of suggesting, from a theoretical perspective, how this model might be operationalized in such a scenario.

Clearly, it is necessary for the swift trust model to be evaluated in an empirical setting and it is initially considered that this could be explored in two phases. The first might be undertaken in an experimental setting in, for example, a university setting in which a class of students are set a suitable exercise to be conducted in groups. In parallel with their task to resolve the particular leadership challenge, individual members of the group would be extracted and interviewed using the five elements of the swift trust model as the broad agenda. Indeed, if circumstances permit,

Table 1. Implications of the swift trust model

Elements of the Swift Trust Model	Implications for Individual HFRN members	Implications for Leaders of the HFRN	Implications for The HFRN	Implications for Organizations of HFRN members
Third Party Information	• Be inclusive also of individuals where prior contact has not been established • Look beyond the reputation of other individual's organizations	• Establish prior common experiences (e.g. training) to facilitate trust among HFRN members • Organize get-togethers at an early stage and exchange background information about organizations and the expertise of individuals	• Draw on third party information on the expertise of individuals to determine the mix of expertise • Establish (virtual) social networks for the HFRN	• Use secondments / secondees to facilitate the sharing of knowledge of each other's organizations representatives • Include experiences and evaluations of individuals in roster databases • Advertise the competence of individual
Dispositional Trust	• Acknowledge different dispositions to trust that reflect individual character traits that may, in turn, reflect cultural norms.			
Rule	• Adhere to "common" rules of the HFRN even if unorthodox for the organization of the individual • Avoid maverick behaviour	• Establish rules and procedures for meetings • Establish common performance indicators that need to be reported to the HFRN	• Establish communication rules • Establish pipeline and indicator visibility across organizations	• Establish prior collaboration mechanisms and cross-organizational teams facilitates their common deployment • Standardize processes across organizations
Category	• Look beyond stereotypes of individuals and organizations	• Recognize categories and stereotypes and discuss them openly to mitigate their effects	• Meet regularly also socially as to overcome categories and stereotypes	• Establish common training to counteract categorization and stereotyping • Select individuals with an open mindset beyond the mission of the organization
Role	• Learn from each other on the job • Undergo certification schemes for humanitarian logisticians	• Establish a common understanding of logistics, related activities and processes	• Focus on commonalities (common pipeline) rather than differences	• Constantly improve the competence level of staff • Recognize certifications and logistics education

additional realism could be injected by having some members of the group operating in a "virtual" way, and the resulting levels of trust between the core team and its virtual members compared.

Evaluating the swift trust framework in an operational setting would be the next stage. This, unfortunately, would require the advent of a significant rapid onset disaster in which multi-sourced team (including the Logistics Cluster) are deployed. With the approval of their host humanitarian organization, it is anticipated that researchers could undertake a number of interviews with key personnel both in the field and at the headquarters which were, again, based around investigation of the impact of the five elements of the swift trust model. Self-evidently, such research would require particularly careful management so as to be timely, and yet not interfere with the vital life-saving work of the organization itself.

Alternatively, the swift trust model could be tested through a survey. Instead of exploring the actual challenges of developing and maintaining swift trust, a survey could contribute to the evaluation of the interdependencies of trust elements, and the mediating effects of perceived risk and the communication environment. However, these very interdependencies pose the challenge to develop independent scales for each element of swift trust.

A further, and clearly more complex strand of research, be that of understanding, from the perspective of the humanitarian logistician, how one might distinguish a successful organization (i.e. that in which a high level of inter-personal trust exists) from an unsuccessful one. In the case of the latter, not only would it be instructive to understand the nature of the perceived failings, but also the implications for the beneficiaries. It is anticipated that such an analysis, which reflects the reverse of the obvious line of enquiry, would help to triangulate the practical development of this model.

Finally, and although some research is currently being conducted in this area, greater clarity is needed over the skills and attributes that make a "good" humanitarian logistician and, in particular, how a linkage between these and logistics performance can be demonstrated.

In drawing up the research agenda outlined above, it is important to acknowledge the complexity of the subject as a whole. The difficulty lies in designing a suitable approach to the next stages of the research that will enable valid conclusions

to be drawn in the face of the many variables within the model.

CONCLUSION

Any particular disaster relief operation includes a number of logisticians from various organizations (and, hence, organizational types and cultures) who come together to form one or more HFRNs. Such HFRN(s) will have the common aim of alleviating the suffering of those affected by the disaster and are likely to show all of the characteristics described above:

- their network is established rapidly;
- they come from different communities, in fact different organizations, countries and cultures;
- they work together in a shared conversation space, with a need to coordinate their activities;
- in which they plan, commit to, and execute actions;
- to fulfill a large, urgent mission. (HFN Research Group, 2006)

But, whilst they may share a common high level goal, the ability of members of an HFRN to work together will have far-reaching consequences and, ultimately, it will have an impact on the success or failure of the disaster response.

Trust, both inter-personal and inter-organizational, has been argued to have positive consequences for the success of a relationship, even to the extent of reducing transaction costs (cf. Laaksonen *et al.*, 2009). Unsurprisingly, therefore, the supply chain collaboration literature draws on trust as a key success factor, although this literature is generally focused in the challenges associated with the development and maintenance of long-term relationships (Skjøtt-Larsen *et al.*, 2003; Barratt, 2004; Fawcett *et al.*, 2008).

However, this chapter has concentrated on the less well researched issues surrounding inter-personal trust in HFRNs and, in particular, those created in the aftermath of a disaster. In doing so, it has drawn on the work of Meyerson *et al.* (1996) who offered the model of swift trust, and the aim of the chapter was to consider how this approach might be operational zed in a disaster relief context. In order to achieve this, each of the elements of swift trust model (see Hung *et al.*, 2004, and Figure 1) has been discussed from the perspective of the leader of such an HFRN.

The swift trust model does, indeed, appear to be highly applicable to this scenario, and its consideration leads to a number of important conclusions. Firstly, it shows that the central and habitual routes to trust that supply network management traditionally considers are, indeed, important aspects in the collaboration between humanitarian organizations and their global suppliers and logistics service providers.

Secondly, in terms of the elements of the swift trust model, the third party information about humanitarian organizations and the individuals they send in response to a disaster, are important aspects in the development of trust in a HFRN. Information on which individuals have been part of a previous successful operation can help in the formation of (parts of) inter-organizational teams that can be co-deployed to a similar operation. Thus, it is in the interest of humanitarian organizations to provide information about the individuals they send to a particular disaster area both to the leader of their own team and also to other humanitarian organizations. This will help to facilitate the individual-individual interaction on the ground even in absence of historical encounters.

Turning to the challenge of dispositional trust, it is clear that selecting individuals on the basis of this particular trait would be impractical and unethical – even if it were it possible to undertake this in anything approaching a rigorous (scientific) way. More important, therefore, is the development of common rules such as standard operating procedures and common forms, to help ensure the inter-operability of logisticians from different humanitarian organizations. The creation of a common set of needs assessment templates represents a good example of where such an approach would pay dividends in terms of efficiency and effectiveness. Similarly, developing standard rules of communication would appear to be important in the context of trust building as these would help enable maverick behavior to be recognized and managed at an early stage.

The categorization of individuals by their colleagues (which is frequently an unconscious process) is clearly more often an impediment rather than an enabler to trust. But, in the same way as for the dispositional element, the key would appear for the HFRN leadership to recognize the potential difficulties here and to ensure that the members of their team and the wider network are fully aware of these challenges. In this sense, a simple awareness of the problem may be sufficient to reduce its negative impact.

The swift trust model suggests that members of an HFRN use their knowledge of an individual's role as a proxy for their competence (and, hence, trustworthiness). Whilst this may be a convenient approach in terms of reducing an individual's cognitive workload, it is undermined by the systemic weakness reflected by the absence of a universally held picture of what skills and attributes makes for a "good" humanitarian logistician. Furthermore, unlike military personnel or fire fighters, there is currently no common training for humanitarian logisticians of different organizations. There would, therefore, appear to be an onus on the broader humanitarian community to play its part by developing pan-humanitarian professionalization mechanisms such as a suite if standard courses (and associated certification) for humanitarian logisticians, together with a register of those who are appropriately qualified. It is recognized that this is not a simpler task, not least because of the concomitant requirement for agreement on, for example, what qualities are needed to be a successful humanitarian logistician. Such an approach also implies the existence of some form of audit function and, indeed, agreement amongst humanitarian organizations that they should be subject to such a regime. Nevertheless, the fact that these are clearly tricky issues is not, of itself, a reason for failing to embark on such a course of action. In this respect, the recent joint training efforts sponsored by the Logistics Cluster and the Humanitarian Logistics Association are to be applauded. Nevertheless, it is clear that much significantly effort is required before appropriate internationally recognized (and audited) certification programmers are developed.

So what can one conclude from the discussion within this chapter? From a theoretical perspective, the model of swift trust would appear to have considerable relevance to the management of post-disaster HFRNs, not least as it provides the leadership of such HFRNs with an agenda that they should bear in mind as the network forms up and develops. In doing so, however, it underlines the fact that the leadership of a given HFRN can, with the best will in the world, only go so far. As with many other aspects of the relief supply chain challenge, they key would appear to be the need for the community as a whole to recognize their broader strategic responsibilities that surmount their specific organizational mandate. The challenge of preparing for, and responding to, disasters will, without question, become more complex and difficult as the century proceeds – in turn, therefore, the community as a whole must learn respond in a way that truly maximizes their ability to support those affected.

REFERENCES

ALNAP. (2008). *Cyclone Nargis: Lessons for operational agencies*. Retrieved July 1, 2010, from www.alnap.org/publications/pdfs/ALNAP-LessonsCycloneNargis.pdf

Bal, J., & Teo, P. K. (2001). Implementing virtual teamworking: Part 2 – A literature review. *Logistics Information Management, 14*(1), 208–222. doi:10.1108/09576050110390248

Barratt, M. (2004). Understanding the meaning of collaboration in the supply chain. *Supply Chain Management: An International Journal, 9*(1), 30–42. doi:10.1108/13598540410517566

Ben-Shalom, U., Lehrer, Z., & Ben-Ari, E. (2005). Cohesion during military operations. *Armed Forces and Society, 32*(1), 63–79. doi:10.1177/0095327X05277888

Butcher, T., Claes, B., & Grant, D. (2008). Supply chain work organisation: Can structuration theory offer new solutions? In *Proceedings of the Logistics Research Network (LRN) Conference,* Liverpool, UK, 10-12 Sept.

Christopher, M. (2011). *Logistics and supply chain managemen* 4th ed. London, UK: Financial Times Prentice Hall.

Coppola, N. W., Hiltz, S. R., & Rotter, N. G. (2004). Building trust in virtual teams. *IEEE Transactions on Professional Communication, 47*(2), 95–104. doi:10.1109/TPC.2004.828203

Das, T. K., & Teng, B.-S. (2001). Trust, control, and risk in strategic alliances: An integrated framework. *Organization Studies, 22*(2), 251–283. doi:10.1177/0170840601222004

Denning, P. J. (2006). Hastily formed networks. *Communications of the ACM, 49*(4), 15–20. doi:10.1145/1121949.1121966

Drabek, T. T. (1985). Managing the emergency response. *Public Administration Review, 45,* 85–92. doi:10.2307/3135002

Fawcett, S. E., Magnan, G. M., & McCarter, M. W. (2008). Benefits, barriers, and bridges to effective supply chain management. *Supply Chain Management: an International Journal, 13*(1), 35–48. doi:10.1108/13598540810850300

Fitzgerald, S. P. (2004). The collaborative capacity framework: from local teams to global alliances. *Advances in Interdisciplinary Studies of Work Teams, 10,* 161–201. doi:10.1016/S1572-0977(04)10007-1

Fynes, B., Voss, C., & de Búrca, S. (2005). The impact of supply chain relationship quality on quality performance. *International Journal of Production Economics, 96,* 339–354. doi:10.1016/j.ijpe.2004.05.008

Gattorna, J. (2006). *Living supply chains.* Harlow, UK: Financial Times Prentice Hall.

Greenberg, P. S., Greenberg, R. H., & Antonucci, Y. L. (2007). Creating and sustaining trust in virtual teams. *Business Horizons, 50*(4), 525–533. doi:10.1016/j.bushor.2007.02.005

Grey, C., & Garsten, C. (2001). Trust, control and post-bureaucracy. *Organization Studies, 22*(2), 229–250. doi:10.1177/0170840601222003

Gulati, R. (1995). Does familiarity breed trust? The implications of repeated ties for contractual choice in alliances. *Academy of Management Journal, 38*(1), 85–112. doi:10.2307/256729

Handy, C. (1995). Trust and the virtual organization. *Harvard Business Review, 73*(3), 40–50.

Head, C. (2000). *Sound doctrine: A tactical primer.* New York, NY: Lantern Books.

HFN (Hastily Formed Networks) Research Group. (2006). *HFN defined.* Retrieved 28 May, 2009, from www.hfncenter.org/cms/node/117

Hung, Y.-T. C., Dennis, A. R., & Robert, L. (2004). Trust in virtual teams: Towards an integrative model of trust formation. In *Proceedings of the 37th Hawaii International Conference on Systems Sciences,* Track 1, Vol. 1.

Järvenpää, S. L., Knoll, K., & Leidner, D. E. (1998). Is anybody out there? Antecedents of trust in global virtual teams. *Journal of Management Information Systems, 14*(4), 29–64.

Järvenpää, S. L., & Leidner, D. E. (1999). Communication and trust in virtual teams. *Organization Science, 103,* 791–815. doi:10.1287/orsc.10.6.791

Jones, G. R., & George, J. M. (1998). The experience and evolution of trust: Implications for cooperation and teamwork. *Academy of Management Review, 23*(3), 531–546.

Knights, D., Noble, F., Vurdubakis, T., & Willmott, H. (2001). Chasing shadows: Control, virtuality and the production of trust. *Organization Studies, 22*(2), 311–336. doi:10.1177/0170840601222006

Kovács, G., & Spens, K. (2008). Humanitarian logistics revisited. In Arlbjørn, J. S., Halldórsson, A., Jahre, M., & Spens, K. (Eds.), *Northern lights in logistics and supply chain management* (pp. 217–232). Copenhagen, Denmark: CBS Press.

Kovács, G., & Spens, K. M. (2010). (forthcoming). Knowledge sharing in relief supply chains. *International Journal of Networking and Virtual Organisations.* doi:10.1504/IJNVO.2010.031219

Kramer, R. M. (1999). Trust and distrust in organizations: Emerging perspectives, enduring questions. *Annual Review of Psychology, 50,* 569–598. doi:10.1146/annurev.psych.50.1.569

Laaksonen, T., Jarimo, T., & Kulmala, H. I. (2009). Cooperative strategies in customer-supplier relationships: The role of interfirm trust. *International Journal of Production Economics, 120*(1), 79–87. doi:10.1016/j.ijpe.2008.07.029

Majchrzak, A., Järvenpää, S. L., & Hollingshead, A. B. (2007). Coordinating Expertise Among Emergent Groups Responding to Disasters. *Organization Science, 18*(1), 147–161. doi:10.1287/orsc.1060.0228

McKnight, D. H., Cummings, L. L., & Chervany, N. L. (1998). Initial trust formation in new organizational relationships. *Academy of Management Review*, 23(3), 473–490.

Mentzer, J. T., DeWitt, W., Keebler, J. S., Min, S., Nix, N. W., Smith, C. D., & Zacharia, Z. G. (2001). Defining supply chain management. *Journal of Business Logistics*, 22(2), 1–25. doi:10.1002/j.2158-1592.2001.tb00001.x

Meyerson, D., Weick, K. E., & Kramer, R. M. (1996). Swift trust and temporary groups. In Kramer, R. M., & Tyler, T. R. (Eds.), *Trust in organizations: Frontiers of theory and research* (pp. 166–195). Thousand Oaks, CA: Sage Publications Inc.

NRCNA. (National Research Council of the National Academies. (2006). *Facing hazards and disasters: Understanding human dimensions*. Committee on Disaster Research in the Social Sciences: Future Challenges and Opportunities, Division of Earth and Life Studies. Washington DC: The National Academies Press.

Oloruntoba, R., & Gray, R. (2006). Humanitarian aid: An agile supply chain? *Supply Chain Management: An International Journal*, 11(2), 115–120. doi:10.1108/13598540610652492

Oloruntoba, R., & Gray, R. (2009). Customer service in emergency relief chains. *International Journal of Physical Distribution and Logistics Management*, 39(6), 486–505. doi:10.1108/09600030910985839

Roberts, A. (2001). *NGOs: New Gods overseas: The World in 2001* (p. 73) London, UK: The Economist Publications.

Rousseau, D. M., Sitkin, S. B., Burt, R. S., & Camerer, C. (1998). Not so different after all: A cross discipline view of trust. *Academy of Management Review*, 23(3), 393–404. doi:10.5465/AMR.1998.926617

Sarkis, J., Talluri, S., & Gunasekaran, A. (2007). A strategic model for agile virtual enterprise partner selection. *International Journal of Operations & Production Management*, 27(11), 1213–1234. doi:10.1108/01443570710830601

Scheper, E., Parakrama, A., & Patel, S. (2006). Impact of tsunami response on local and national capacities. London, UK: Tsunami Evaluation Coalition (TEC). Retrieved September 4, 2008, from http://www.tsunami-evaluation.org/NR/rdonlyres/8E8FF268-51F0-4367-A797-F031C0B51D21/0/capacities_final_report.pdf

Skjøtt-Larsen, T., Thernøe, C., & Andresen, C. (2003). Supply chain collaboration. Theoretical perspectives and empirical evidence. *International Journal of Physical Distribution and Logistics Management*, 33(6), 531–549. doi:10.1108/09600030310492788

Smith, W., & Dowell, J. (2000). A case study of co-ordinative decision-making in disaster management. *Ergonomics*, 43(8), 1153–1166. doi:10.1080/00140130050084923

Sphere. (2004). *Humanitarian charter and minimum standards in disaster response*. The Sphere Project. Retrieved September 4, 2008, from http://www.sphereproject.org/handbook/pages/navbook.htm?param1=0

Stocking, D. B. (2010). *Natural disasters: How can we improve?* Seminar at the Royal Geographic Society. Retrieved 25 May, 2010, from http://www.21stcenturychallenges.org/challenges/25-may-natural-disasters-how-can-we-improve/

Tatham, P. H., Kovács, G., & Larson, P. D. (2010). What skills and attributes are needed by humanitarian logisticians - A perspective drawn from international disaster relief agencies. *Proceedings of the 21st Production and Operations Management Society (POMS) Annual Conference*, Vancouver, May 7-10, 2010.

The Lancet. (2010). Editorial. Growth of aid and the decline of humanitarianism. *Lancet, 375*(9711), 253. doi:10.1016/S0140-6736(10)60110-9

Thomas, A. (2003). Why logistics? *Forced Migration Review, 18*, 4.

Uhr, C., & Ekman, O. (2008). Trust among decision makers and its consequences to emergency response operations. *Journal of Emergency Management, 6*(3), 21–37.

UNJLC (United Nations Joint Logistics Centre). (2008). *Logistics cluster Myanmar: Cyclone Nargis emergency response* 10th May – 10th August 2008. End of Mission Report. Retrieved October 28, 2008, from http://www.logcluster.org/logistics-cluster/meeting/global-logistics-cluster-meeting-3-4-octobre-2008

Uzzi, B. (1997). Social structure and competition in interfirm networks: The paradox of embeddedness. *Administrative Science Quarterly, 42*(1), 35–67. doi:10.2307/2393808

van Wassenhove, L. N. (2006). Humanitarian aid logistics: Supply chain management in high gear. *The Journal of the Operational Research Society, 57*(5), 475–589. doi:10.1057/palgrave.jors.2602125

Völz, C. (2005). Humanitarian coordination in Indonesia: An NGO viewpoint. *Forced Migration Review*, Special Issue, July, 26-27.

Walker, P., & Russ, C. (2010). Professionalising the humanitarian sector. *ELRHA*. Retrieved July 3, 2010, from http://www.elrha.org/professionalisation

Weick, K. E. (1988). Enacted sensemaking in crisis situations. *Journal of Management Studies, 25*(4), 305–317. doi:10.1111/j.1467-6486.1988.tb00039.x

Weick, K. E. (1993). The collapse of sensemaking in organizations: The Mann Gulch disaster. *Administrative Science Quarterly, 38*(4), 628–652. doi:10.2307/2393339

Zolin, R. (2002). *Swift trust in hastily formed networks*. The Hastily Formed Networks Research Group. Retrieved September 4, 2008, from http://www.hfncenter.org/cms/files/swift-trustinHFN10-03-02.pdf

ADDITIONAL READING

Bachmann, R., & Zaheer, A. (Eds.), *Handbook of Trust Research*. Cheltenham, UK: Edward Elgar.

Bal, J., Wilding, R., & Gundry, J. (1999). Virtual teaming in the agile supply chain. *International Journal of Logistics Management, 10*(2), 71–82. doi:10.1108/09574099910806003

Christopher, M., Peck, H., & Towill, D. (2006). A taxonomy for selecting global supply chain strategies. *International Journal of Logistics Management, 17*(2), 277–287. doi:10.1108/09574090610689998

Douglas, M. (1978). Cultural Bias. *Royal Anthropological Society of Great Britain, Occasional Paper, No. 35.*

Ellickson, R. C. (1986). Adverse possession and perpetuities law: two dents in the libertarian model of property rights. *Washington University Law Quarterly, 64*, 723–738.

Fukuyama, F. (1995). *Trust: The Social Virtues and the Creation of Prosperity*. New York: Free Press.

Goranson, H. T. (1999). *The Agile Virtual Enterprise: Cases, Metrics, Tools*. Westport, CT: Quorum Books.

Jahre, M., & Heigh, I. (2008). Does the current constraints in funding promote failure in humanitarian supply chains? *Supply Chain Forum, 9*(2), 44–54.

Kasper-Fuehrer, E. C., & Ashkanasy, N. M. (2001). Communicating trustworthiness and building trust in interorganizational virtual organizations. *Journal of Management, 27*, 235–254. doi:10.1016/S0149-2063(01)00090-3

Lewicki, R. L., McAllister, D. J., & Bies, R. J. (1998). Trust and distrust: New relationships and realities. *Academy of Management Review, 23*(3), 438–458.

Maguire, A., Phillips, N., & Hardy, C. (2001). When 'silence = death', keep talking: trust, control and the discursive construction of identity in the Canadian HIV/AIDS treatment domain. *Organization Studies, 22*(2), 285–310. doi:10.1177/0170840601222005

Stoddard, A. (2003). Humanitarian NGOs: challenges and trends. *HPG Briefing, 12 (July), 1-4.* Retrieved 16 Sep, 2008, from http://www.odi.org.uk/HPG/papers/hpgbrief12.pdf.

Tatham, P. H., & Spens, K. (2008). The developing humanitarian logistics knowledge management system – a proposed taxonomy. *Proceedings of POMS*, San Diego, 9-12 May.

Telford, J., & Cosgrave, J. (2007). The international humanitarian system and the 2004 Indian Ocean earthquake and tsunamis. *Disasters, 31*(1), 1–28. doi:10.1111/j.1467-7717.2007.00337.x

Zack, P. J., & Knack, A. (2001). Trust and growth. *The Economic Journal, 111*(April), 295–321. doi:10.1111/1468-0297.00609

KEY TERMS AND DEFINITIONS

Hastily Formed Relief Network: HFRNs are co-located teams in disaster areas with intersectoral partnerships that link humanitarian organizations to governments, local communities, businesses and the military, in order to form the humanitarian aid supply network. Members of a hastily formed relief network share the mission of bringing disaster relief but do not subscribe to common goals and do not share a common history or future beyond the short term deployment.

Trust: Is present when one party has a fundamental belief that the other can be relied upon to fulfill their obligations with integrity, and will act in the best interests of the other. Hastily formed network: A HFN is a network of people established rapidly, from different communities, working together in a shared conversation space, in which they plan, commit to, and execute actions (HFN Research Group, 2006).

Chapter 11
A Study of Barriers to Greening the Relief Supply Chain

Joseph Sarkis
Clark University, USA

Karen M. Spens
HUMLOG Institute, Hanken School of Economics, Finland

Gyöngyi Kovács
HUMLOG Institute, Hanken School of Economics, Finland

ABSTRACT

Relief supply chain (SC) management is a relatively unexplored field. In this field, practitioners have shown some interest in greening practices, but little practical or academic literature exists to help provide insights into combining the two fields. Adoption of green SC principles in the relief SC requires a systematic study of existing barriers in order to remove these barriers and allow introduction of green practices. The aim of this chapter is to explore barriers to implementation of green practices in the relief SC. Expert opinions and literature from humanitarian logistics and green supply chain management are used to establish a list of barriers and to propose a categorization of barriers. Further research to evaluate the relationships and importance of these barrier factors is identified.

INTRODUCTION

Supply chain management requires the planning, design, implementation, coordination and maintenance of various flows across many boundaries. Supply chain social responsibility has received increased interest over the past decade. One aspect of socially responsible supply chains includes their application and management for humanitarian purposes. Another important dimension of the socially responsible supply chains is the greening aspect, or green supply chain management. Separately these two topical disciplines have seen a paralleled growth and interest by practitioners and researchers. Even though numerous researchers have been investigating sustainable supply chains, where the term sustainable includes social, economic, and environmental dimensions, it is surprising that the intersection of these inchoate fields has yet to be carefully examined. Therefore, in this chapter

DOI: 10.4018/978-1-60960-824-8.ch011

the two fields, green supply chain management (GSCM) and relief supply chain management are integrated through a study on barriers to greening the relief supply chain.

Relief supply chain management is a relatively new area of investigation which is typically associated with unexpected disasters that require immediate action. As stated by Kovács and Spens (2007), recent humanitarian logistics literature also focuses on disaster relief, yet, most relief supply efforts can be attributed to longer term effects, especially those situations which result from war and famine (Hoerz, 1997). One good example is refugee camps, which may last for years and need effective long-term and short-term operational planning. The burden on the environment is of such a magnitude that comments such as "the UNHCR [aka The UN Refugee Agency] has destroyed our environment" have been cited (Hoerz, 1997). A general list of potential environmental impacts associated with water and related activities in a camp situation include the

• Depletion of the source as a result of unsustainable extraction or collection of water.
• Contamination of the local water due to improper disposal of waste water and human-waste, faulty design and operation/maintenance of the piped water network, excessive extraction of groundwater and other related activities in the camp.
• Impacts to local environment due to construction and operation of water supply system intensity and magnitude of which would largely depend on the nature and size of the project and the sensitivity of the local ecosystem.
• Inappropriate drainage, soil and water conservation measures as well as poor water management in irrigation systems may lead to erosion, floods, groundwater contamination and soil salinization.

• Camps or settlements close to open streams or over unconfined aquifers may cause downstream contamination.

Inside Haiti, more than two months after the 2010 earthquake, it was reported by the BBC that some areas had yet to receive humanitarian supplies. When some of these supplies arrived in a community not too far from the epicenter of the 2010 earthquake, they came in many forms. One set of supplies included military rations (Meal, Ready to Eat, MRE). These rations were culturally inappropriate due to their 'individual' nature and containing non-traditional, to Haitians, foods. It was expected that many of these MRE would be discarded. Important from an environmental perspective, is that these MREs came in hard plastic containers. It was observed in the BBC radio that piles of green plastic, hazardous and not easily disposable, existed as over 250,000 MREs were delivered to Haitians. Given this type of scenario, one suggestion to reduce the impact of camps on the environment has been to involve refugees in the battle against this environmental destruction, something which has been successfully deployed for example by the UN. An example is provided by the Sherkole camp in the western Highlands in Ethiopia where environmental education and awareness is being put into place to help fight climate change (UNHCR, 2009). The other type of situation to which humanitarian logistics is usually linked is disaster relief operations. Literature suggests that up to 80 percent of the costs involved in relief operations are in fact related to logistics. Not only are the costs of the disaster relief operations high, the impact on the environment is also severe as in many cases the urgency of the situation enforces the use of environmentally unsound transportation modes and means. Thereby, one aspect of the potential conflict between efficiency in humanitarian logistics and environmentally sound green supply chains is in the selection of modes of transportation for material delivery. If delivery time is a major concern then air transport may be the most time efficient approach. Yet, air transport is between 5-20 times worse in terms of fuel and air emissions than train or road transport and even greater than water transport. This situation is one example of the conflict where planning for relief supply chains and distribution may mitigate the inefficient transportation mode selection problem.

The Office of Coordination for Humanitarian Affairs (OCHA) also promotes preplanning (OCHA, 2007), as they state that environmental considerations should be integrated into physical planning and shelter from the very start of an emergency. They point out that location and layout of refugee camps, provisions made for emergency shelter, and the use of local resources for construction and fuel can have a major negative environmental impact. Planning may incorporate a more efficient design of relief supply chains where distributed management or relief supplies and improved communication networks may actually help impoverished areas socially and reduce the ecological footprint of relief supply chains. We posit that these win-win situations can be more effectively implemented with joint consideration of greening relief supply chains. Another humanitarian event that showed the implications of materials logistics on the environment are occurrences of flooding. For example, when the Red River overflowed along the Minnesota-North Dakota border, millions of sandbags were used to provide protection against the flooding. The sand to fill the sandbags had to travel 30-40 miles to the site where the bags were filled and stacked. After the flooding occurred, the sandbags could not be left there. Communities than take it upon themselves to dispose of the sandbags. The issue of what to do with the sand from these sandbags has environmental implications. Sending the sand back 30-40 miles to its source was not a feasible alternative. Current plans included recycling the sand as part of road construction material. Future considerations to help with materials management have yet to be determined, but more effective environmental plans are needed.

The chapter will seek to introduce a brief state-of-the research review of issues in each of these fields. We will then identify the various elements of both these fields and discuss how barriers to implementation (organizational and otherwise) may lead to difficulty in their joint implementation. Expert opinions will be used to identify barriers setting a foundation for further research on how these barriers may be mitigated.

BACKGROUND AND LITERATURE ON GREEN AND RELIEF SUPPLY CHAINS

Green Supply Chain Management

This field has evolved over the years from focusing on descriptive works and elements of good practice, e.g. building awareness, to basic theoretical understanding of mechanisms on identifying how to improve the relationships and practices in green supply chain management for competitive and environmental improvements. The green supply chain is defined as supply chain management that has been utilized to mitigate the impact of industrial supply chain activities on the environment.

The supply chain can be described from at least four flows and relationships perspectives, upstream, downstream, internal organizational activities and the closing of the supply chain loop, or reverse logistics. Upstream activities, flows and relationships would include purchasing and procurement topics. Included amongst these topics might be outsourcing, vendor auditing, management and selection, supplier collaboration and supplier development. Internal organizational supply chain activities are generally related to the traditional production and operations management topics of an organization. Managing the flows, relationships and resources inside the boundaries of a stand-alone unit or organization, the enterprise, is the scope of this dimension. Such activities may include research and design, quality, inventory, materials, and technology management within an organization could each influence environmental characteristics of internal organizational processes. The next juncture of a supply chain focuses upon the outbound and downstream relationships and flows. Activities and functions here may include outbound logistics and transportation, marketing, distribution, packaging, and warehousing. The flows are utilized by downstream customers who may be commercial or individual consumers. Closing of the supply chain loop supply chain activities focus on end-of-life materials will eventually be consumed back into the system with recycling, remanufacturing, reclamation, and reverse logistics all part of this concept.

Evaluating all aspects of the supply chain from an environmental perspective a goal of greening the supply chain. The extensive internal and external relationships that need to be managed within and between organizations is not a trivial task, with additional complexities of managing environmental dimensions in addition to standard business dimensions. These environmental dimensions include various organizational greening practices that may affect standard supply chain management activities. For example, the use of life cycle analysis and design for the environment require close cooperation and collaboration with suppliers and customers. The many activities of the supply chain can be put into practices such as eco-design, supplier relationships, customer cooperation, internal practices, and investment recovery. These practices map to the four areas of traditional supply chain management.

The need for greening supply chains arises from a variety of pressures faced by the organization including regulatory, competitive, and community/public pressures. All of these pressures can be related to the competitive and economic sustainability of the organization. The response to these pressures varies, but can be categorized into reactive and proactive responses. The organization may gain various benefits from the responses

ranging from very tangible results, e.g. reducing costs and generating revenues, to less tangible results including building reputation and image or having the legitimate right to operate. Yet, the responses are not without their limitations. For example, even if an organization faces significant pressure to green their supply chain, without appropriate coordination, information, and resources the likelihood of the greening implementation is limited.

Relief Supply Chain Management

Humanitarian logistics has during the last few years gained increasing interest in the logistics research community. Until recently only a few articles had been published, however, much due to the poor logistics performance during the South East Asian earthquake and tsunami in 2004, the need for research in the field was recognized. Evidence also points at an increasing number of natural and man-made emergencies worldwide, nevertheless, there is still relatively little work published aimed at improving the understanding of the nature of supply chain management (SCM) in crisis conditions (Pettit & Beresford, 2009). The existing literature in the field has also been quite skewed towards examining immediate disaster response (Kovács & Spens, 2008), whereas the other phases of the disaster response cycle, i.e. mitigation, preparedness and recovery have received less attention. The latest publications in the field of humanitarian logistics indicate a growth in the interest for the other stages as well (see e.g. Richey, 2009).

According to a literature review conducted recently by Richey (2009), key topics in supply chain disaster and crisis management-related supply chain strategy and logistics operations include: agility, risk management/insurance issues, humanitarian issues, inventory management, facility location, collaboration/networks, and multi-level partner/non-partner integration. There is also an issue of whether supply chain management in the business sector can be compared to supply chain management and logistics in the humanitarian arena. They also describe the unique characteristics of the disaster relief environment and compare and contrast humanitarian relief chains and commercial supply chains (Balcik & Beamon 2008). A major conclusion from the literature is that many of the tools and techniques, despite the challenges posed by the differences, used in the commercial arena can be applied in the humanitarian arena. There are, however, also major differences that separate relief SCM and logistics from its commercial counterparts. The major difference being that relief supply chains operate in an unpredictable, dynamic and chaotic environment (Balcik & Beamon, 2008). In summary, the dominating characteristics that bring additional complexity and unique challenges to relief chain design and management are:

- unpredictability of demand, in terms of timing, location, type, and size,
- suddenly-occurring demand in very large amounts and short lead times for a wide variety of supplies,
- high stakes associated with adequate and timely delivery,
- lack of resources (supply, people, technology, transportation capacity, and money).
- uncertainty
- communications
- degraded infrastructure
- human resources
- earmarking of funds

In addition there are other characteristics of the relief environment that differentiate this area from the commercial area, e.g. the amount of and type of actors in the complex aid supply network and funding constraints (Kovács & Spens, 2007; Balcik & Beamon, 2008). The actors in the relief supply chain involve a diverse group of direct stakeholders including military forces, donor and recipient governments, commercial actors

(corporations provide goods and services) and ultimately the end users; the beneficiaries. The actor structure and the number of actors in a relief operation, the complex relations that evolve around them, their different and often conflicting goals and demands lead to problems in setting and prioritizing goals for the whole supply chain or supply network. Issues such as donor pressure and other stakeholder involvement might affect resource allocation and competition for funding might inhibit coordination efforts between non-governmental organizations (NGOs) and other network actors. In many instances, donors tend to fund NGOs for specific missions or activities relating to their own agendas, which often might not contribute to infrastructure (Balcik & Beamon, 2008). This behavior again encourages the NGOs to focus on operational disaster relief activities rather than disaster preparedness which would reduce expenses or make relief activities more effective over the long-term. So, for example, although access to timely and accurate information is vital for relief organizations, information systems are yet not well-established in the relief supply chain (Balcik & Beamon, 2008).

Even though logistics is central to disaster response activities the aid sector has viewed logistics more as a necessary expense rather than an important strategic component of their work (Beamon & Kotleba, 2006). Only recently have humanitarian relief organizations started to understand the criticality and importance of relief chain management to the success of disaster relief operations. In addition, relief supply chain management also entails longer-term considerations e.g. the provision of food aid during complex emergencies as well as supplying refugee camps, of which some become rather permanent settlements. In these situations the defining characteristics ascribed to relief supply chains are no longer valid. Therefore in longer-term situations, environmental concerns and green practices should be more easily incorporated in relief supply chains in order to mitigate the impacts of the aid operations. Barriers to implementation of green supply chain practices should therefore be investigated in order to help understand and overcome them, both with regards to longer-term aid efforts as well as shorter-term emergency relief efforts. In this chapter, however, the focus is on short-term emergency relief efforts as a first step in exploring the barriers to greening the relief supply chain as the emergency phase is the critical moment at which environmental degradation may be confined or limited and activities undertaken at the early stage of the operation are usually far more cost-effective than those taken later (OCHA, 2007)

BARRIERS TO INTRODUCING NEW PRACTICES AND PROCEDURES

In the supply chain literature, potential barriers or resisting forces have been investigated. Barriers to strategic supply management derive from both the nature of the organization itself and the people that compose the organization (Fawcett *et al.* 2008). Other barriers to SCM fall under managerial complexity (information system and technological incompatibility, inadequate measurement systems, and conflicting organizational structures and culture) or misalignments in allying firms' processes, structures, and culture. Interestingly, the single greatest barrier that Fawcett *et al.* (2008) identified in the survey they conducted was information systems. However, when they conducted interviews, human nature was found to be the primary barrier to successful SC collaboration. Unsurprisingly, they also concluded that people are change averse and prefer to maintain the status quo. Human and organizational behavior was at the root of nearly each of the major barriers (i.e. organizational culture and structure, functional conflicts, lack of managerial commitment, conflicting and non-transparent processes, policies, and procedures, performance measurement, information sharing, lack of trust, resource constraints, and complexity of SC networks).

Similar barriers investigation has occurred in other supply chain and operations topics. For example a study on barriers to implementation of agile manufacturing found 11 barriers typically facing agility and agile manufacturing adoption. The barriers included lack of top management support and commitment and fear of organizational change (Hasan *et al.,* 2007). Similar barriers were found in various organizational programmatic studies such as implementation of ERP systems (deVries, 2007). Organizational culture and training of employees were also found to be substantial barriers in these settings. In recent research in the field of green supply chain management concludes that there are different categories of barriers, and names 11 categories of barriers or. In summary, the operations and supply chain literature indicates that the following categories of barriers exist (Sarkis, 2009):

- Informational barriers
- Political barriers
- Proximal barriers
- Organizational
- Economic barriers
- Operational barriers
- Temporal barriers
- Technological barriers
- Cultural barriers

These categories of barriers provide a good basis for exploring barriers in other fields. The barriers are general enough to provide a broad categorization, but there is room to also explore whether there are additional categories that need to be added.

METHODOLOGY

Using a workshop in Canada on humanitarian logistics, logistics experts representing different aid organizations were asked to participate in our study on barriers to greening the relief supply chain. Climate change-related disasters were already mentioned at the workshop, and we briefed the experts that the question in our research would be on the ecological dimension of humanitarian logistics and supply chain management. Altogether 6 organizations (out of 9 asked) participated of which one was not represented at the workshop. The organizations that took part in the study represent different types of aid organizations; UN related (World Food Program, WFP), large non-governmental international aid organization (World Vision International), secular organization (Canadian Foodgrains Bank), private sector aid initiative (American Logistics Aid Network, ALAN), a logistics professionals initiative (The World Organization for Relief Logistics Development, World), and a government funded organization (National Emergency Supply Agency, NESA). The logistics experts from these aforementioned organizations were in the first round asked to list 10 or more barriers through electronic correspondence. The experts all replied by providing at least 10 barriers, in some cases more. Usually the barriers were provided in the form of statements such as *"low awareness of technology to reduce carbon footprint"*.

The next round consisted of sending out an email to the same experts that included a list of barriers discovered in round 1. This list was compiled based on an analysis of the barriers (usually provided in the form of statements) received in round one. The first step of the analysis involved grouping the barriers. Thereafter the barriers were assigned to the most appropriate categories available from the literature review. If the statements did not fit the category, a remark was made on the statement that indicated a misfit. The misfit usually meant that the category the barrier had been assigned to needed some further refinement or a broadening of the category. In Table 1 an example is provided of the grouping of barriers and their categorization into more general categories.

Table 1 shows the grouping and categorization of "informational" and "organizational" barriers.

The lack of information inhibits the adoption and implementation of new practices, interestingly enough, the lack of information and communication is also stated as one of the most important features of a non-resilient supply chain (Sheffi, 2005). The statements provided by the experts (see column 3 Relief SC barriers) indicated that there was a low awareness of technologies to reduce carbon outputs; low awareness of green operations; lack of awareness and knowledge how to handle different materials and lack of communication. All these barriers were included under the general category "informational" based on the earlier review of literature. Organizational barriers mentioned by the experts were for example lack of top management support, lack of training and education in green issues, poor coordination and the resistance of organizations to change. In earlier literature these barriers have also been mentioned as resisting forces to introducing new practices and procedures in supply chains. In a similar manner all different statements provided by the experts were categorize. As seen in Table 1, many of the initial barriers (statements) were similar in nature therefore the final step of the analysis involved interpretation of the statements and the forming of more general barriers in line with earlier literature.

In a third round, a table that showed the categorizations, the more general barriers as well as examples of the initial statements was sent out to the experts for review. The experts were told that they could add barriers at this point or also change the categorization of the barriers. However, all experts agreed to the categorization and no further barriers were added in the third round, thereby confirming the initial results of the analysis.

BARRIERS TO GREENING THE RELIEF SUPPLY CHAIN

By using literature on resisting forces to adoption of manufacturing and implementation of SC

Table 1. Categorizing barriers

Informational Barriers: General	Informational Barriers in the Relief SC
• Inappropriate measurement [HSS, WR] • Communications [R] • Lack of information [V] • Lack of communication with other NGOs, GOs [KS] • Inadequate knowledge, inadequate data [S]	• Low awareness of technologies to reduce carbon outputs • Low awareness of low carbon / green operations • Low awareness of management approaches to reduce carbon footprint • Lack of awareness and knowledge how to handle different materials • Lack of communication
Organizational Barriers: General	**Organizational Barriers in the Relief SC**
• Lack of top management support and commitment [HSS, WR] • Fear of and resistance to organizational change, resistance of the workforce [HSS, WR] • Poor partnership (supply chain) formation and management [HSS] • Insufficient training, education and rewards system [HSS, R, KS, WR] • Poor planning [WR] • Performance standards, measurement [R] • Stakeholder influence on inventory system [V] • Lack of co-ordination with other humanitarian organizations [KS]	• Lack of top management support • Use of spontaneous volunteers who may not have full training • Low skill levels among field worker • Jobs/tasks are given out without the same degree of oversight that would be present in a non emergency situation • Many organizations have a set way in which they work. Changing this structure can be tough, but also beneficial. • Material required -- not all food aid and other aid are packaged in environmentally friendly materials. So, even if the supply chain is green, there will still be environmental damage being caused after-the-fact with the litter of these materials. • Lack of green training, education & experience (many still need logistics training and may not have knowledge/training on green methods) • Lack of education and training • Donor response and public/media reaction • Poor preparedness activities - better logistics capacity assessments would identify opportunities for greening relief supply interventions • Level of coordination between UN/NGOs etc. • Poor inter agency coordination - duplication of effort and therefore excess carbon output and excess waste / environmental impact • Poor donor coordination • Lack of coordination • Lack of attitude and willingness to operate in green way • Performance measurement • Lack of performance measurement systems • Little time to plan and make decisions.

[HSS]Hasan *et al.* (2004), [WR]Whalen & Rahim (2004), [R]Rodman (2004), [V]de Vries (2007), [KS]Kovács & Spens (2009), [S]Sarkis (2009)

techniques and tools and expert opinions, barriers to greening the relief supply chain were found. Table 2 summarizes the results of this expert round in combination with the findings from the literature review.

A closer look at Table 2 indicates that the barriers that were found to be relevant for relief supply chains in fact resemble the general barriers found in literature. As no emphasis has yet been put on trying to sort or rank the barriers, no statements on the internal hierarchy of the barriers can be made. If we assume that the amount of statements that could be assigned to a category

indicate the importance of this category, clearly the four categories; organizational, technological, cultural and economic, stand out as the most important ones. One of the primary barriers to successful SC collaboration (Fawcett *et al.*, 2008) was human nature due to the fact that people are change averse and prefer to stick to status quo. In our study, this aspect did not stand out, however, further research is needed in order to establish the importance of the noted categories and barriers.

The analysis also shows that even though the general categories fit well with the barriers

Table 2. Barriers to greening the relief supply chain

General categories	Barriers to greening the relief SC
Informational	Lack of information Inadequate knowledge Lack of communication
Political	Political limitations Lack of policies
Organizational/Interorganizational	Poor SC partnership management Insufficient training and education Stakeholder influence Lack of coordination Last mile considerations Lack of top management support Poor planning Lack of performance measurement systems
Temporal	Uncertainty of time of event Little time to plan and make decisions Unpredictable demand Urgency
Technological	Unavailability of appropriate technology Degraded infrastructure Lack of transport infrastructure
Cultural -organizational culture	Goal to help people, environment second at best Lack of attitude and willingness to operate in a green way
Economic	Funding Lack of supplies, equipment Lack of resources Inadequate human resources
Operational	Structures and /or processes not in place

provided in the relief supply chain setting, some barriers indicated a refinement of the category. For example the category culture in relief supply chains does not only relate to organizational culture as e.g. in the study by de Vries (2007). Culture is defined both in terms of organizational culture as well as in terms of the local culture in a country where disaster has struck. Cultural issues play an important role when supplies are chosen for the beneficiaries, for example food items are very culturally bound and are usually specified according to the receiving country preferences. Another important distinction in the barriers is the "temporal" barrier. The temporal barrier indicates an urgency of the disaster response, which is typical for relief supply chains. In business supply chains, urgency is usually not the determining factor and does not override environmental

matters. However, in relief supply chains, the need to save lives in the immediate aftermaths of a disaster usually overrides other issues. The unpredictability that relief supply chains operate in is also included in the "temporal" category. Since the immediacy and temporal factors are a major difference separating relief supply chains from business supply chains (Balcik & Beamon, 2008), this category may also turn out to be one of the most important ones.

In summary, our initial analysis indicates that there are general barriers to greening the relief supply chain that resemble those barriers found in earlier SC literature. The main differences to for-profit supply chains arise from the context that these supply chains operate in where unpredictability, uncertainty and even security play an important role. In this context, other issues than environmental issues play an important role, maybe the most important role, as lives are at stake. In order to overcome such barriers, planning becomes even more important, that is if we preplan, saving lives might not come at the expense of the environment. However, what is needed is a more thorough review and additional research on the importance of the barriers to be able to address them.

CONCLUDING DISCUSSION

The global demand for humanitarian assistance, which is already considerable, is likely to grow in the coming decade, and to see a major increase in our lifetimes. The biggest single cause will be climate change and the increased incidence and severity of extreme weather events associated with it. Indeed, we are beginning to feel the effects. What we are already witnessing is not an aberration but rather a 'curtain raiser' on the future. These events are what I call the 'new normal'. The number of recorded disasters has doubled from approximately 200 to over 400 per year over the past two decades. Nine of out every 10 disasters are now climate-related. Last year, my office at the UN issued an unprecedented 15 funding appeals for sudden natural disasters, five more than the previous annual record. 14 of them were climate-related. Holmes, 2008, p.4

This citation from the Under-Secretary-General for Humanitarian Affairs and Emergency Relief Coordinator underlines the severity of climate change impact on humanitarian assistance. Humanitarian assistance is expected to grow substantially due to climate change. Logistics costs in relief supply chains are substantial, and thereby the field of relief supply chain management will also increase in importance. The need to introduce green supply chain principles and practices in the relief supply chain is evident, however, no studies have so far addressed the greening of the relief supply chain, a void that this chapter is starting to fill. This chapter aims to explore barriers to implementation of green SC practices in the relief SC. Expert interviews and literature from the fields of humanitarian logistics and green supply chain management are used to establish a list of barriers and to propose a categorization of the barriers. The barriers that were found in this study revealed that earlier literature on barriers or resistance to change in supply chains well describe the barriers that exist in the relief supply chain. However, similar to the differences that exist between business logistics and humanitarian logistics, some barriers are specific to this field, unpredictability and urgency standing out as the most important ones. Nevertheless, this chapter is only a first step in an investigation of barriers that exist when seeking to green the relief supply chain. More research is needed in order to validate the results and in order to rank the barriers in order of importance. Interpretative structural modeling (ISM) is proposed as the next step to integrate expert opinion and help structure the relationships amongst the barriers. Once this structure is understood, prioritization and addressing the barriers becomes clearer. By doing so, the most important barriers can be addressed and future studies can focus on how to overcome the barriers and hopefully aid in greening the relief supply chains.

REFERENCES

Balcik, B., & Beamon, B. M. (2008). Facility location in humanitarian relief. *International Journal of Logistics Research and Applications*, *11*(2), 101–121. doi:10.1080/13675560701561789

de Vries, J. (2007). Diagnosing inventory management systems: An empirical evaluation of a conceptual approach. *International Journal of Production Economics*, *108*(1-2), 63–73. doi:10.1016/j.ijpe.2006.12.003

Fawcett, S. E., Magnan, G. M., & McCarter, M. W. (2008). Benefits, barriers, and bridges to effective supply chain management. *Supply Chain Management: An International Journal, 13*(1), 35–48. doi:10.1108/13598540810850300

Hasan, M. A., Shankar, R., & Sarkis, J. (2007). A study of barriers to agile manufacturing. *International Journal of Agile Systems and Management, 2*(1), 1–22.

Hoerz, T. (1997). *The environment of refugee camps.* Retrieved January 6, 2010, from http://www.fmreview.org/HTMLcontent/rpn185.htm

Holmes, J. (2008). The need for collaboration. *Forced Migration Review, 31*(4). Retrieved January 8, 2010, from http://www.fmreview.org/FMRpdfs/FMR31/FMR31.pdf

Kovács, G., & Spens, K. (2009). Identifying challenges in humanitarian logistics. *International Journal of Physical Distribution and Logistics Management, 39*(6), 506–528. doi:10.1108/09600030910985848

Kovács, G., & Spens, K. M. (2007). Humanitarian logistics in disaster relief operations. *International Journal of Physical Distribution and Logistics Management, 29*(12), 801–819.

Maspero, E. L., & Ittmann, H. W. (2008). The rise of humanitarian logistics. *Proceedings of the 27th Southern African Transport Conference,* 7-11 July, Pretoria, South Africa. Retrieved December 8, 2009, from http://repository.up.ac.za/upspace/bitstream/2263/6251/1/Ittmann%2027.pdf

OCHA. (2007). *Environment.* Retrieved January 10, 2010, from *http://74.125.93.132/search?q=cache%3AhLpcfdsUq1AJ%3Aochaonline.un.org%2FOchaLinkClick.aspx%3Flink%3Docha%26docId%3D1091518+refugee+camp+environment&hl=en&gl=us*

Pettit, S., & Beresford, A. (2009). Critical success factors in the context of humanitarian aid supply chains. *International Journal of Physical Distribution & Logistics Management, 39*(6), 450–468. doi:10.1108/09600030910985811

Richey, R. G. Jr. (2009). The supply chain crisis and disaster pyramid: A theoretical framework for understanding preparedness and recovery. *International Journal of Physical Distribution & Logistics Management, 39*(7), 619–628. doi:10.1108/09600030910996288

Russell, T. E. (2005). *The humanitarian relief supply chain: Analysis of the 2004 South East Asia earthquake and tsunami.* Masters thesis at MIT, Massachusetts. Retrieved December 8, 2009, from http://dspace.mit.edu/bitstream/handle/1721.1/33352/62412847.pdf?sequence=1

Sarkis, J. (2009). *A boundaries and flows perspective of green supply chain management.* GPMI Working Paper 2009-07. George Perkins Marsh Institute, Worcester, MA. Retrieved January 4, 2010, from http://www.clarku.edu/departments/marsh/news/WP2009-07.pdf

Sheffi, Y. (2005). *The resilient enterprise. Overcoming vulnerability for competitive advantage.* Cambridge, MA: The MIT Press.

UNHCR. (2009). *Ethiopia/Green refugee camp.* Retrieved January 10, 2010, from http://www.unmultimedia.org/tv/unifeed/d/14105.html

Van Wassenhove, L. N. (2006). Humanitarian aid logistics: Supply chain management in high gear. *The Journal of the Operational Research Society, 57*(5), 475–489. doi:10.1057/palgrave.jors.2602125

ADDITIONAL READING

Supply-Chain Council. (2008). *Supply-Chain Operations Reference Model v9.0.* Supply-Chain Council, Pittsburgh, USA. www.supply-chain.org.

KEY TERMS AND DEFINITIONS

Barrier: Resisting force, arising from the nature of the organization and people that compose the organization.

Green Supply Chain Management: Evaluation of all aspects of the supply chain from an environmental perspective.

Chapter 12
Disaster Impact and Country Logistics Performance

Ira Haavisto
Hanken School of Economics, Finland

ABSTRACT

The study in this chapter seeks to answer the question whether a country's logistics performance has a correlation with the impacts of a disaster; impact being measured in average amount of affected, the average amount of deaths, the average amount of injured in a disaster or the average amount of economic damage. This is a quantitative study where the EM-DATs disaster data is analyzed through correlation analysis against the World Bank's logistics performance index (LPI). The findings do not show a significant relationship between countries LPI and the average number of deaths or injured persons in a disaster. A positive correlation between the variable LPI and the variable economic damage can be found. A negative correlation between the LPI and the average amount of affected can be found for countries with an average ranking LPI. Countries with low LPI and high disaster occurrence are further identified. Findings encourage the identified countries to take into consideration their logistics performance when planning and carrying out humanitarian response operations. Results also encourage humanitarian organizations to pay attention to the receiving countries' logistics performance in planning and carrying out humanitarian response operations.

INTRODUCTION

Each country or area has a different logistics performance in e.g. transporting goods. Accord-

DOI: 10.4018/978-1-60960-824-8.ch012

ing to Hausman *et al.* (2005) it can take 93 days to export a 20-foot full container load (FCL) of cotton apparel in Kazakhstan while in Sweden it takes only 6 days. A country's logistics performance affects the country's trade competitiveness (Arvis *et al.* 2010), but is there an effect as well on a country's ability to transport goods in the event of a disaster? A country's capacities to handle the effects of an event are fundamental in the determination whether an event is a disaster or not. It's fundamental since the effects of an event are not determined as a major disaster as long as a system or a nation has the capabilities to cope with the effects of the event. (Kovács and Spens 2007). A country's logistics performance is therefore important knowledge for humanitarian organizations in order for them to know when and if their assistance is needed.

A country's own capacity to handle a disaster is seen to affect the impact of the disasters (Beresford and Pettit 2009). In the first 72 hours the affected country is most likely responsible for handling the effects of the disaster singlehandedly. In the case of an event where immediate relief is required, humanitarian organizations have in average the aim to reach the affected area within 72 hours of the disaster occurrence. And, when humanitarian organizations arrive on location, they still rely heavily on the resources in the country: can supplies and other resources be found locally and what kind of infrastructure is in place in the affected country? In recent years, humanitarian organizations' planning and preparing for disasters has improved (McEntire 1999) but an increased capacity building for country's own planning has not been seen.

Countries' logistics performances vary and there are several different measurements in use for determining performance. There are variations in the level of infrastructure and large variations as well in country specific policies and procedures which in the commercial sector affect the trade competitiveness (Hausman *et al.* 2005). A country's trade competiveness has in empirical studies been found to have a statistical link with the county's logistics performance. The link has been found between transport cost and trade flows, and between the quality of the infrastructure and transport costs (Hausman *et al.* 2005; Limao and Venables 2005). The logistics performance of a

country could be likened to the timeliness and cost in a humanitarian response operation in a similar manner it is linked to trade competitiveness. The logistics performance in a country might even have a larger significance for the humanitarian sector than for the commercial, since a disaster is determined by time and place uncertainty and the outcome of the operations is measured in lives (Kovács and Spens, 2007). In a relief operation the logistics performance of the affected country might therefore be crucial in successfully accessing and aiding the ones affected by a disaster.

IS THERE A CORRELATION BETWEEN DISASTER IMPACT AND COUNTRY LOGISTICS PERFORMANCE?

This study compares countries' logistics performance with the impact of occurred disasters. The aim is to analyze whether a country's logistics performance has a correlation with the average amount of affected population, the average amount of deaths, the average amount of injured or the average amount of economic damage per disaster.

The secondary aim of the study is to identify countries where disasters are re-occurring or where a high number of people are affected by the disasters and where the logistics performance is low. For example, between 1990 and 1998, approximately 94 per cent of major natural disasters and more than 97 per cent of all natural disaster-related deaths occurred in developing countries (World Bank 2001). Developing countries also have in average a lower logistics performance than developed countries when calculated in relation to income per capita (Arvis *et al.* 2010). In this study three hypotheses are tested through correlation analysis.

Hypothesis 1: There is a negative correlation between country logistics performance and disaster impact (low country logistics performance correlates with high disaster impact; high country logistics performance correlates with low disaster impact.)

Hypothesis 2: The correlation between country logistics performance and disaster impact differ depending on the level of logistics performance (high, medium, low).

Hypothesis 3: The correlation between country logistics performance and disaster impact differ depending on used disaster impact measurement (affected, dead, injured, economic damage).

The first hypothesis states that there would be a negative correlation between the two variables, which would mean that low performance indicator correlates with high disaster impact and high logistics performance. The second hypothesis states that depending on where in the logistics performance ranking the country is, the correlation might differ. E.g. if the country is in the group of low performers the correlation might be very significant, but in the group of high performers there might not be a significant correlation at all. The third hypothesis again states that depending on which measurement for disaster impact is used the correlation might differ. This means that there might be a correlation detected between the logistics performance and the total number of affect but e.g. not between logistics performance and number of injured.

The practical indication with this study is to see whether a low logistics performance correlates with the impact of a disaster. The practical implication of the study is as well to indicate which areas might be disaster prone and have a low logistics performance and therefore might need a larger focus on country preparedness and humanitarian organizational preparedness to tackle the impacts of a disaster.

WHY IS LOGISTICS IMPORTANT IN DISASTER RESPONSE?

Logistics performance has been proven to have an impact on trade competitiveness but why is it important in disaster response? The commercial supply chain can be seen as a process of managing the flow of goods, information and finances from the source to the final customer. Similarly to commercial logistics operations, logistics in disaster response struggle with conflicting interests of stakeholders and with unpredictable demand. There are differences between humanitarian logistics and commercial logistics. The most essential difference can be seen as the motivation to improve the logistics operations (Kovács and Spens 2007). The motivation for private companies comes from being monitored and measured by profitability, but in the case of humanitarian logistics the output of the performance could be measured in human lives.

Disaster response is characterized by numerous factors of uncertainty which do not exist in the commercial sector. In most cases, the beneficiaries, their location and their needs are unknown. A relief operation is therefore characterized by demand uncertainties in the form of location, type and volume (Beamon and Balcik 2008). This uncertainty and unpredictability leads according to Beresford and Pettit (2009) to the relief operations being reactive rather than proactive, which would mean that the response operations are seldom prepared for. Beamon and Balcik (2008) further argue that the unpredictability of a disaster makes the planning and preparing even more important. The lack of preparedness in the humanitarian sector can be due to the unpredictability of the event, or due to the affected countries not having capacity for disaster preparedness and/or organizations not having funds allocated for planning. Organizations rarely have funds for planning and preparing since donors are hesitant to provide funds in advance to humanitarian organization in the fear of them "spending the money on heavy administration", instead of saving lives. Countries again, especially developing countries might also not have resources for disaster prevention and preparedness. Bringing to mind that emergencies often take place in less developed areas with poor infrastructure (Jennings *et al.* 2000; Beresford and Pettit 2009). Kovács and Spens (2009) emphasize that it is crucial for a humanitarian operation to know what the preparedness level of the pre-disaster area is. Part of the area preparedness level can be viewed as terms of infrastructure such as road network, access points, electrical grid and medical centers in the region. Because of the uncertain characteristics of a disaster, logistics performance and especially country preparedness and organization preparedness seem crucial in humanitarian response.

Preparedness

Preparedness means planning how to respond when an emergency or disaster occurs. The most important part of the preparing is to get to know the area, recognize the risks and and plan how to respond in that particular setting. Planning for a disaster is often focused on developing the capacity to respond. Disaster preparedness focused on developing the capacity to respond quickly and appropriately to a disaster is according to Perry (2007) the foundation of all relief activities. Preparing or planning can save lives and minimize the damage of a disaster. The impact of a disaster can be reduced by setting up warning systems and by effective disaster management. To respond properly, an authority must have a plan for response, trained personnel to respond, and the necessary resources with which to respond (Oloruntoba 2005; McEntire 2002).

Certain disaster prone areas are very prepared to meet a disaster. Iceland, Japan and New Zealand are good examples of high mitigation. The areas are prone to earthquakes, but have a good possibility of predicting an upcoming event and have a high level of local preparedness. In other areas such as on the African continent there is a pattern

of slow on-set disasters and lack of preparedness for these sorts of disasters. Slow onset disasters are disasters such as famine, drought and poverty; while rapid onset disasters can be hurricanes, earthquakes and tornados (van Wassenhove 2006). The characteristic for a slow onset disaster is that they sneak up slowly while the rapid onset ones are often unpredictable. One might think that countries and organizations would have more time to plan and prepare for a slow onset disaster but since the disaster often develop for a long period of time the same drama and media attention is seldom involved in slow onset disasters as for a sudden onset-one. There is not too much humanitarian logistics research on preparedness or response for disasters on the African continent. The majority of all research in humanitarian logistics concerning disasters and disaster preparedness has been conducted in Asia (Beresford and Petitt 2009; Kovács and Spens 2009). This is logical in the sense that 60 percent of the world's disasters take place in Asia. Amin and Goldstein (2008) claim that there is a strong relationship between vulnerability, poverty and natural disasters. The statistics do not show a higher level of disasters occurring in developing countries but the impact if measured in deaths is higher. 11 percent of the people being exposed to natural disasters live in developing countries but the disasters occurring in developing countries account for 53 percent of the recorded deaths.

In spite of where in the world a disaster occurs, preparedness plays a crucial role in the possibilities to respond to the disaster. Part of being prepared is good logistics performance in the country or the area of the disaster and availability of resources such as infrastructure, supplies. Part of being prepared as a humanitarian organization is to have the knowledge of a country's logistics performance. Even if the logistics performance is seen as an important factor for disaster preparedness, does the performance have a correlation with the disaster response and the impact of a disaster? In the following sections we will discuss countries' logistics performance, the possibility to measure it.

Logistics Performance

In research, logistics performance is often referred to as the logistics performance of a company, organizations, a supply chain or a supply chain network. In this study we are however looking at country specific logistics performance. There is a limited amount of research conducted on country specific logistics performance, but there are indicators developed to measure that performance. Logistics performance indicators and different indexes are used to indicate what a country´s logistics performance is. There are several different types of performance indicators and several different factors that are included in the calculation. Logistic performance indicators are often indicators that are calculations of different factors that influence the logistics performance in the area. Indicators of time, indicators of costs and indicators of complexity and risk factors can be included (Hausman *et al.* 2005).

Existing indexes that measure the logistics performance that are currently used in research and by practitioners are e.g. The Port Infrastructure Index, The Port Efficiency Index, The Transport Cost Index (Clark *et al.* 2004), Cargo Handling Restriction Index (Fink *et al.* 2002), The European Freight Forwarding Index, Global Competitiveness Index and Global Enabling Trade (Karamperidis *et al.* 2010). They all measure different logistics actions, and are used for different purposes and studies. The Port Infrastructure Index e.g. measures the ratio between the number of ports per country in relation to the country's surface and population.

The logistics indicator chosen to be used in this study is published by the World Bank 2010 and constitutes of data gathered in 2009. The indicator is one of the broader logistics performance indicators since it includes measures for; customs, infrastructure, international shipment, logistics competence, tracking and tracing and domestic

logistics. As discussed in the introduction the logistics performance in a country is seen to have direct link with trade competitiveness, efficient logistics play an important role in the worldwide flow of goods and services. Dollar *et al.* (2004) state for example that firms in countries with better logistics have a higher probability of attracting foreign direct investment. Logistics inefficiencies again harm the competitiveness through their effect on both time and cost. These affects have been the trigger for the World Bank to develop the Logistics Performance Indicator, where countries' logistics performance is evaluated and ranked. The World Banks Logistics performance index (LPI) is studied to be directly linked to important economic outcomes, such as trade expansion and growth (Arvis *et al.* 2010). The aim of ranking the logistics performance of countries is according to the World Bank (2010) for the indicators to serve as a catalyst for domestic policy reform.

There are several ways to measure country logistics performance. All above mentioned indexes have their own purpose and should be used accordingly. The World Banks LPI is one of the broadest indicators and is therefore chosen to be used in this study. But what about measuring the other variable in this study; disaster impact? There is no disaster impact index available, but there are several different measurements which can indicate the disaster impact. The following section will discuss the different sources of disaster impact data.

Disaster Impact

All disasters, in spite of classification or type, have a common denominator: the severe impact they have on people's lives, properties and the environment (Shaluf 2007). But how can that impact on peoples' lives be measured? An industrial accident might have a high economic impact while a predicted hurricane might drive people to evacuate and an unpredicted earthquake might cause injuries and casualties.

Researchers have tried to quantify a disasters impact on the population and on a nation's development. The damage caused by a disaster is seen to not only have an immediate impact on peoples' lives but also a long term negative impact on a country's economic development. The immediate impact is by Sharma (2010) be separated into direct impact; such as physical and human capital and indirect impact; on capital flow and on production, consumption, income and employment.

To be able to measure the impact of a disaster one must recognize the different characteristics and types of disasters. Disasters can be divided into several categories according to their characteristics. Disasters can be divided into disaster sub-groups such as geophysical, meteorological, hydrological, climatologically and biological disasters (EM-DAT, 2010). In humanitarian logistics literature the most common distinction between disasters is; man-made disaster and natural disaster (van Wassenhove 2006; Kovács and Spens 2009) or as mentioned in the introduction slow onset disasters (famine, drought and poverty) and rapid onset disasters (hurricanes, earthquakes and tornados). The predicted impact of a disaster can be measured in magnitude for example on the Righter scale for earthquakes, or the Saffir-Simpson Hurricane Damage Intensity Scale for hurricanes. For an industrial accident such as for example an oil spill, the environmental impact is often measured. The most common way of measuring the impact of a disaster is to measure post-even variables.

There exists a variation of databases that gather data on disasters and on disaster impact. These are; EM-DAT Database, NatCat Database, Sigma Database, Disaster Database Project and regional or local disaster databases. The EM-DAT database is maintained by The Centre for Research on the Epidemiology of Disasters. The database is meant to be used to for example to determine whether an event is a disaster or not and whether the impact of the event requires aid. The information has been gathered since 1900 (EM-DAT, 2010). The NatCat database again gathers post-event data on

property damage and whether persons are injured or dead. The database is maintained by Munich Reinsurance Company since 1970 for insurance purposes (Guha-Sapis and Below 2002). The Sigma Database was set up for similar purposes as the NatCat database. In this study the EM-DAT database data is used because of the broadness of data included in the database, and because the data is fully available to the public. The EM-DAT has though been criticized for only gathering data on the direct affects of a disaster (Sharma, 2010). The direct affects do not necessarily take in consideration the the full scale of the disaster impact, such as longer term social and economic impact as the impact on the structure of economy, on social and behavioral considerations, and on political and institutional factors.

This study seeks to find out if there is a similar link between LPI and disaster impact as there is between the LPIs and trade competitiveness. In the following section the method of the study and the used variables are discussed.

Research Design

Statistical data is used to analyze disaster impact and the logistics performance through a correlation analysis. The analysis is used to seek a correlation between the logistics performance in a country and the disaster impact. Three hypotheses where formed; there is a negative correlation between logistics performance and disaster impact, the correlations differs depending on the level of logistics performance (high, medium, low), the correlation differs depending on used disaster impact measurement (affected, dead, injured, economic damage).

The data used to measure the country logistics performance is the World Bank's LPI and the data used to measure the disaster impact is the EM-DAT disaster data.

When looking at the LPIs and the EM-DAT disaster occurrence data it must be kept in mind that the World Bank only calculates the LPI for

154 countries while EM-DAT gathers disaster data for all countries of the world. Because this study is looking at both data sources the countries that do not have a World Bank calculated LPI are not included in the study. Countries that are left out but who do have a high disasters occurrence rate are countries such as for example Central African Republic and Burundi. Furthermore because of the irregularity of disasters, data regarding disaster occurrence was used from a longer period of time (2000-2009) while the most recent data (2010) regarding the country's logistics performance was used.

A correlation analysis is conducted for all variables. Correlation analysis can be used to indicate a predictive relationship and can suggest possible causal relationships. Both Pearson correlation and Spearman's rank correlation is conducted. Pearson's correlation seeks to find linear correlations when Spearman's rank correlation measure if and to what extent one variable increases or decreases compared to another variable. The Spearman's rank correlation do not requires that the relationship is a linear relationship.

The variable that is chosen to indicate the country's logistics capacity is the World Banks LPI. The variables chosen to express the disaster impact are; the average number of affected, the average number of deaths, the average number of injured and the average economic damage. The average for the variables is calculated as total divided by occurred disasters.

Correlation analyses are as well conducted for low, average and high LPI groups separately. Further a correlation analysis for each LPI group and each disaster impact variable is conducted separately. The data is divided into these 3 LPI groups by dividing the total amount of 154 countries into equally sized groups (51 countries in the low and average LPI groups and 52 countries in the high LPI group).

In the following section the data sources are discussed in more detail. Thereafter the calculations and the results are presented.

The World Bank Logistics Performance Index (LPI)

The World Bank first started with the biannual Logistics Performance Indicators ranking in 2005. The data constitutes of detailed level data on time and cost to move a typical 20-foot container from the port of entry to a populous or commercially active city in the country. Measured activities are trade document processing, approvals needed for import and export transactions, customs clearance, technical clearance, inland transport, terminal handling and container security measures (Hausman *et al.*, 2005). The data was collected through a detailed questionnaire distributed to experienced logistics practitioners, mostly freight forwarders.

The data that is used in this study is from the World Banks logistics ranking from 2010. The LPI is rated from 1 (=worst) to 5 (=best). The 2010 LPI is a snapshot of the logistics performance in 154 countries. The World Bank LPI of 2010 constitutes of data regarding a country's; *Customs, Infrastructure, International shipments, Logistics capacity and competence, Tracking and tracing, and Timeliness* (Arvis *et al.* 2010). In disaster situations one of the indicators might be more important than others. In a sudden onset disaster "timeliness" might become more crucial than the other indicators and if the country eases up on their customs requirements due to the disaster, customs as an indicator might not be significant at all.

The LPI is proven to indicate a country's trade competitiveness but is it relevant in humanitarian response? Humanitarian response operations have many of the same challenges in moving relief and aid supplies to and from an affected area as a private company with trade in the area might have. There are differences as well. In a disaster situation countries for example tend to ease up on their customs regulations and entry into the country with relief supplies might be much quicker than with commercial goods. Also the LPI as calculated by the World Bank are only looking at the timeliness with the final destination in the country assumed to be a commercial hub, while relief supplies often need to be transported to remote areas. Since the disasters might occur in a remote area, the countries inland capacity becomes more crucial, than the capacity to import and export goods to commercial hubs.

Emergency Data Base Disaster Occurrence Data

EM-DAT classifies the disasters into following disaster types: complex, drought, earthquake, epidemic, extreme temperature, flood, industrial accident, insect infestation, mass movement dry, mass movement wet, miscellaneous accident, storm, transport accident, volcano and wildfire. In this study all disaster types are included. For disasters to be entered into the database one out of four disaster criteria must be met. The criteria are: ten or more people killed, hundred or more people reported affected, a declaration of emergency or call for international assistance. As can be seen in Table 1, the EM-DAT database includes data on number of deaths, number of people affected by disaster, number of people injured in the disasters and economic damage. The number

of deaths is by the EM-DAT defined as a person who is confirmed or presumed dead. Injured is a person who suffers from physical injuries, trauma or illness and requires medical treatment. A person who is affected by the disaster is one who needs immediate assistance; he or she can be a displaced or evacuated person. In year 2003 there was for example 37 million uprooted people in the world (Shaluf 2007). The estimated damage is given in US dollars and there is no standard procedure on how to determine the figure for the economic impact.

THE RELATIONSHIP BETWEEN LOGISTICS PERFORMANCE AND DISASTER IMPACT

High Disaster Occurrence and Low Logistics Performance

The country with most disasters occurring in average per year between the 2000 and 2009 is China with an average of 87.56 defined disasters per year (see table 1). China has a high LPI and a high risk of earthquake. Some of the world's most

Table 1. Top 10 disaster occurrence countries

Country	LPI	Average # disasters/ year	Average economic damage / disaster	Average # injured/ disaster	Average # affected/ disaster	Average # deaths/ disaster
China	3.49	88	230 645*	914*	1 489 054*	142*
India	3.12	44	62 115*	497*	1 554 490*	168*
US	3.86	34	1 132 407*	21	68 599*	18
Nigeria	2.59**	31	43	8	2 351	44
Indonesia	2.76**	28	50 496	619*	47 249	740*
Philippines	3.14	21	11 594	42	261 619*	61
Russian fed.	2.61**	19	26 433	25	11 840	27
Iran Isl. Rep.	2.57**	17	28 747	200*	253 931*	203*
Bangladesh	2.74**	17	39 489	522*	487 660*	93
Pakistan	2.53**	15	58 523	1 027*	148 311*	594*

(**. Lower LPI than total average; *. Higher than total average)

deathly earthquakes have taken place in China. One of the latest was the Sichuan earthquake in year 2008. The effects of the disaster were severe, with 68,712 casualties and an estimated economic damage of US$120 billion (Sharma 2010). If the countries are prepared and the nature of the disaster is known, affected populations can be trained to respond to the disaster or to avoid the disaster altogether (Kovács and Spens 2009). In the case if the Sichuan earthquake the Chinese government could undertake the response operation by using soldiers from the National army and they could in a fairly short time access and aid the areas of the disaster.

In number of occurred disasters between 2000 and 2009, China is followed by India (average of 44 disasters/ year) and United States (average of 34 disasters/ year). All top three disaster occurrence countries have a higher than average LPI. These countries are also densely populated. China has a total population of 1.340 billion and 139 people per square kilometer, India has a total population of 1.180 billion and a population density of 361 persons per square kilometer and the United States has a total population of 310 million and a population density of 31 persons per square kilometer (UN 2010). What moreover needs to be kept in mind when looking at table 1, is that the disaster occurrence data in this study takes in consideration all types of disasters, not only natural disasters but also e.g. industrial accidents. The study entails all types of disasters since there can be similar requirements in the response operation and the disaster impact regardless of what sort the disaster is. The figures might look quite different if only looking at natural disasters since the top three countries are industrialized and have a high occurrence of industrial accidents and transport accidents as well as natural disasters.

Table 1 also shows that out of the 10 countries with most disasters occurring per year, 6 countries have a below average (2.86) LPI. Low logistics performer countries with a high disaster occur-

rence rate are Nigeria, Indonesia, Philippines, Russia, Iran, Bangladesh and Pakistan.

The table also shows that the disaster impact data seems to vary quite a bit. With the highest average amount of affect per disaster in India with 1.55 million affected in average per disaster, China with 1.48 million and Bangladesh with 0.48 million affected. Out of the disasters that have occurred between 2000- 2009 what further can be stated is that United States seems to have a high amount of economic damage, with Hurricane Katrina (2005) probably as one of the main factors. The sole economic damage from the hurricane is estimated as US$125 billion (Amin and Goldstein 2008). Indonesia had a high number of its population affected in the tsunami of 2004 (Banomyong *et al.* 2009, Perry 2007, Régnier *et al.* 2008). Pakistan again has been hit with re-occurring earthquakes both in year 2005 (Kashmir) with 80 000 deaths and 2008 (Balochistan) which left 150 000 people homeless. (EM-DAT 2010)

Figure 1 shows the average amount of affected per disaster for all (154) countries as well as the LPI for these countries. The countries are sorted in order of the variable average amount of affected per disaster. What can be seen in the figure is that there seems to be a slight increasing trend line for LPI compared to the decreasing trend line for the average amount of affected. That would indicate that when the countries are sorted from high to low average amount of affected, the LPI seems to decrease to a minor extent. This would mean that there could be relationship where the higher the amount of affected per disaster the lower the LPI is.

The countries that stand out as such with a low LPI and high disaster occurrence or high disaster impact can be seen in Table 2. The table shows the lowest ranking logistics performers. Somalia, Nepal and Sudan are such countries that have a low LPI and a higher than average disaster occurrence rate. Sudan also has a higher than average number of people affected by disasters. Sudan has been distressed by a civil war since 1983

Figure 1. Trend line (all 154 countries) for LPI and the disaster impact variable average affected

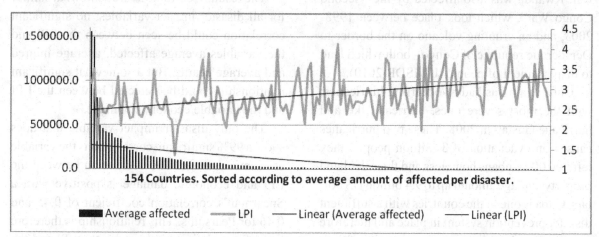

leaving over two million dead and over four million displaced (Beamon and Kotleba 2006). The four million displaced people in Sudan are accounted for in the variable affected per disaster.

Out of the countries with a low logistics performance Eritrea, Rwanda and Cuba do not have a very high disaster occurrence rate, but they do have a high number of people who in average were affected by the disasters that did occur. Eritrea's population was affected by disasters since it was in war with Ethiopia 1993-2000 when

the population was affected by famine and displacement. Eritrea was also hit by drought in 2008 (Devereux 2009). Rwanda again has had a limited amount of natural disaster in the years 2000-2009, there are some landslides, flooding, drought, epidemics, road accidents, forest fires and social conflicts. The reason that Rwanda has such a large amount of people affected by disaster (see Table 2) is probably due to the displacement and resettlement that has taken place still between 2000 and 2009 as a result of the 1994 genocide and

Table 2. Bottom 10 logistics performers (2010) and disaster occurrence (2000-2009)

Country	LPI	Average # disasters / year	Average economic damage / disaster	Average # injured / disaster	Average # affected / disaster	Average # deaths / disaster
Somalia	1.34	6.1*	1 818	9	10 541	58
Eritrea	1.70	0.6	-	9	801 409*	11
Sierra Leone	1.97	1.9	-	4	1 383	56
Namibia	2.02	1.9	499	-	48 939	20
Rwanda	2.04	2.1	-	48	102 444*	22
Cuba	2.07	3.2	259 349*	10	339 292*	7
Guinea-Bissau	2.10	1.4	-	0.8	13 344	57
Iraq	2.11	2.7	54	44	3 256	64
Nepal	2.20	6.6*	1 164	20	44 643	62
Sudan	2.21	7.9*	6 845	15	112 447*	54

(*. Higher than average)

war. Rwanda was also affected by the "Second Congo War", which took place between 1998-2003, and an erupting volcano on the border to Democratic republic of Congo - both which lead to displacement of people (UNISDR 2010).

Cuba again has had re-occurring hurricanes; with the most severe ones, hurricane Ike and hurricane Gustav in 2008. These two hurricanes caused an evacuation of 3 million people, they left 200 000 Cubans homeless and the calculated property damage amount up to 9.4 billion US dollars. Cuba is one of the countries with a sufficient disaster prevention system in place and therefore the disasters that hit the country, in spite of their magnitude, seldom cause a large loss of lives (Keyser and Smith 2009).

Logistics Performance Indicators Correlation with Disaster Impact Variables

To test the hypothesis of the study, correlation analyses were conducted for the variables. The variable LPI was seen as the dependent variable since the data is normally distributed and the disaster impact variables as independent ones. Spearman's correlation and Pearson's correlation was carried out.

Potential outliers could be the countries with a high population density. Countries such as China, India and United States could already in the previous chapter been seen to stand out in the descriptive statistic with having a high disaster occurrence and a high amount of people affected by the disasters (see Table 1).

The results (see Table 3) are somewhat similar for all disaster impact variables; no significant correlation could be seen between the LPI and the variables average affected, average injured and average deaths. But a somewhat significant relationship could be detected between the LPI and the variable economic damage.

The only disaster impact variable that does show a 99% significant correlation is the variable economic damage. The correlation between the LPI and economic damage is positive with a Spearman's correlation coefficient of 0.45 and 0.46 for Pearson's. This relationship is therefore linear The relationship between the LPI and the economic damage is positive which indicates that the countries with a higher logistics performance tend to as well have a larger economic damage measured in dollars per disaster exactly opposite to what the first hypothesis states. The found relationship could be explained by the fact that the higher logistics performers tend to have a higher GDP and therefore these countries tend to have more economic resources that can be damaged in a disaster. The first hypothesis stating that there could be a negative correlation between LPI and disaster impact variables is proven false for all variables.

Countries with Low, Average or High LPI

When correlations are calculated for countries with a low (ranking: 1.38- 2.57) LPI, the same results cannot be detected (see Table 4). There is no significant relationship between the variable LPI and the variable economic damage. This seems

Table 3. Correlations between LPI and disaster impact variables

	LPI	Average Affected	Average Injured	Average Death	Average Damage
Spearman's Correlation	1.00	0.04	0.01	0.09	**0.45***
Pearson's Correlation	1.00	0.01	0.03	0.07	**0.46***

(*p < 0.01)

to indicate that the LPI and economic damage correlation that could be seen in the correlation analysis for all countries cannot be detected when looking at countries with low indicators. This might be due to the fact that low LPI countries tend to have a lower economic development level and could perhaps have less economic resource that can get damaged by the impacts of a disaster.

For countries with an average (ranking: 2.58-2.99) LPI there can neither be found a significant relationship between a country's LPI and the economic damage as could be seen for all countries. There can though be found a 95% significant negative correlation (Spearman's correlation coefficient -0.347) between LPI and the variable average total affected. That would mean that in the countries with a higher LPI there seems to be less average affected per disaster. The relationship is not linear since Pearson's correlation shows no significant correlation. The results from Spearman's correlation would prove the first hypothesis correct meaning that for such countries that have an average LPI, there seems to be a negative correlation between the LPI and the average affected. The higher the LPI the less number of people seem to be affected in average per disaster.

Countries with a high (3.02-4.11) LPI show the same correlation with a 99% significance, between

LPI and the variable economic damage as could be detected in the correlation for all countries. For the countries with a high LPI there seems to be correlation, the higher the LPI, the larger the average economic damage per disaster. This relationship is linear.

FUTURE RESEARCH DIRECTIONS

Logistics performance is determined in this study to somewhat correlate with disaster impact and it would be valuable to furthered analyze this statement, since the direct negative correlation that was sought after with the first hypothesis could not be directly proven. Further research should test the hypothesis with a broader set of variables e.g. taking in account population density and country economic development. The study could as well be conducted by only taking in consideration natural disasters, since some types of disasters such as e.g. industrial accidents might have a different impact on the area and population and the humanitarian response operations might differ to some extent form the response to a natural disaster. Further a regression analysis could be conducted with the LPI as a dependent variable and all the variables in the study and the above mentioned variables

Table 4. Correlations for low, average and high LPI

	Low LPI	Average Affected	Average Injured	Average Deaths	Average Damage
Spearman's Correlation	1	-0.052	-0.149	-0.143	0.288
Pearson's Correlation	1	-0.083	-0.197	0.017	0.050
	Average LPI	Average Affected	Average Injured	Average Deaths	Average Damage
Spearman's Correlation	1	-0.347**	0.029	-0.044	0.093
Pearson's Correlation	1	0.037	0.032	0.016	0.000
	High LPI	Average Affected	Average Injured	Average Deaths	Average Damage
Spearman's Correlation	1	-0.097	-0.073	0.128	0.476*
Pearson's Correlation	1	0.039	-0.194	0.167	0.512*

(* $p < 0.01$, ** $p < 0.05$)

accounted for as independent variables. It would also be valuable to measure whether the level of preparedness in a country correlates with the disaster impact. Could there be found a correlation between the level of preparedness in a country and the disaster impact variables average amount of affected, average amount of deaths and average amount of economic damage?

Further research could also be conducted on the logistics performance indicators and whether they can serve as measurement for country logistics performance. Logistics performance in humanitarian settings could as well be further studied.

CONCLUDING DISCUSSION

This study aimed to analyze the relationship between the logistics performance in a country and the disaster impact. The primary aim was to analyze the relationship between countries' logistics performance and the disaster impact, this relationship showed no significant relationship for all countries between the LPI and variables; average amount of affected, average amount of injured and average amount of deaths. A significant relationship between the LPI and average amount of affected could be found when only analyzing countries with average LPI. For these countries there is a significant negative relationship, which indicates that countries with a higher LPI tend to have a lower amount of affected per disaster. Countries in the group of average LPIs with an LPI on the lower side of that group therefore tend to have more people affected per disaster then countries with a higher LPI in that group. This result aligns with the first hypothesis of the study.

The analysis did as well show a relationship between the LPI and the variable average economic damage. The relationship is positive which indicates that the countries with a higher logistics performance might have more resources that can be and are damage in the course of disaster. This correlation could with a high significance be detected

for countries with high LPI as well. For countries with low or average LPI there seemed to be no correlation between the logistics performance and the disaster impact variable economic damage. Amin and Goldstein (2008) though claim that the economic burden is proportionally much higher in poor countries, which are also the countries with low logistics performance (Arvis *et al.*, 2010). So if the variable economic damage would be proportioned against a country's economic development the results might be different. The economic loss for example from Hurricane Katrina amount to only 0,1 percent of United States gross domestic product. While the UNISDR (2004) calculates that the economic losses due to disasters occurred in the past twenty years in developing countries account for between 134 and 378 percent of the countries' gross domestic product.

The results show no direct correlation between a country's logistics performance and the other disaster impact variables (deaths, affected, injured) when looking at all countries. But when analyzing only the countries with average LPIs an interesting relationship between the LPI and the average amount of affected could be detected. In this group of countries, the level of logistics performance and the disaster impact correlate with 95 percent significance according to Spearman coefficient. The results indicated that the higher the logistics performance the lesser the disaster impact and the lower the logistics performance the higher the disaster impact measured as number of people affected by a disaster. None of these relationships could be seen when analyzing the low LPI countries which also tend to be the countries with the highest disaster occurrence. For these countries it seems like there are other variables that are crucial to take inconsideration in disaster response. It might also be that for the countries with a low LPI, humanitarian organizations take a larger role in the disaster response than they do for the countries with an average LPI and therefore the relationship with the countries logistics performance and average amount of affected cannot be seen.

As the third hypothesis stated, the correlation differ depending on which disaster impact variable was used. For the variables average number of injured and average number of death there was no correlation with the LPI found. For the disaster impact variable economic damage a correlation could be detected for all countries and for the disaster impact variable total amount of affected a correlation was found for the group of average LPI countries. The disaster impact can differ depending on what type of disaster is in question and where the disaster occurs. For example in the case of a slow onset disaster people might have time to prepare and even evacuate and therefore the number of injured or dead might not be high, but there could be a high number of people who are affected. On other hand, in the case of a rapid onset disaster in a highly populated and economically developed area the economic damage could ascend.

The second and third hypothesis could be confirmed since the results differed depending on the group of LPI (low, average and high) and depending on which disaster impact variable was used.

The secondary aim was to identify areas with a high disaster occurrence rate and low logistics performance. Countries with a high disaster occurrence rate and a low logistics performance level are: Somalia, Nepal and Sudan. Countries with low logistics performance and a higher than average numbers of people affected per disaster are Eritrea, Rwanda, Cuba and Sudan. All these countries have been hit by both natural and complex emergencies between 2000 and 2009. The population might have been affected by war or political disturbance (Somalia, Eritrea, Rwanda, Sudan) or natural disasters (Nepal and Cuba). Since the LPI is calculated with measurements for e.g. customs, it is important to keep in mind when analyzing the results for example for Cuba that the inland transportation has a fairly high level of performance while the customs cause a lengthy wait for exports and imports to Cuba and therefore Cuba's ranking as a low logistics performance might be a bit misleading.

As practical implication, increased disaster preparedness in areas with high disaster occurrence and low logistics performance is suggested. Country preparedness is crucial in those areas where the disasters are re-occurring. While country preparedness could be emphasized in areas where natural disasters re-occur, humanitarian organizations should focus a high level of preparedness to handle the aftermath of disasters in such areas that are conflict prone and that have high disaster occurrence and low logistics performance. The results from this study can be used by countries to identify themselves as disaster prone and as low logistics performers. Humanitarian organizations again could use the logistics performance indicators in their planning so that their humanitarian logistics planning (e.g. lead time and transportation planning) could account for the differences in the logistics performance in affected countries.

REFERENCES

Amin, S., & Goldstein, M. (2008). *Data against natural disasters: Establishing effective systems for relief, recovery, and reconstruction*. World Bank Report.

Arvis, J.-F., Mustra, M. A., Ojala, L., Shepherd, B., & Saslavsky, D. (2010). *Connecting to compete 2010, trade logistics performance index and its indicators*. Washington, DC: The World Bank.

Banomyong, R., Beresford, A., & Pettit, S. (2009). Logistics relief response model: The case of Thailand's tsunami affected area. *International Journal of Services Technology and Management, 12*(4). doi:10.1504/IJSTM.2009.025816

Beamon, B., & Balcik, B. (2008). Performance measurement in humanitarian relief chains. *International Journal of Public Sector Management, 21*(1). doi:10.1108/09513550810846087

Beamon, B., & Kotleba, S. (2006). Inventory management support systems for emergency humanitarian relief operations in South Sudan. *The International Journal of Logistics Management, 17*(2), 187–212. doi:10.1108/09574090610689952

Beresford, A., & Pettit, S. (2009). Critical success factors in the context of humanitarian aid supply chains. *International Journal of Physical Distribution & Logistics Management, 39*(6), 450–468. doi:10.1108/09600030910985811

Beresford, A., & Pettit, S. (2009). Emergency logistics and risk mitigation in Thailand following the Asian tsunami. *International Journal of Risk Assessment and Management, 13*(1), 7–21. doi:10.1504/IJRAM.2009.026387

Clark, X., Dollar, D., & Micco, A. (2004). Port efficiency, maritime transport costs, and bilateral trade. *Journal of Development Economics, 75*, 417–450. doi:10.1016/j.jdeveco.2004.06.005

Devereux, S. (2009). Why does famine persist in Africa? *Food Security, 1*, 25–35. doi:10.1007/s12571-008-0005-8

Dollar, D., Hallward-Driemeier, M., & Mengistae, T. (2004). *Investment climate and international integration.* Policy Research Working Paper 3323, Washington, DC: World Bank.

EM-DAT. (2010). *EM-DAT – Emergency events database.* Centre for Research on the Epidemiology of Disasters (CRED), Universite´ Catholique de Louvain, Louvain-La-Neuve. Retrieved September 2010, from www.emdat.be

Fink, C., Matoo, A., & Neagu, I. C. (2002). Trade in international maritime services: How much does policy matter? [Oxford University Press.]. *The World Bank Economic Review, 16*(1), 81–108. doi:10.1093/wber/16.1.81

Guha-Sapis, D., & Below, R. (2002). *The quality and accuracy of disaster data: A comparative analyses of three global data sets. WHO Centre for Research on the Epidemiology of Disasters University of Louvain School of Medicine.* Brussels, Belgium: ProVention Publications.

Hausman, W., Lee, H., & Subramanian, U. (2005). *Global logistics indicators, supply chain metrics, and bilateral trade patterns.* Final Draft, World Bank report.

Karamperidis, S., Mangan, J., & Jackson, E. (2010). Developing an index of maritime costs and connectivity for the UK. In the *Proceedings of LRN 2010*, (p. 354).

Keyser, J., & Smith, W. (2009). *Disaster relief management in Cuba, why Cuba's disaster relief model is worth careful study.* International policy report, Center for International Policy, May.

Kovács, G., & Spens, K. (2007). Humanitarian logistics in disaster relief operations. *International Journal of Physical Distribution and Logistics Management, 37*(2), 99–114. doi:10.1108/09600030710734820

Kovács, G., & Spens, K. (2009). Identifying challenges in humanitarian logistics. *International Journal of Physical Distribution and Logistics Management, 39*(6), 506–528. doi:10.1108/09600030910985848

Limao, N., & Venable, A. J. (2001). Infrastucture, geographical disadvantages, transport costs and trade. *The World Bank Economic Review, 15*, 451–479. doi:10.1093/wber/15.3.451

McEntire, D. A. (2002). Coordinating multi-organizational responses to disaster: Lessons from the March 28, 2000 Fort Worth tornado. *Disaster Prevention and Management: An International Journal, 11*(5), 369–379. doi:10.1108/09653560210453416

Oloruntoba, R. (2005). A wave of destruction and the waves of relief: Issues, challenges and strategies. *Disaster Prevention and Management: An International Journal, 14*(4), 506–521. doi:10.1108/09653560510618348

Perry, M. (2007). Natural disaster management planning: A study of logistics managers responding to the tsunami. *International Journal of Physical Distribution and Logistics Management, 37*(5), 409–433. doi:10.1108/09600030710758455

Régnier, P., Neri, B., Scuteri, S., & Miniati, S. (2008). From emergency relief to livelihood recovery: Lessons learned from post-tsunami experiences in Indonesia and India. *Disaster Prevention and Management, 17*(3), 410–429. doi:10.1108/09653560810887329

Shaluf, I. M. (2007). An overview on disasters. *Disaster Prevention and Management, 16*(5), 687–703. doi:10.1108/09653560710837000

Sharma, K. (2010). *Socio-economic aspects of disaster's impact: An assessment of databases and methodologies*. Economic Growth Centre, Division of Economics School of Humanities and Social Sciences Nanyang Technological University, Working Paper, No 01.

UN. (2010). *Population statistics*. Retrieved October, 2010, from http://unstats.un.org/unsd/demographic/default.htm

UNISDR. (2010). *Disaster risk reduction and prevention in Rwanda*. Retrieved October, 2010, from http://www.unisdr.org/eng/country-inform/reports/Rwanda-report.pdf

van Wassenhove, L. N. (2006). Humanitarian aid logistics: Supply chain management in high gear. *The Journal of the Operational Research Society, 57*(5), 475. doi:10.1057/palgrave.jors.2602125

World Bank. (2010). *Trade logistics and facilitation, logistics performance index*. Retrieved September, 2010, from http://web.worldbank.org

ADDITIONAL READING

Altay, N., & Green, W. G. (2006). OR/MS research in disaster operations management. *European Journal of Operational Research, 175*(1), 475–493. doi:10.1016/j.ejor.2005.05.016

Hale, T., Moberg, C.R., (2005), Improving supply chain disaster preparedness -A decision process for secure site location, *International Journal of Physical Distribution & Logistics Management,* Vol. 35 No. 3.

McEntire, D. A. (2002). Coordinating multi-organizational responses to disaster: lessons from the March 28, 2000 Forth Worth tornado. *Disaster Prevention and Management: An International Journal, 11*(5), 369–379. doi:10.1108/09653560210453416

Raleigh, C., & Urdal, H. (2007). Climate change, environmental degradation and armed conflict. *Political Geography, 26,* 674. doi:10.1016/j.polgeo.2007.06.005

Roh, S., & Pettit, S. (2009), "Pre-positioning networks for humanitarian aid logistics", *14th Annual Logistics Research Network Conference, 9th – 11th September 2009, Cardiff.*

Tatham, P. (2009) " An investigation into the suitability of the use of unmanned aerial vehicle systems (UAVS) to support the initial needs assessment process in rapid onset humanitarian disasters", *Int. J. Risk Assessment and Management,* Vol. 13, No1.

Tatham, P., & Spens, K. (2008), The developing humanitarian logistics knowledge management system - a proposed taxonomy. *Proceedings of the 19th Annual Conference of the Production and Operations Management Society,* POMS 2008, La Jolla/CA, USA, paper 008-0047.

KEY TERMS AND DEFINITIONS

Disaster Impact: The direct or indirect impact that an event classified as a disaster have on people's lives, properties and the environment.

Disaster Occurrence: Amount of disaster occurred in a particular time. The EM-DAT disasters are entered in the database by event and by country.

Humanitarian Response Operation: Responding (a country or an organization responding) to an event or circumstance where humanitarian assistance is needed, both humanitarian relief operations and development aid programs.

Logistics Performance Index (LPI): The by the World Bank measured LPI a multidimensional assessment of logistics performance, rated on a scale from one (worst) to five (best).

Compilation of References

Agence France Presse. (2010). *Haïti le Canada va rouvrir l'aéroport de Jacmel*. Retrieved 10th March, 2010, from http://www.romandie.com /infos/news2/100120160031 .a9e10wyg.asp

Agndal, H. (2004). *Internationalisation as a process of strategy and change*. Thesis (PhD). Jönköping International Business School.

Aldworth, P. (Ed.). (2009). *Lloyd's maritime atlas of world ports and shipping places* (25th ed.). London, UK: Informa Professional.

Allen, G. W. (2007). Reconstruction and development on Operation Herrick 5 utilising local contractors. *The Royal Engineers Journal, 121*(2), 84–89.

ALNAP. (2008). *Cyclone Nargis: Lessons for operational agencies*. Retrieved July 1, 2010, from www.alnap.org/publications/pdfs/ALNAPLessonsCycloneNargis.pdf

Altay, N. (2008). Issues in disaster relief logistics . In Gal-el-Hak, M. (Ed.), *Large-scale disasters: Prediction, control, and mitigation* (pp. 121–146). New York, NY: Cambridge University Press.

Altiok, T., & Melamed, B. (2007). *Simulation modeling and analysis with arena*. Burlington, MA: Elsevier.

Amin, S., & Goldstein, M. (2008). *Data against natural disasters: Establishing effective systems for relief, recovery, and reconstruction*. World Bank Report.

Anderson, M. B., & Woodrow, P. J. (1998). *Rising from the ashes. Development strategies in times of disaster*. Boulder, CO: Lynne Rienner Publishers.

Anderson, J., & Katz, J. M. (2010). *US forces scale back Haiti earthquake relief role*. Retrieved 10th March, 2010, from http://www.breitbart.com/ article.php?id=D9DS9B880 &show_article=1

Anonymous. (8 January 2010). UN aid agencies will not abandon Somalia despite insecurity, says official. *UN News Centre*. Retrieved 12 February, 2010, from http://www.un.org/apps/news/story.asp?NewsID=33433&Cr=somali&Cr1

Antill, P. (2001). *Military involvement in humanitarian aid operations*. HistoryofWar.org

Arvis, J.-F., Mustra, M. A., Ojala, L., Shepherd, B., & Saslavsky, D. (2010). *Connecting to compete 2010, trade logistics performance index and its indicators*. Washington, DC: The World Bank.

Associated Press. (2010). *Post-earthquake rubble removal to take three years: Haitian president*. Retrieved 10th March, 2010, from http://dcnonl.com/ article/id37658

Austin, T. (2009). Tobruk's proud mission. *Navy News*. Retrieved 12 February, 2010, from www.defence.gov.au/news/navynews

Axelsson, B., Rozemeijer, F., & Wynstra, F. (2005). The case for change . In Davis, J. A. (Ed.), *Developing sourcing capabilities* (pp. 3–13). West Sussex, UK: John Wiley & Sons Ltd.

Bacon, L. M. (2010). *Carl Vinson, other ships headed to Haiti*. Retrieved 10th March, 2010, from http://www.navytimes.com/ news/2010/01/navy_vinson _haiti_update_011310w/

Bal, J., & Teo, P. K. (2001). Implementing virtual teamworking: Part 2 – A literature review. *Logistics Information Management*, *14*(1), 208–222. doi:10.1108/09576050110390248

Balcik, B., & Beamon, B. M. (2008). Facility location in humanitarian relief. *International Journal of Logistics Research and Applications*, *11*(2), 101–121. doi:10.1080/13675560701561789

Balcik, B., Iravani, S., & Smilowitz, K. (2010in press). A review of equity in nonprofit and public sector: A vehicle routing perspective . In Cochran, J. J. (Ed.), *Wiley encyclopedia of operations research and management science*. John Wiley & Sons.

Bammel, J. L., & Rodman, W. K. (2006/2007). Humanitarian logistics: A guide to operational and tactical logistics in humanitarian emergencies. *Air Force Journal of Logistics, 30/31*(4/1), 1-42.

Banomyong, R., Beresford, A. K. C., & Pettit, S. (2009). Supply chain relief response model: The case of Thailand's tsunami affected area. *International Journal of Services Technology and Management*, *12*(4), 414–429. doi:10.1504/IJSTM.2009.025816

Banomyong, R., Beresford, A., & Pettit, S. (2009). Logistics relief response model: The case of Thailand's tsunami affected area. *International Journal of Services Technology and Management*, *12*(4), 414–429. doi:10.1504/IJSTM.2009.025816

Banomyong, R., Beresford, A., & Pettit, S. (2009). Logistics relief response model: The case of Thailand's tsunami affected area. *International Journal of Services Technology and Management*, *12*(4). doi:10.1504/IJSTM.2009.025816

Barbarosoglu, G., & Arda, Y. (2004). A two-stage stochastic programming framework for transportation planning in disaster response. *The Journal of the Operational Research Society*, *55*(1), 43–53. doi:10.1057/palgrave.jors.2601652

Barbarosoglu, G., Ozdamar, L., & Cevik, A. (2002). An interactive approach for hierarchical analysis of helicopter logistics in disaster relief operations. *European Journal of Operational Research*, *140*(1), 118–133. doi:10.1016/S0377-2217(01)00222-3

Barratt, M. (2004). Understanding the meaning of collaboration in the supply chain. *Supply Chain Management: An International Journal*, *9*(1), 30–42. doi:10.1108/13598540410517566

Barringer, F., & Longman, J. (2005). Police and owners begin to challenge looters. *New York Times*.

Barry, J., & Jefferys, A. (2002). *A bridge too far: Aid agencies and the military humanitarian response*. Humanitarian Practice Network (HPN) Paper, No. 37.

BBC. (2010a). Haiti will not die, President Rene Preval insists. *BBC News*. Retrieved 10th March, 2010, from http://news.bbc.co.uk/1/hi/world/americas/8511997.stm

BBC. (2010b). Haiti quake victims' bodies 'piled up by roads. *BBC News*. Retrieved 18th January, 2010, from http://news.bbc.co.uk/2/hi /uk_news/england/devon/8465916.stm.

BBC. (2010c). Haiti quake victim rescue operation declared over. *BBC News*. Retrieved 23rd January, 2010, from http://news.bbc.co.uk/1/hi /world/americas/8476474.stm.

BBC. (2010d). Haiti aid effort one month after earthquake. *BBC News*. Retrieved 30th March, 2010, from http://news.bbc.co.uk/1/hi /world/americas/8509333.stm

Beamon, B., & Balcik, B. (2008). Performance measurement in humanitarian relief chains. *International Journal of Public Sector Management*, *21*(1). doi:10.1108/09513550810846087

Beamon, B., & Kotleba, S. (2006). Inventory management support systems for emergency humanitarian relief operations in South Sudan. *The International Journal of Logistics Management*, *17*(2), 187–212. doi:10.1108/09574090610689952

Beauregard, A. (1998). *Civil-military cooperation in joint humanitarian operations: A case analysis of Somalia, the former Yugoslavia and Rwanda*. Waterloo, Canada: Ploughshares Monitor.

Beausang, F. (2003). *Is there a development case for United Nations-business partnerships*. LSE Working Paper Series. ISSN: 1470-2320

Ben-Shalom, U., Lehrer, Z., & Ben-Ari, E. (2005). Cohesion during military operations. *Armed Forces and Society, 32*(1), 63–79. doi:10.1177/0095327X05277888

Beresford, A. K. C., & Pettit, S. (2009). Emergency supply chain and risk mitigation in Thailand following the Asian tsunami. *International Journal of Risk Assessment and Management, 13*(1), 7–21. doi:10.1504/IJRAM.2009.026387

Beresford, A., & Pettit, S. (2009). Critical success factors in the context of humanitarian aid supply chains. *International Journal of Physical Distribution & Logistics Management, 39*(6), 450–468. doi:10.1108/09600030910985811

Beresford, A., & Pettit, S. (2009). Emergency logistics and risk mitigation in Thailand following the Asian tsunami. *International Journal of Risk Assessment and Management, 13*(1), 7–21. doi:10.1504/IJRAM.2009.026387

Beresford, A. K. C., & Pettit, S. (2007, July). *Disaster management and risk mitigation in Thailand following the Asian Tsunami.* Paper presented at the International Conference on Supply Chain Management, Bangkok.

Beresford, A., & Pettit, S. (2007). Disaster management and mitigation: A case study of logistics problems in Thailand following the Asian Tsunami. In Á. Halldórsson & G. Stefánsson, (Eds.), *Proceedings of the 19th Annual Conference for Nordic Researchers in Logistics, NOFOMA 2007, Reykjavík, Iceland* (pp.121-136).

Binder, A., & Witte, J. M. (2007). *Business engagement in humanitarian relief: Key trends and policy implications.* London, UK: Humanitarian Policy Group, Overseas Development Institute.

Boadle, A. (2010). *U.S. military says Haiti airport jam easing.* Retrieved 30th March, 2010, from http://www.reuters.com/article/ idUSTRE60H00020100118

Borton, J. (1993). Recent trends in international relief system. *Disasters, 17*(3), 187–201. doi:10.1111/j.1467-7717.1993.tb00493.x

Boutros-Ghali, B. (1992). *An agenda for peace: Preventive diplomacy, peacemaking and peace-keeping (A/47/277-S/24111).* New York: Secretary-General, United Nations.

Brannigan, M. (2010). *Haiti seaport damage complicates relief efforts.* Retrieved 10th March, 2010, from http://www.miamiherald.com/ news/breaking-news/story/1426067.html

Brazier, D. (2009). Heavy lifting for United Nations peacekeeping: Strategic deployment stocks . *Logistics and Transport Focus, 11*(11), 37–40.

Briscoe, G., & Dainty, A. (2005). Construction supply chain integration: An elusive goal? *Supply Chain Management: An International Journal, 10*(4), 319–326. doi:10.1108/13598540510612794

Brocades-Zaalberg, T. (2005). *Soldiers and civil power: Supporting or substituting civil authorities in peace support operations during the 1990s. Amsterdam.* Amsterdam: University.

Buckley, P. J., & Casson, M. (1976). *The future of the multinational enterprise.* New York, NY: Holmes & Meier.

Butcher, T., Claes, B., & Grant, D. (2008). Supply chain work organisation: Can structuration theory offer new solutions? In *Proceedings of the Logistics Research Network (LRN) Conference,* Liverpool, UK, 10-12 Sept.

Canadian Press. (2010). *Canada stops Haitian evacuation flights, death toll set to jump.* Retrieved 28th February, 2010, from http://www.google.com/ hostednews/canadianpress /article/ALeqM5jMrV3 QsuxEtZVBiLmFwVcL _XgMnQ

Carter, W. N. (1999). *Disaster management: A disaster management handbook.* Manila, Philippines: Asian Development Bank.

China Communication News Net. (2008). *The Ministry of Transport issued 4 water-road transport routes.* Retrieved 20th June, 2010, from http://www.zgjtb.com/101179 /101182/101215/32931.html

China.com. (2008). *Up to July 20 2008, the Sichuan Wenchuan earthquake has caused 69,197 deaths, 374,176 people injured, and 18,222 people were missing.* Retrieved 20th June, 2008, from http://www.china.com.cn/news/ zhuanti/wxdz/2008-07/20/content_16038392.htm

Chinanews. (2008a). *The first international aid relief supplies has arrived in Chengdu*. Retrieved 20th June, 2008, from http://www.chinanews.com.cn /gn/news/2008/05-14/1250176.shtml

Chinanews. (2008b). *Japan military will send the relief supplies to the disaster area*. Retrieved 20th June, 2008, from http://bjyouth.ynet.com/ view.jsp?oid=41090285

Christopher, M. (2011). *Logistics and supply chain managemen* 4th ed. London, UK: Financial Times Prentice Hall.

CIC (Centre on International Cooperation). (2006). *Annual review of global peace operations, 2006*. Boulder, CO: Lynne Rienner Publishers.

Clark, X., Dollar, D., & Micco, A. (2004). Port efficiency, maritime transport costs, and bilateral trade. *Journal of Development Economics*, 75, 417–450. doi:10.1016/j.jdeveco.2004.06.005

CNN. (2010a). *Haiti pier opens, road laid into Port-au-Prince*. Retrieved 20th March, 2010, from http://www.cnn.com/2010/ WORLD/americas/01/21/ haiti.earthquake/index.html?hpt=T2

CNN. (2010b). *Massive food distribution begins in quake-ravaged Haitian capital*. Retrieved 20th March, 2010, from http://edition.cnn.com/2010/ WORLD/americas/01/31/haiti.food.aid/

Connaughton, R. (1996). *Military support and protection for humanitarian assistance: Rwanda, April-December 1994*. Occasional Paper No.18, Camberly, UK: Strategic and Combat Institute.

Coppola, N. W., Hiltz, S. R., & Rotter, N. G. (2004). Building trust in virtual teams. *IEEE Transactions on Professional Communication*, 47(2), 95–104. doi:10.1109/TPC.2004.828203

Cowell, A., & Otterman, S. (2010). *Relief groups seek alternative routes to get aid moving*. Retrieved 20th March, 2010, from http://www.nytimes.com/ 2010/01/16/world/americas/ 16relief.html?hp

Cox, A. (2004). The art of the possible: Relationship management in power regimes and supply chains. *Supply Chain Management: An International Journal*, 9(5), 346–356. doi:10.1108/13598540410560739

Croft, S., & Treacher, T. (1995). Aspects of intervention in the South. In Dorman, A. M., & Otte, T. G. (Eds.), *Military intervention: From gunboat diplomacy to humanitarian intervention*. Dartmouth, NH: Dartmouth Publishing.

Cross, P. (2004). *The hidden power of social networks*. Boston, MA: Harvard Business School.

Currey, C. J. (2003). *A new model for military/nongovernmental relations in post-conflict operations*. Carlisle. PA: U.S. Army War College.

Dainty, A. R. J., Millett, S. J., & Briscoe, G. H. (2001). New perspectives on construction supply chain integration. *Supply Chain Management: an International Journal*, 6(4), 163–173. doi:10.1108/13598540110402700

Das, T. K., & Teng, B.-S. (2001). Trust, control, and risk in strategic alliances: An integrated framework. *Organization Studies*, 22(2), 251–283. doi:10.1177/0170840601222004

De Conning, C. (2007). Civil-military coordination practices and approaches within United Nations peace operations. *Journal of Military and Strategic Studies*, 10(1).

de Ville de Goyet, C. (2008). *The use of a logistics support system in Guatemala and Haiti*. Washington, DC: World Bank.

de Vries, J. (2007). Diagnosing inventory management systems: An empirical evaluation of a conceptual approach. *International Journal of Production Economics*, 108(1-2), 63–73. doi:10.1016/j.ijpe.2006.12.003

DEFRA. (2009). *2008 guidelines to Defra's greenhouse gas conversion factors- Methodology paper for transport emission factors*. Retrieved from www.defra.gov.uk

Denning, P. J. (2006). Hastily formed networks. *Communications of the ACM*, 49(4), 15–20. doi:10.1145/1121949.1121966

Department of Peacekeeping Operations & Department of Field Support. (2009). *A new partnership agenda: Charting a new horizon for UN peacekeeping*. New York.

Devereux, S. (2009). Why does famine persist in Africa? *Food Security*, 1, 25–35. doi:10.1007/s12571-008-0005-8

DiarioLibre.com. (2010). *LF viaja a Haití, acuerda con Préval plan para mitigar daños.* Retrieved 18th March, 2010, from http://www.diariolibre.com/ noticias_det.php?id=230910

Diaz, R. (2010). *Dominican Republic: Helping neighboring Haiti after earthquake.* Retrieved 16th March, 2010, from http://globalvoicesonline.org/ 2010/01/14/dominican-republic -helping-neighboring- haiti -after-earthquake/

Disaster Management Training Programme. (1993). *Logistics* (1st ed.). New York, NY: United Nations Development Programme/Department of Humanitarian Affairs.

Doel, M. T. (1995). Military assistance in humanitarian aid operations: Impossible paradox or inevitable development? *Royal United Services Institute Journal, 140*(5), 26–32.

Dollar, D., Hallward-Driemeier, M., & Mengistae, T. (2004). *Investment climate and international integration.* Policy Research Working Paper 3323, Washington, DC: World Bank.

Dolmetsch, C. (2010). *UN urges Haiti coordination as supplies flood airport.* Retrieved 16th March, 2010, from http://www.businessweek.com/ news/2010-01-22/un-urges- haiti-relief-coordination-as- supplies- flood-airport.html.

Drabek, T. T. (1985). Managing the emergency response. *Public Administration Review, 45,* 85–92. doi:10.2307/3135002

Duffey, T. (2000). Cultural issues in contemporary peacekeeping. *International Peacekeeping, 7*(1), 142–168.

Dunning, J. H. (1980). Towards an eclectic theory of international production: Some empirical tests. *Journal of International Business Studies, 11*(1), 9–31. doi:10.1057/palgrave.jibs.8490593

Dwyer, F. R., Schurr, P. H., & Oh, S. (1987). Developing buyer-seller relationships. *Journal of Marketing, 51,* 11–27. doi:10.2307/1251126

EAR. (2002). *Kosovo housing reconstruction programme 2000-2001.* Evaluation Report, Programming, Coordination and Evaluation Division Evaluation Unit, September 2002.

EAR. (2003). *FYROM housing reconstruction programme 2001-2003.* Evaluation Report, Programming, Coordination and Evaluation Division Evaluation Unit, August 2003.

Ebersole, J. M. (1995). Mohonk criteria for humanitarian assistance in complex emergencies. *Disaster Prevention and Management, 4*(3), 14–24. doi:10.1108/09653569510088032

Edwards, K., & Mathews, L. (2009). Fires and floods: What a month for 1AOSS. *Combat Support Spring,* 12-14.

Eisenhardt, K. (1989). Building theory from case study research. *Academy of Management Review, 14*(4), 532–550.

Ellram, L. M. (1996). The use of case study method in logistics research. *Journal of Business Logistics, 17*(2), 93–138.

EM-DAT. (2008). *Emergency events database-Université Catholique de Louvain.* Retrieved January 29, 2009, from http://www.emdat.be/Database/terms.html

EM-DAT. (2010). *EM-DAT–Emergency events database.* Centre for Research on the Epidemiology of Disasters (CRED), Universite´ Catholique de Louvain, Louvain-La-Neuve. Retrieved September 2010, from www.emdat.be

English, C. R. I. (2010). *Chinese team offers aid in Haiti.* Retrieved 20th March, 2010, from http://english.cri.cn/6909/ 2010/01/15/45s542729.htm

Eriksson, P. (2000). Civil-military co-ordination in peace support operations – An impossible necessity? *The Journal of Humanitarian Assistance.*

Ernst, R. (2003). The academic side of commercial logistics and the importance of this special issue. *Forced Migration Review, 18,* 5.

Evans, J., Treadgold, A., & Mavondo, F. (2000). Psychic distance and the performance of international retailers- A suggested theoretical framework. *International Marketing Review, 17*(4/5), 373–391. doi:10.1108/02651330010339905

Expatica. (2010). *Expatica, Dutch aid ship arrives in Haiti.* Retrieved 16th March, 2010, from http://www.expatica.com/nl/ news/dutch-rss-news/dutch-aid- ship-arrives-in-haiti_20254.html

Fawcett, P., McLeish, R., & Ogden, I. (1992). *Logistics management*. London, UK: Pitman Publishing.

Fawcett, S. E., Magnan, G. M., & McCarter, M. W. (2008). Benefits, barriers, and bridges to effective supply chain management. *Supply Chain Management: An International Journal, 13*(1), 35–48. doi:10.1108/13598540810850300

Fearne, A., & Fowler, N. (2006). Efficiency versus effectiveness in construction supply chains: The dangers of lean thinking in isolation. *Supply Chain Management: An International Journal, 11*(4), 283–287. doi:10.1108/13598540610671725

Ferks, G., & Klem, B. (2006). *Conditioning peace among protagonists: A study into the use of peace conditionalities in the Sri Lankan peace process*. Netherlands Institute of International Relations, Clingendael Institute, Conflict Research Unit.

Fink, C., Matoo, A., & Neagu, I. C. (2002). Trade in international maritime services: How much does policy matter? [Oxford University Press.]. *The World Bank Economic Review, 16*(1), 81–108. doi:10.1093/wber/16.1.81

Fitzgerald, S. P. (2004). The collaborative capacity framework: from local teams to global alliances. *Advances in Interdisciplinary Studies of Work Teams, 10*, 161–201. doi:10.1016/S1572-0977(04)10007-1

Forman, S., & Parhad, R. (1997). *Paying for essentials: Resources for humanitarian assistance*. Paper prepared for meeting at Pocantico Conference Centre of the Rockefeller Brothers Fund. New York.

Försvarsmakten. (2005). *Grundsyn Log Fu*. Stockholm, Sweden: Försvarsmakten.

Foxton, P. D. (1994). *Powering war, modern land force logistics*. London, UK: Brassey's Ltd.

Franke, V. C., & Warnecke, A. (2009). Building peace: An inventory of UN peace missions since the end of the Cold War. *International Peacekeeping, 16*(3), 407–436. doi:10.1080/13533310903036467

Fuentes, G. (2010). *Bunker Hill en route to help Haiti mission*. Retrieved 20th February, 2010, from http://www.navytimes.com/news/2010/01/navy_bunkerhill_011610/

Fynes, B., Voss, C., & de Búrca, S. (2005). The impact of supply chain relationship quality on quality performance. *International Journal of Production Economics, 96*, 339–354. doi:10.1016/j.ijpe.2004.05.008

Gadde, L.-E., & Snehota, I. (2000). Making the most of supplier relationships. *Industrial Marketing Management, 29*, 305–316. doi:10.1016/S0019-8501(00)00109-7

Gaoyan, C. (2008). *Strategy assessments of China army in Sichuan earthquake relief*. Retrieved 20th June, 2008, from http://military.china.com/zh_cn /critical3/27/20080604/14886960.html

Garamone, J. (2010). *Top navy doc predicts long USNS comfort deployment*. Retrieved 20th March, 2010, from http://www.defense.gov/news /newsarticle.aspx?id=57565

Gattorna, J. (2006). *Living supply chains*. Harlow, UK: Financial Times Prentice Hall.

Ghezán, G., Mateos, M., & Vileri, L. (2002). Impact of supermarkets and fast-food chains on horticulture supply chains in Argentina. *Development Policy Review, 20*(4), 389–408. doi:10.1111/1467-7679.00179

Gill, T., Leveillee, J., & Fleck, D. (2006). *The rule of law in peace operations*. General Report of the seventeenth Congress of the International Society for Military Law and the Law of War, 16-21 May, Scheveningen, Holland.

Goddard, T. (2005). Corporate citizenship and community relations. Contributing to the challenges of aid discourse. *Business and Society Review, 110*(3), 269–296. doi:10.1111/j.0045-3609.2005.00016.x

Gooley, T. B. (1999). In time of crisis, logistics is on the job. *Logistics Management and Distribution Report, 38*, 82–86.

Gordon, S. (2001). Understanding the priorities for civil-military co-operation (CIMIC). *The Journal of Humanitarian Assistance*.

Gourlay, C. (2000). Partners apart: Managing civil-military co-operation in humanitarian interventions. *Disarmament Forum, 3*, 33–44.

Greenberg, P. S., Greenberg, R. H., & Antonucci, Y. L. (2007). Creating and sustaining trust in virtual teams. *Business Horizons, 50*(4), 525–533. doi:10.1016/j.bushor.2007.02.005

Greet, N. (2009). ADF experience on humanitarian operations: A new idea? *Security Challenges, 4*(2), 45–63.

Grey, C., & Garsten, C. (2001). Trust, control and post-bureaucracy. *Organization Studies, 22*(2), 229–250. doi:10.1177/0170840601222003

Guha-Sapis, D., & Below, R. (2002). *The quality and accuracy of disaster data: A comparative analyses of three global data sets. WHO Centre for Research on the Epidemiology of Disasters University of Louvain School of Medicine.* Brussels, Belgium: ProVention Publications.

Guillon, J. (2010). *In Haiti, the Jacmel cathedral clock stopped at 5:37 pm.* Retrieved 20th March, 2010, from http://www.mysinchew.com /node/34251

Gulati, R. (1995). Does familiarity breed trust? The implications of repeated ties for contractual choice in alliances. *Academy of Management Journal, 38*(1), 85–112. doi:10.2307/256729

Gullander, S., & Larsson, A. (2000, May). *Outsourcing and location – Comparing industrial parks in the automotive industry and contract manufacturing in the electronics industry.* Paper presented at the Conference on New Tracks on Swedish Economic Research in Europe, Mölle, Sweden.

Haas, J. E., Kates, R. W., & Bowden, M. (1977). *Reconstruction following disaster.* Cambridge, MA: MIT press.

Haghani, A., & Oh, S. C. (1996). Formulation and solution of a multi-commodity, multi-modal network flow model for disaster relief operations. *Transportation Research Part A, Policy and Practice, 30*(3), 231–250. doi:10.1016/0965-8564(95)00020-8

Handfield, R. B., & Melnyk, S. A. (1998). The scientific theory-building process: A primer using the case of TQM. *Journal of Operations Management, 16*(4), 321–339. doi:10.1016/S0272-6963(98)00017-5

Handy, C. (1995). Trust and the virtual organization. *Harvard Business Review, 73*(3), 40–50.

Harr, J. (2009, January 5). Lives of the Saints. *New Yorker (New York, N.Y.),* 47–59.

Hasan, M. A., Shankar, R., & Sarkis, J. (2007). A study of barriers to agile manufacturing. *International Journal of Agile Systems and Management, 2*(1), 1–22.

Hausman, W., Lee, H., & Subramanian, U. (2005). *Global logistics indicators, supply chain metrics, and bilateral trade patterns.* Final Draft, World Bank report.

Head, C. (2000). *Sound doctrine: A tactical primer.* New York, NY: Lantern Books.

Heaslip, G. (2010, 19 January). Civil military coordination. *Irish Times,* p. 13.

Heaslip, G., Mangan, J., & Lalwani, C. (2007a). *Humanitarian supply chains, the Irish defence forces and NGOs – A cultural collision or a meeting of minds.* CCHLI International Humanitarian Logistic Symposium, Cranfield, United Kingdom, November 2007.

Heaslip, G., Mangan, J., & Lalwani, C. (2007b). *Integrating military and non governmental organisation (NGO) objectives in the humanitarian supply chain: A proposed framework.* Logistics Research Network, Hull, United Kingdom, September 2007.

Heaslip, G., Mangan, J., & Lalwani, C. (2008a). *Strengthening partnerships in humanitarian supply chain.* Nordic Logistics Research Network (NOFOMA), Helsinki, Finland, June 2008.

Henderson, J. H. (2007). *Logistics in support of disaster relief.* Bloomington, IN: AuthorHouse.

HFN (Hastily Formed Networks) Research Group. (2006). *HFN defined.* Retrieved 28 May, 2009, from www.hfncenter.org/cms/node/117

Hines, P., Holweg, M., & Rich, N. (2004). Learning to evolve: A review of contemporary lean thinking. *International Journal of Operations & Production Management, 24*(9/10).

Hoerz, T. (1997). *The environment of refugee camps.* Retrieved January 6, 2010, from http://www.fmreview.org/HTMLcontent/rpn185.htm

Hofmann, C. A., & Hudson, L. (2009). *Military responses to natural disasters:last resort or inevitable trend?*Humanitarian Exchange Magazine.

Holmes, J. (2008). The need for collaboration. *Forced Migration Review, 31*(4). Retrieved January 8, 2010, from http://www.fmreview.org/FMRpdfs/FMR31/FMR31.pdf

Howden, M. (2009). *How humanitarian logistics Information Systems can improve humanitarian supply chains: A view from the field.* Paper presented at the Proceedings of the 6th International ISCRAM Conference, Gothenburg, Sweden.

Hung, Y.-T. C., Dennis, A. R., & Robert, L. (2004). Trust in virtual teams: Towards an integrative model of trust formation. In *Proceedings of the 37th Hawaii International Conference on Systems Sciences*, Track 1, Vol. 1.

IASC. (2006). *Logistics cluster - About the logistics cluster.* Retrieved 12 February, 2010, from http://www.logcluster.org/about/logistics-cluster/

IAWG. (2009). Retrieved from http://www.iawg.gov/

ICESAR. (2010). *The Icelandic urban SAR team has landed at Haiti.* Retrieved 20th March, 2010, from http://www.icesar.com/

ICRC. (2010). *Haiti earthquake: Reaching victims outside the capital.* Retrieved 20th March, 2010, from http://www.icrc.org/web/eng/ siteeng0.nsf/html/haiti- earthquake-update-190110

India PRWire. (2010). *Statement from Digicel on Haiti earthquake.* Retrieved on 10th March, 2010, from http://www.indiaprwire.com/ pressrelease/telecommunications /2010011441347.htm

Jackson, J. M. E. (2009). Desalination technology increases naval capabilities, meets humanitarian needs. *The Military Engineer, 101*(662), 36.

Jackson, H. C. (2009). *Number of hungry Americans increases.* USDA: Food Manufacturing. http://www.food-manufacturing.com /scripts/Products-USDA-Number -Of-Hungry-Americans.asp

Jahre, M., & Spens, K. (2007). Buy global or go local – That's the question. In P. Tatham, (Ed.), *Proceedings of the International Humanitarian Logistics Symposium*, Faringdon, UK.

James, A. (1997). Humanitarian aid operations and peace-keeping . In Belgrad, E. A., & Nachmias, N. (Eds.), *The politics of international humanitarian aid operations.* Westport, CT: Praeger.

Järvenpää, S. L., Knoll, K., & Leidner, D. E. (1998). Is anybody out there? Antecedents of trust in global virtual teams. *Journal of Management Information Systems, 14*(4), 29–64.

Järvenpää, S. L., & Leidner, D. E. (1999). Communication and trust in virtual teams. *Organization Science, 103*, 791–815. doi:10.1287/orsc.10.6.791

Jelinek, P., & Burns, R. (2010). 10,000 troops on scene by Monday. *Navy Times.* Retrieved 30th March, 2010, from http://www.navytimes.com/news/2010/01/ap_military_haiti _update_011510/

Jennings, E., Beresford, A. K. C., & Banomyong, R. (2000). *Emergency relief supply chain: A disaster response model.* Department of Maritime Studies and International Transport. (Cardiff University Occasional Paper No. 64).

Jennings, E., Beresford, A. K. C., & Pettit, S. J. (2002). Emergency relief logistics: A disaster response model. In *Proceedings of the Logistics Research Network Conference* (pp. 121–128).

Johanson, J., & Vahlne, J.-E. (1977). The internationalization process of the firm - A model of knowledge development and increasing foreign market commitments. *Journal of International Business Studies, 8*(1), 23–32. doi:10.1057/palgrave.jibs.8490676

Johanson, J., & Vahlne, J.-E. (1990). The mechanism of internationalization. *International Marketing Review, 7*(4), 11–24. doi:10.1108/02651339010137414

Johanson, J., & Vahlne, J.-E. (2003). Business relationship learning and commitment in the internationalization process. *Journal of International Entrepreneurship, 1*(1), 83–101. doi:10.1023/A:1023219207042

Johanson, J., & Vahlne, J.-E. (2009). The Uppsala internationalization process model revisited: From liability of foreignness to liability of outsidership. *Journal of International Business Studies, 40*(9), 1411–1431. doi:10.1057/jibs.2009.24

Jones, G. R., & George, J. M. (1998). The experience and evolution of trust: Implications for cooperation and teamwork. *Academy of Management Review, 23*(3), 531–546.

Jones, B., & Stoddard, A. (2003). *External review of the inter-agency standing committee.* New York, NY: Centre on International Cooperation, December 2003.

JTA. (2010). *Israeli medical, rescue workers help Haitians.* Retrieved 30th March, 2010, from http://www.jta.org/news/article /2010/01/17/1010200/israeli- medical-rescue-workers-help-haitians

Justinger, L. (2009). USACE deploys response teams following Pacific tsunami. *The Military Engineer, 101*(662), 22–23.

Kaldor, M. (1999). *New and old wars, organized violence in a global era.* Cambridge, UK: Polity Press.

Kaldor, M. (2003). Civil society and accountability. *Journal of Human Development, 4*(1), 5–27. doi:10.1080/1464988032000051469

Kamphuis, B. (2005). Economic policy for building peace . In Junne, G., & Verkoren, W. (Eds.), *Postconflict development-Meeting new challenges* (pp. 185–210). Boulder, CO: Lynne Rienner Publishers Inc.

Kane, T. M. (2001). *Military logistics and strategic performance.* London, UK: Frank Cass Publishers.

Karamperidis, S., Mangan, J., & Jackson, E. (2010). Developing an index of maritime costs and connectivity for the UK. In the *Proceedings of LRN 2010*, (p. 354).

Keating, C., Weschler, J., Ward, C. A., Claude, A., & Fernandes, F. R. (2006). *Twenty days in August: The security council sets massive new challenges for UN peacekeeping.* New York, NY: United Nations Security Council.

Kent, R. C. (1987). *Anatomy of disaster relief: The international network in action.* London, UK: Pinter.

Keyser, J., & Smith, W. (2009). *Disaster relief management in Cuba, why Cuba's disaster relief model is worth careful study.* International policy report, Center for International Policy, May.

King, L. (2010). *Hampton roads, The Carl Vinson departs Haiti.* Retrieved 30th March, 2010, from http://hamptonroads.com/2010/02 /carl-vinson-departs-haiti

Kirby, J. (2003). Supply chain challenges: Building relationships. *Harvard Business Review, 81*(7), 64–73.

Kleindorfer, P. R., & Saad, G. H. (2005). Managing disruption risks in supply chains. *Production and Operations Management, 14*(1), 53–68. doi:10.1111/j.1937-5956.2005.tb00009.x

Knights, D., Noble, F., Vurdubakis, T., & Willmott, H. (2001). Chasing shadows: Control, virtuality and the production of trust. *Organization Studies, 22*(2), 311–336. doi:10.1177/0170840601222006

Korac, M. (2006). Gender, conflict and peace-building: Lessons from the conflict in the former Yugoslavia. *Women's Studies International Forum, 29*, 510–520. doi:10.1016/j.wsif.2006.07.008

Kovács, G., & Tatham, P. (2009). Humanitarian logistics performance in the light of gender. *International Journal of Productivity and Performance Management, 58*(2), 174–187. doi:10.1108/17410400910928752

Kovács, G., & Tatham, P. (2009). Responding to disruption in the supply network – From dormant to action. *Journal of Business Logistics, 30*(2), 215–228. doi:10.1002/j.2158-1592.2009.tb00121.x

Kovács, G., & Spens, K. M. (2010). (forthcoming). Knowledge sharing in relief supply chains. *International Journal of Networking and Virtual Organisations.* doi:10.1504/IJNVO.2010.031219

Kovács, G., & Spens, K. (2007). Humanitarian logistics in disaster relief operations. *International Journal of Physical Distribution and Logistics Management, 37*(2), 99–114. doi:10.1108/09600030710734820

Kovács, G., & Spens, K. (2009). Identifying challenges in humanitarian logistics. *International Journal of Physical Distribution and Logistics Management, 39*(6), 506–528. doi:10.1108/09600030910985848

Kovács, G., & Spens, K. (2008). Humanitarian logistics revisited . In Arlbjørn, J. S., Halldórsson, A., Jahre, M., & Spens, K. (Eds.), *Northern lights in logistics and supply chain management* (pp. 217–232). Copenhagen, Denmark: CBS Press.

Kramer, R. M. (1999). Trust and distrust in organizations: Emerging perspectives, enduring questions. *Annual Review of Psychology, 50*, 569–598. doi:10.1146/annurev.psych.50.1.569

Ku, B. (2008). *Relief operations in Wenchuan.* Retrieved 10th June, 2008, from http://blog.sina.com.cn/s/blog_4bb4c26301009e9h.html

Laaksonen, T., Jarimo, T., & Kulmala, H. I. (2009). Cooperative strategies in customer-supplier relationships: The role of interfirm trust. *International Journal of Production Economics, 120*(1), 79–87. doi:10.1016/j.ijpe.2008.07.029

Lambert, D. M., & Knemeyer, M. A. (2004). We're in this together. *Harvard Business Review, 82*(12), 2–9.

Lambert, D., Emmelhanz, M., & Gardner, J. (1996). So you think you want a partner? *Marketing Management, 5*(2), 25–41.

Lambert, D., & Knemeyer, M. (2004). We're in this together. *Harvard Business Review, 82*(12), 114–122.

Lambert, D. M., Cooper, M. C., & Pagh, J. D. (1998). Supply chain management: Implementation issues and research opportunities. *The International Journal of Logistics Management, 9*(2), 1–19. doi:10.1108/09574099810805807

Larson, P. D., & Halldorsson, A. (2004). Logistics versus supply chain management: An international survey. *International Journal of Logistics: Research and Applications, 7*(1), 17–31. doi:10.1080/13675560310001619240

Laurence, T. (1999). *Humanitarian assistance and peacekeeping: An uneasy alliance.* London, UK: The Royal United Services Institute for Defence Studies.

Lee, H. (2005). Triple A supply chain. *Harvard Business Review, 82*(10), 102–112.

Lett, D. (2010). *Canada earns its wings,* Retrieved 30th March, 2010, from http://www.winnipegfreepress.com/opinion/columnists/Canada-earns-its- wings-83147562.html

Limao, N., & Venable, A. J. (2001). Infrastucture, geographical disadvantages, transport costs and trade. *The World Bank Economic Review, 15*, 451–479. doi:10.1093/wber/15.3.451

Long, D. (1997). Logistics for disaster relief: Engineering on the run. *IIE Solutions, 29*(6), 26–29.

Long, D. C., & Wood, D. F. (1995). The logistics of famine relief. *Journal of Business Logistics, 16*(1), 213–229.

Lund, M. (2003). *What kind of peace is being built? Assessing the record of post-conflict peacebuilding, charting future directions.* Ottawa, Canada: International Development Research Centre.

Mackey, R. (2010). *Latest updates on rescue and recovery in Haiti.* Retrieved 30th March, 2010, from http://thelede.blogs.nytimes.com/ 2010/01/15/latest-updates-on- rescue-and-recovery-in-haiti/?hp

Mackinlay, J. (Ed.). (1996). *A guide to peace support operations.* Providence, RI: The Thomas J. Watson Jr. Institute, Brown University.

Macrae, J., & Leader, N. (2000). The politics of coherence: The UK government's approach to linking political and humanitarian responses to complex political emergencies. *HPG Research in Focus, 1*, 1–3.

Macrae, J. (2002). Analysis and synthesis . In Macrae, J. (Ed.), *The new humanitarianisms: A review of trends in global humanitarian action. Report to the Humanitarian Policy Group.* London, UK: Overseas Development Institute.

Majchrzak, A., Järvenpää, S. L., & Hollingshead, A. B. (2007). Coordinating Expertise Among Emergent Groups Responding to Disasters. *Organization Science, 18*(1), 147–161. doi:10.1287/orsc.1060.0228

Maon, F., Lindgreen, A., & Vanhamme, J. (2009). Developing supply chains in disaster relief operations through cross-sector socially oriented collaborations. *Supply Chain Management: An International Journal, 14*(2), 149–164. doi:10.1108/13598540910942019

Markowski, S., Hall, P., & Wylie, R. (Eds.). (2010). *Defence procurement and industry policy: A small country perspective.* Abingdon, UK: Routledge.

Maslow, A. H., & Stephens, D. C. (Eds.). (2000). *The Maslow business reader.* New York, NY: John Wiley & Sons Inc.

Maspero, E. L., & Ittmann, H. W. (2008). The rise of humanitarian logistics. *Proceedings of the 27th Southern African Transport Conference*, 7-11 July, Pretoria, South Africa. Retrieved December 8, 2009, from http://repository.up.ac.za/upspace/bitstream/2263/6251/1/Ittmann%2027.pdf

McClintock, A. (1997). Global cases in logistics and supply chain management . In Taylor, D. H. (Ed.), *The logistics of third-world relief operations* (pp. 354–369). London, UK: International Thomson Business Press.

McCutcheon, S. D., Handfield, R., McLachlin, R., & Samson, D. (2002). Effective case research in operations management: A process perspective. *Journal of Operations Management, 20*, 419–433. doi:10.1016/S0272-6963(02)00022-0

McEntire, D. A. (2002). Coordinating multi-organizational responses to disaster: Lessons from the March 28, 2000 Fort Worth tornado. *Disaster Prevention and Management: An International Journal, 11*(5), 369–379. doi:10.1108/09653560210453416

McHugh, G. (2009, 29 October). Ready to help Padang assist. *Navy News*. Retrieved from www.defence.gov.au/news/navynews

McKnight, D. H., Cummings, L. L., & Chervany, N. L. (1998). Initial trust formation in new organizational relationships. *Academy of Management Review, 23*(3), 473–490.

McLachlin, R., Larson, P. D., & Khan, S. (2009). Not-for-profit supply chains in interrupted environments: The case of a faith-based humanitarian relief organization. *Management Research News, 32*(11), 1050–1064. doi:10.1108/01409170910998282

Mentzer, J. T., DeWitt, W., Keebler, J. S., Min, S., Nix, N. W., Smith, C. D., & Zacharia, Z. G. (2001). Defining supply chain management. *Journal of Business Logistics, 22*(2), 1–25. doi:10.1002/j.2158-1592.2001.tb00001.x

Meredith, J. (1998). Building operations management theory through case and field research. *Journal of Operations Management, 16*, 441–454. doi:10.1016/S0272-6963(98)00023-0

Meyerson, D., Weick, K. E., & Kramer, R. M. (1996). Swift trust and temporary groups . In Kramer, R. M., & Tyler, T. R. (Eds.), *Trust in organizations: Frontiers of theory and research* (pp. 166–195). Thousand Oaks, CA: Sage Publications Inc.

Moody, F. (2001). Emergency relief logistics: A faster way across the global divide. *Logistics Quarterly, 7*(2). Retrieved on March 9, 2007, from http://www.lq.ca/issues/summer2001/articles/article07.html.

Mooney, C. Z. (1997). *Monte Carlo simulation*. Sage University Paper series on Quantitative Applications in the Social Sciences, 07-116, Thousand Oaks, CA: Sage.

Moore, D. M., & Antill, R. D. (2002). Opportunities and challenges in logistics for humanitarian aid operations: A role for UK Armed Forces? *Proceedings of the Logistics Research Network Conference*, ILT, Plymouth, September.

Moxham, C., & Boaden, R. (2007). The impact of performance measurement in the voluntary sector. *International Journal of Operations & Production Management, 27*(8), 826–845. doi:10.1108/01443570710763796

Moyo, D. (2009). *Dead aid*. London, UK: Allen Lane.

MSNBC. (2010). *Haiti aid bottleneck is easing up*. Retrieved 30th March, 2010, from http://www.msnbc.msn.com/id /34915151/ns/world _news-haiti_earthquake/

Munslow, B., & O'Dempsey, T. (2009). Loosing soft power in hard places: humanitarianism after the US invasion of Iraq. *Progress in Development Studies, 9*(1), 3–13. doi:10.1177/146499340800900102

Murray, V. (2006). Introduction: What's so special about managing nonprofit organizations? In Murray, V. (Ed.), *The management of nonprofit and charitable organizations in Canada*. Markham, Ontario: LexisNexis-Butterworths.

NATO. (2004). *AJP-9 NATO civil-military co-operation (CIMIC) doctrine*. Retrieved 10 April, 2007, from http://www.nato.int/ims/docu/AJP-9.pdf

NATO. (2007). *NATO logistics handbook*. Retrieved July 15, 2009, from www.nato-otan.org/docu/logi-en/logistics_hndbk_2007-en.pdf

Natsios, A. S. (1995a). The international humanitarian response system. *Parameters, 25*, 68–81.

Natsios, A. S. (1995b). NGOs and the UN System in complex humanitarian emergencies: Conflict or co-operation? *Third World Quarterly, 16*(3), 405–419. doi:10.1080/01436599550035979

Navy, U. S. (2010a). *Maritime force serves as cornerstone of relief operations in Haiti*. Retrieved 20th March, 2010, from http://www.navy.mil/search/ display.asp?story_id=50696

Navy, U. S. (2010b). *Vinson helicopters perform medical evacuations*: *Sea Base on the way*. Retrieved 20th March, 2010, from http://www.navy.mil/search/ display.asp?story_id=50582

Navy, U. S. (2010c). *USS Normandy arrives off coast of Port-Au-Prince*. Retrieved 20th March, 2010, from http://www.navy.mil/search/ display.asp?story_id=50593

News, J. D. (2010). *22nd MEU departs for Haiti*. Retrieved 20th March, 2010, from http://www.jdnews.com/news/ uss-71828-equipment-leave.html

Nordin, F. (2008). Linkages between service sourcing decisions and competitive advantage: A review, propositions, and illustrating cases. *International Journal of Production Economics, 114*(1), 40–55. doi:10.1016/j.ijpe.2007.09.007

NRCNA. (National Research Council of the National Academies. (2006). *Facing hazards and disasters: Understanding human dimensions*. Committee on Disaster Research in the Social Sciences: Future Challenges and Opportunities, Division of Earth and Life Studies. Washington DC: The National Academies Press.

OCHA. (2003). *Guidelines on the use of military and civil defence assets to support United Nations humanitarian activities in complex emergencies*. Geneva, Switzerland: OCHA.

OCHA. (2009). *Reference guide of OCHA's strategic framework 2010-2013*. New York, NY: United Nations.

OCHA. (2010). *Annual plan and budget*. Geneva, Switzerland: United Nations.

OCHA. (2007). *Environment*. Retrieved January 10, 2010, from *http://74.125.93.132/search?q=cache%3AhLpcfds Uq1AJ%3Aochaonline.un.org%2FOchaLinkClick.aspx% 3Flink%3Docha%26docId%3D1091518+refugee+camp +environment&hl=en&gl=us*

OCHA. (2010a). Haiti earthquake situation reports #1 & 2. Office of the Coordination for Humanitarian Affairs, New York, January 12-13, 2010. Retrieved from http:// www.reliefweb.int

OCHA. (2010b). Haiti earthquake situation reports #3 & 4. Office of the Coordination for Humanitarian Affairs, New York, January 12-13, 2010. Retrieved from http:// www.reliefweb.int

OECD. (2005). *Paris declaration on aid effectiveness*. Retrieved on September 12, 2007, from www.oecd.org

OED. (2000). *Concise Oxford English dictionary*. Oxford, UK: Oxford University Press.

O'Grady, S., & Lane, H. W. (1996). The psychic distance paradox. *Journal of International Business Studies, 27*(2), 309–333. doi:10.1057/palgrave.jibs.8490137

Oloruntoba, R., & Gray, R. (2009). Customer service in emergency relief chains. *International Journal of Physical Distribution and Logistics Management, 39*(6), 486–505. doi:10.1108/09600030910985839

Oloruntoba, R., & Gray, R. (2006). Humanitarian aid: An agile supply chain? *Supply Chain Management: An International Journal, 11*(2), 115–120. doi:10.1108/13598540610652492

Oloruntoba, R. (2005). A wave of destruction and the waves of relief: Issues, challenges and strategies. *Disaster Prevention and Management: An International Journal, 14*(4), 506–521. doi:10.1108/09653560510618348

Olson, L., & Gregorian, H. (2007). Interagency and civil-military coordination: Lessons from a survey of Afghanistan and Liberia. *Journal of Military and Strategic Studies, 10*(1). Retrieved on June 26, 2008, from http://www.jmss.org/2007/2007fall/articles/olson-gregorian.pdf

Ozdamar, L., Ekinci, E., & Kucukyazici, B. (2004). Emergency logistics planning in natural disasters. *Annals of Operations Research, 129*, 217–245. doi:10.1023/B:ANOR.0000030690.27939.39

Padgett, T. (2010). *With the military in Haiti: Breaking the supply logjam*. Retrieved 20th March, 2010, from http://www.time.com/time/specials/ packages/ article/0,28804,1953379 _1953494,00.html

Pagonis, W. G. (1992). *Moving mountains, lessons in leadership and logistics from the Gulf War*. Boston, MA: Harvard Business School Press.

Pan American Health Organization. (2001). *Humanitarian supply management and logistics in the health sector*. Washington, DC: World Health Organization. Emergency Preparedness and Disaster Relief Program, Department of Emergency and Humanitarian Action, Sustainable Development and Healthy Environments.

Pan American Health Organization. (2000). *Manual logistical management of humanitarian supply*. Washington, DC: PAHO.

Pande, R. K., & Pande, R. (2007). Resettlement and rehabilitation issues in Uttaranchal (India) with reference to natural disasters. *Disaster Prevention and Management, 16*(3), 361–369. doi:10.1108/09653560710758314

Pardasani, M. (2006). Tsunami reconstruction and redevelopment in the Maldives. A case study of community participation and social action. *Disaster Prevention and Management, 15*(1), 79–91. doi:10.1108/09653560610654257

Peck, H. (2006). Reconciling supply chain vulnerability with risk and supply chain management. *International Journal of Logistics Research and Applications, 9*(2), 127–142.

Penrose, E. (1959). *The theory of the growth of the firm*. London, UK: Basil Blackwell.

Perry, M. (2007). Natural disaster management planning: A study of logistics managers responding to the tsunami. *International Journal of Physical Distribution and Logistics Management, 37*(5), 409–433. doi:10.1108/09600030710758455

Pettit, S. J., & Beresford, A. K. C. (2005). Emergency relief logistics: An evaluation of military, non-military and composite response models. *International Journal of Logistics: Research and Applications, 8*(4), 313–331.

Pettit, S., & Beresford, A. K. C. (2005). Emergency relief logistics: An evaluation of military, non-military and composite response models. *International Journal of Logistics: Research and Applications, 8*(4), 313–331.

Pettit, S., & Beresford, A. K. C. (2009). Critical success factors in the context of humanitarian aid supply chains. *International Journal of Physical Distribution and Logistics Management, 39*(6), 450–468. doi:10.1108/09600030910985811

Pettit, S. J., & Beresford, A. K. C. (2006). Emergency relief logistics: An evaluation of military, non-military, and composite response models. *International Journal of Logistics: Research and Applications, 8*(4), 313–331.

Pettit, S. J., & Beresford, A. K. C. (2005). Emergency relief logistics: An evaluation of military, non military and composite response models. *International Journal of Logistics: Research and Applications, 8*(4), 313–332.

Port World. (2009). *Distance calculator*. Retrieved May 2009, from http://www.portworld.com/map

Porter, M. E. (1980). *Competitive strategy*. New York, NY: Free Press.

PRC. (2008). The relief supplies arrived in Chengdu successfully and will be delivered to the disaster areas as soon as possible. Retrieved 15th July, 2008, from http://www.gov.cn/jrzg/2008-05 /17/content_980034.htm

Pugh, M. (2001). *Civil-military relations in peace support operations: Hegemony or emancipation?* Plymouth, UK: University of Plymouth.

Red Cross. (2010). *Haiti earthquake appeal*. Retrieved 30th March, 2010, from http://www.redcross.org.uk/donatesection.asp?id=102168

Régnier, P., Neri, B., Scuteri, S., & Miniati, S. (2008). From emergency relief to livelihood recovery: Lessons learned from post-tsunami experiences in Indonesia and India. *Disaster Prevention and Management, 17*(3), 410–429. doi:10.1108/09653560810887329

Rhoads, C. (2010). *Earthquake sets back Haiti's efforts to improve telecommunications*. Retrieved 30th March, 2010, from http://online.wsj.com/article/SB10001424052748703657604575005453223257096.html

Richey, R. G. Jr. (2009). The supply chain crisis and disaster pyramid: A theoretical framework for understanding preparedness and recovery. *International Journal of Physical Distribution & Logistics Management, 39*(7), 619–628. doi:10.1108/09600030910996288

Rietjens, S. J. H., Voordijk, H., & De Boer, S. J. (2007). Co-ordinating humanitarian operations in peace support missions. *Disaster Prevention and Management, 16*(1), 56–69. doi:10.1108/09653560710729811

Rietjens, S. J. H. (2006). *Civil-military cooperation in response to a complex emergency: Just another drill?* Doctoral Dissertation, University of Twente, Enschede, the Netherlands.

Roberts, A. (2001). *NGOs: New Gods overseas: The World in 2001* (p. 73) London, UK: The Economist Publications.

Rodman, W. K. (2004). *Supply chain management in humanitarian relief logistics.* MSc Thesis, Air Force Institute of Technology, Wright-Patterson Air Force Base, Ohio, US.

Rousseau, D. M., Sitkin, S. B., Burt, R. S., & Camerer, C. (1998). Not so different after all: A cross discipline view of trust. *Academy of Management Review, 23*(3), 393–404. doi:10.5465/AMR.1998.926617

Russell, T. E. (2005). *The humanitarian relief supply chain: Analysis of the 2004 South East Asia earthquake and tsunami.* Masters thesis at MIT, Massachusetts. Retrieved December 8, 2009, from http://dspace.mit.edu/bitstream/handle/1721.1/33352/62412847.pdf?sequence=1

Saad, M., Jones, M., & James, P. (2002). A review of the progress towards the adoption of supply chain management (SCM) relationships in construction. *European Journal of Purchasing and Supply Management, 8*(3), 173–183. doi:10.1016/S0969-7012(02)00007-2

Samii, R. (2008). *Leveraging logistics partnerships: Lessons from humanitarian organizations.* Erasmus Research Institute of Management.

Samii, R., & Van Wassenhove, L. (2003). *The United Nations Joint Logistics Centre (UNJLC): The genesis of a humanitarian relief coordination platform.* INSEAD case study. Retrieved from www.ecch.com/humanitariancases

Sarkis, J., Talluri, S., & Gunasekaran, A. (2007). A strategic model for agile virtual enterprise partner selection. *International Journal of Operations & Production Management, 27*(11), 1213–1234. doi:10.1108/01443570710830601

Sarkis, J. (2009). *A boundaries and flows perspective of green supply chain management.* GPMI Working Paper 2009-07. George Perkins Marsh Institute, Worcester, MA. Retrieved January 4, 2010, from http://www.clarku.edu/departments/marsh/news/WP2009-07.pdf

Scan Global Logistics. (2009). *Distance calculator.* Retrieved May 2009, from www.scangl.com

Schary, P. B., & Skjøtt-Larsen, T. (2004). *Managing the global supply chain* (3rd ed.). Copenhagen, Denmark: Copenhagen Business School Press.

Schein, E. (1996). Three cultures of management: The key to organizational learning. *Sloan Management Review,* (Fall): 9–20.

Scheper, E., Parakrama, A., & Patel, S. (2006). Impact of tsunami response on local and national capacities. London, UK: Tsunami Evaluation Coalition (TEC). Retrieved September 4, 2008, from http://www.tsunami-evaluation.org/NR/rdonlyres/8E8FF268-51F0-4367-A797-F031C0B51D21/0/capacities_final_report.pdf

Schulz, S. (2009). *Disaster relief logistics. Benefits of and impediments to cooperation between humanitarian organizations.* Haupt Berne.

Schulz, S. F., & Heigh, I. (2009). Logistics performance management in action within a humanitarian organization. *Management Research News, 32*(11), 1038–1049. doi:10.1108/01409170910998273

Seiple, C. (1996). *The US military/NGO relationship in humanitarian interventions.* Carlisle Barracks, PA: Peacekeeping Institute Centre for Strategic Leadership, U.S. Army War College.

Shaluf, I. M. (2007). An overview on disasters. *Disaster Prevention and Management, 16*(5), 687–703. doi:10.1108/09653560710837000

Sharma, K. (2010). *Socio-economic aspects of disaster's impact: An assessment of databases and methodologies.* Economic Growth Centre, Division of Economics School of Humanities and Social Sciences Nanyang Technological University, Working Paper, No 01.

Sheffi, Y. (2005). *The resilient enterprise. Overcoming vulnerability for competitive advantage.* Cambridge, MA: The MIT Press.

Siegel, A. (2003). Why the military think that aid workers are over-paid and under-stretched. *Humanitarian Affairs Review*, *1*, 52–55.

Siegler, M. (2010). *Twitter strikes deal to bring free SMS tweets to Haiti.* Retrieved 20th March, 2010, from http://www.washingtonpost.com /wp-dyn/content/article/2010/02/ 23/AR2010022 300234.html

Simmins, C. (2010). *Two months after the Haitian earthquake.* Retrieved 20th March, 2010, from, http:// northshorejournal.org/ two-months-after-the-haitian-earthquake

Singer, P. W. (2008). *Corporate warriors* (updated ed.). Ithaca, NY: Cornell University Press.

Skjøtt-Larsen, T. (1999). Supply chain management: A new challenge for researchers and managers in logistics. *The International Journal of Logistics Management*, *10*(2), 41–54. doi:10.1108/09574099910805987

Skjøtt-Larsen, T., Thernøe, C., & Andresen, C. (2003). Supply chain collaboration. Theoretical perspectives and empirical evidence. *International Journal of Physical Distribution and Logistics Management*, *33*(6), 531–549. doi:10.1108/09600030310492788

Slevin, P. (2010). *Quake-damaged main port in Port-au-Prince, Haiti, worse off than realized.* Retrieved 30th March, 2010, from http://www.washingtonpost.com /wp-dyn/content/article/2010/01/27/ AR2010012705250.html

Slim, H. (2006). Humanitarianism with borders? NGOs, belligerent military forces and humanitarian action. *The Journal of Humanitarian Assistance*.

Smith, W., & Dowell, J. (2000). A case study of co-ordinative decision-making in disaster management. *Ergonomics*, *43*(8), 1153–1166. doi:10.1080/00140130050084923

Sohunet. (2008). *Premier Wen answered the questions from pressmen in Yingxiu County.* Retrieved 15th July, 2008, from http://news.sohu.com/ 20080903/n259341067.shtml

Sowinski, L. L. (2003). The lean, mean supply chain and its human counterpart. *World Trade*, *16*(6), 18.

Sphere Project. (2004). *The humanitarian charter and minimum standards in disaster response.* Oxford, UK: Oxfam Publishing.

Sphere. (2004). *Humanitarian charter and minimum standards in disaster response.* The Sphere Project. Retrieved September 4, 2008, from http://www.sphereproject.org/ handbook/pages/navbook.htm?param1=0

Spring, S. (2006, September 11). Relief when you need it: Can FedEx, DHL and TNT bring the delivery of emergency aid into the 21st century? *Newsweek International Edition*.

Srinivasan, K. (2009). International conflict and cooperation in the 21ˢᵗ century. *The Round Table*, *98*(400), 37–47. doi:10.1080/00358530802601660

Steele, J. (2010). *Navy destroyer to return after helping out in Haiti.* Retrieved 30th March, 2010, from http://www.signonsandiego.com/ news/2010/feb/03/navy-destroyer-return-after-helping-out-haiti/

Steinle, C., & Schiele, H. (2008). Limits to global sourcing? Strategic consequences of dependency on international suppliers: Cluster theory, resource-based view and case studies. *Journal of Purchasing and Supply Management*, *14*(1), 3–14. doi:10.1016/j.pursup.2008.01.001

Stephenson, M., Jr. (2004). *Making humanitarian relief networks more effective: Exploring the relationships among coordination, trust and sense making.* Paper prepared for Delivery at the National Conference of the Association for Research on Non-Profit Organizations and Voluntary Action (ARNOVA). Los Angeles, California.

Stephenson, R. S. (1993). *Logistics.* Geneva: United Nations Disaster Management Programme, United Nations Development Program (UNDP).

Stocking, D. B. (2010). *Natural disasters: How can we improve?* Seminar at the Royal Geographic Society. Retrieved 25 May, 2010, from http://www.21stcenturychallenges.org/challenges/25-may-natural-disasters-how-can-we-improve/

Stöttinger, B., & Schlegelmilch, B. B. (1998). Explaining export development through psychic distance: Enlightening or elusive. *International Marketing Review*, *15*(5), 357–372. doi:10.1108/02651339810236353

Studer, M. (2001). The ICRC and civil-military relations in armed conflict. *International Review of the Red Cross*, *83*(842), 367–391.

Supply-Chain Council. (2008). *Supply-Chain Operations Reference Model v9.0.* Supply-Chain Council, Pittsburgh, USA. www.supply-chain.org.

Swift, J. S. (1999). Cultural closeness as a facet of cultural affinity: A contribution to the theory of psychic distance. *International Marketing Review, 16*(3), 182–201. doi:10.1108/02651339910274684

Tatham, P., & Kovács, G. (2010). The application of swift trust to humanitarian logistics. *International Journal of Production Economics, 126*(1), 35–45. doi:10.1016/j.ijpe.2009.10.006

Tatham, P. H., Kovács, G., & Larson, P. D. (2010). What skills and attributes are needed by humanitarian logisticians - A perspective drawn from international disaster relief agencies. *Proceedings of the 21st Production and Operations Management Society (POMS) Annual Conference,* Vancouver, May 7-10, 2010.

Tatham, P., & Kovács, G. (2007). The humanitarian supply network in rapid onset disasters. In A. Halldorsson & G. Stefansson (Eds.), *Proceedings of the 19th Annual Conference for Nordic Researchers in Logistics,* (pp. 1059-1074). NOFOMA 2007, Reykjavik, Iceland.

Taylor, D. H. (2005). Value chain analysis: An approach to supply chain improvement in agri-food chains. *The International Journal of Physical Distribution & Logistics Management, 35*(10), 744–761. doi:10.1108/09600030510634599

Taylor, D. H. (2009). An application of value stream management to the improvement of a global supply chain: A case study in the footwear industry. *International Journal of Logistics Research and Applications, 12*(1). doi:10.1080/13675560802141812

Taylor, D., & Pettit, S. (2009). A consideration of the relevance of lean supply chain concepts for humanitarian aid provision. *International Journal of Services Technology and Management, 12*(4), 430–444. doi:10.1504/IJSTM.2009.025817

Teague, P. (2008, September 11). P&G is king of collaboration. *Purchasing,* 46.

The Lancet. (2010). Editorial. Growth of aid and the decline of humanitarianism. *Lancet, 375*(9711), 253. doi:10.1016/S0140-6736(10)60110-9

Thomas, A., & Mizushima, M. (2005). Logistics training: Necessity or luxury? *Forced Migration Review, 22,* 60–61.

Thomas, A. (2003). Fritz Institute: Leveraging private expertise for humanitarian supply chains. *Forced Migration Review, 21,* 64–65.

Thomas, A. (2003). Why logistics? *Forced Migration Review, 18,* 4.

Thomas, A., & Kopzack, L. (2005). *From logistics to supply chain management. The path forward in the humanitarian sector.* Fritz Institute. Retrieved on September 15, 2009, from http://www.fritzinstitute.org/ PDFs/ WhitePaper/ FromLogisticsto.pdf

Times-Picayune. (2010). *New Orleans to Haiti barge initiative seeks donations of cash, goods.* Retrieved 30th March, 2010, from http://www.nola.com/news/ index.ssf/2010/02/new_orleans _to_haiti_ barge_ini.html

Tomasini, R., & Van Wassenhove, L. (2009). *Humanitarian logistics.* Palgrave MacMillan. doi:10.1057/9780230233485

Tomasini, R. M., & Van Wassenhove, L. N. (2004a). Pan-American health organisation's humanitarian supply management system: De-politicization of the humanitarian supply chain by creating accountability. *Journal of Public Procurement, 4*(3), 437–449.

Tomasini, R. M., & Van Wassenhove, L. N. (2004b). The TPG-WFP partnership: Looking for a partner. (INSEAD case study 06/2004-5187).

Tomasini, R., & Van Wassenhove, L. (2007). *UNJLC moving the world: Transport optimization for South Sudan.* INSEAD Case Study. Retrieved from www.ecch.com/ humanitariancases

Trent, R. J., & Monczka, R. M. (2005). Achieving excellence in global sourcing. *MIT Sloan Management Review, 47*(1), 24–32.

Trenton, D., Clark, L., & Rosenberg, C. (2010). *New airfield, more troops to increase delivery of aid, security.* Retrieved 20th March, 2010, from http://www.miamiherald.com/ 2010/01/19/1433097/us- pledges -aid-security-will-improve.html

Tripp, R. S., Amouzegar, M. A., McGarvey, R. G., Bereit, R., & George, D. (2006). *Sense and respond logistics: Integrating prediction, responsiveness, and control capabilities*. RAND Corp.

Tuttle, W. G. T. Jr. (2005). *Defense logistics for the 21st century*. Annapolis, MD: Naval Institute Press.

Tysseland, B. E. (2008). Life cycle cost based procurement decisions: A case study of Norwegian defence procurement projects. *International Journal of Project Management, 26*(4), 366–375. doi:10.1016/j.ijproman.2007.09.005

Uhr, C., & Ekman, O. (2008). Trust among decision makers and its consequences to emergency response operations. *Journal of Emergency Management, 6*(3), 21–37.

UN. (2010). *Population statistics*. Retrieved October, 2010, from http://unstats.un.org/unsd/demographic/default.htm

UNHCR. (1997). *The state of the world's refugees, a humanitarian agenda*. Oxford, UK: Oxford University Press.

UNHCR. (2009). *Ethiopia/Green refugee camp*. Retrieved January 10, 2010, from http://www.unmultimedia.org/tv/unifeed/d/14105.html

UNHCR. (2010). Retrieved from http://www.unhcr.org/pages/49f6d3d26.html

UNISDR. (2010). *Disaster risk reduction and prevention in Rwanda*. Retrieved October, 2010, from http://www.unisdr.org/eng/country-inform/reports/Rwanda-report.pdf

United Nations. (1994). *Guidelines on the use of military and civil defence assets in disaster relief*. Geneva, Switzerland: United Nations.

United Nations ISDR. (2004). *Terminology of disaster risk reduction*. Retrieved 25th July, 2010, from http://www.unisdr.org/ eng/library/lib-terminology -eng%20 home.htm

United Nations. (2005). *United Nations humanitarian CMCoord concept endorsed by the IASC*. Rome, Italy: Inter-Agency Standing Committee (IASC).

UNJLC (United Nations Joint Logistics Centre). (2008). *Logistics cluster Myanmar: Cyclone Nargis emergency response* 10th May – 10th August 2008. End of Mission Report. Retrieved October 28, 2008, from http://www.logcluster.org/logistics-cluster/meeting/global-logistics-cluster-meeting-3-4-octobre-2008

UNJLC. (2008). *UNJLC training material*. Copenhagen.

UNOCHA. (2006). *Guidelines on the use of military and civil defence assets to support United Nations humanitarian activities in complex emergencies*. Brussels United Nations Office for the Coordination of Humanitarian Affairs.

USAID. (2002). *Foreign aid in the national interest: Promoting freedom, security and opportunity* (pp. 1–149). Washington, DC: U.S. Agency for International Development.

Uzzi, B. (1997). Social structure and competition in interfirm networks: The paradox of embeddedness. *Administrative Science Quarterly, 42*(1), 35–67. doi:10.2307/2393808

van der Laan, E. A., de Brito, M. P., & Vergunst, D. A. (2009). Performance measurement in humanitarian supply chains. *International Journal of Risk Assessment and Management, 13*(1), 22–45. doi:10.1504/IJRAM.2009.026388

Van der Laan, E., de Brito, M. P., & Vermaesen, S. (2007). Logistics information and knowledge management issues in humanitarian aid organizations. *Proceedings of the SIMPOI/POMS conference*, Brazil, August 8-10.

Van der Vorst, J. G. A. J., & Beulens, A. J. M. (2002). Identifying sources of uncertainty to generate supply chain redesign strategies. *International Journal of Physical Distribution & Logistics Management, 32*(6), 409–430. doi:10.1108/09600030210437951

van Wassenhove, L. N. (2006). Humanitarian aid logistics: Supply chain management in high gear. *The Journal of the Operational Research Society, 57*(5), 475–489. doi:10.1057/palgrave.jors.2602125

Van Weele, A. J. (2005). *Purchasing & supply chain management* (4th ed.). London, UK: Thomson Learning.

Verkuil, P. R. (2007). *Outsourcing sovereignty*. Cambridge, UK: Cambridge University Press. doi:10.1017/CBO9780511509926

Völz, C. (2005). Humanitarian coordination in Indonesia: An NGO viewpoint. *Forced Migration Review*, Special Issue, July, 26-27.

Vrijhoef, R., & Koskela, L. (2000). The four roles of supply chain management in construction. *European Journal of Purchasing and Supply Management*, 6(3-4), 169–178. doi:10.1016/S0969-7012(00)00013-7

Walker, P., & Russ, C. (2010). Professionalising the humanitarian sector. *ELRHA*. Retrieved July 3, 2010, from http://www.elrha.org/professionalisation

Washington Post. (2010). *US Navy en route to make Haiti seaport usable*. Retrieved 20th March, 2010, from http://www.washingtonpost.com /wp-dyn/content/article/2010/01 /16/AR2010011601601.html

Weick, K. E. (1988). Enacted sensemaking in crisis situations. *Journal of Management Studies*, 25(4), 305–317. doi:10.1111/j.1467-6486.1988.tb00039.x

Weick, K. E. (1993). The collapse of sensemaking in organizations: The Mann Gulch disaster. *Administrative Science Quarterly*, 38(4), 628–652. doi:10.2307/2393339

Weiss, T. G. (1998). Civilian-military interactions and ongoing UN reforms: DHA's past and OCHA's remaining challenges. *International Peacekeeping*, 5(4), 49–70.

Weiss, T. G., & Campbell, K. M. (1991). Military humanitarianism. *Survival*, 33(5), 451–465. doi:10.1080/00396339108442612

Wheatley, G., & Welsch, S. D. (1999). The use and limitations of technology in civil-military interactions. In The Cornwallis Group (Eds.), *Volume IV: Analysis of civil-military interactions*. Nova Scotia: The Lester B. Person Canadian International Peacekeeping Training Centre.

Whipple, J. M., & Russell, D. (2007). Building supply chain collaboration: A typology of collaborative approaches. *International Journal of Logistics Management*, 18(2), 174–196. doi:10.1108/09574090710816922

Whiting, M. C., & Ayala-Ostrom, B. E. (2009). Advocacy to promote logistics in humanitarian aid. *Management Research News*, 32(11), 1081–1089. doi:10.1108/01409170910998309

Whiting, M. (2009). Enhanced civil military cooperation in humanitarian supply chains. In Gattorna, J. (Ed.), *Dynamic supply chain management* (pp. 107–122). Surrey, UK: Gower Publishing.

Wijkman, A., & Timberlake, L. (1988). *Natural disasters: Acts of God or acts of man?* Earthscan Publications.

Williams, P. (2010). *Emergency architects of Canada to aid in Haiti reconstruction effort*. Retrieved 20th March, 2010, from http://dcnonl.com/article/id37551

Womack, J., & Jones, D. (1996). *Lean thinking*. New York, NY: Simon and Schuster.

Wood, R. A. (2010, 13 January). Vinson deploys to respond to Haiti earthquake. *Navy Military News*.

World Bank. (2010). *Trade logistics and facilitation, logistics performance index*. Retrieved September, 2010, from http://web.worldbank.org

Xinhua. (2010). *S.Korea to dispatch peacekeepers to Haiti next week if parliament approves*. Retrieved 20th March, 2010, from http://www.istockanalyst.com/ article/ viewiStockNews/articleid/ 3841113

Xinhuanet. (2008a). *China publicized the deployment of the military in the disaster area first time*. Retrieved from http://www.chinaelections.org/ NewsInfo.asp?NewsID=127999

Xinhuanet. (2008b). *The military mission focuses on the relief operation of remote villages*. Retrieved 15th July, 2008, from http://www.ce.cn/xwzx/gnsz /gdxw/200805/23/ t20080523_ 15598398.shtml

Xinhuanet. (2008c). *1.6 million dollars relief supplies provided by U.S. military has arrived in Chengdu*. Retrieved 15th July, 2008, from http://china.zjol.com.cn/05china / system/2008/05/18/ 009525011.shtml

Xinhuanet. (2008d). *Military relief operation record*. Retrieved 15th July, 2008, from http://news.xinhuanet.com/ mil/2008-05/19/ content_8205680.htm

Xinhuanet. (2008e). *113080 soldiers have been dispatched to the earthquake relief mission*. Retrieved 15th July, 2008, from http://news.sina.com.cn/c/ 2008-05-18/163215566124.shtml

Xinhuanet. (2008f). *People will never forget*. Retrieved 15th July, 2008, from http://www.ce.cn/xwzx/gnsz/gdxw/200808/27/t20080827_16635783.shtml

Xinhuanet. (2008g). *The earthquake relief military continue retracing*. Retrieved 15th July, 2008, from http://news.sohu.com/20080814/n258843155.shtml

Yamamoto, T. (1999). Corporate – NGO partnership: Learning from case studies . In Yamamoto, T., & Gould, K. (Eds.), *From corporate–NGO partnership in Asia-Pacific*. Japan Centre for International Exchange.

Yin, R. K. (1994). *Case study research: Design and methods*. Newbury Park, CA: Sage Publications.

Yin, R. K. (1981). The case study crisis: Some answers. *American Quarterly*, *26*(1), 58–65.

Yin, R. (1989). *Case study research design and methods*, 2nd edition. Applied Research Methods Series, vol. 5. Sage Publications.

Zandee, D. (1999). Civil-military interaction in peace operations. *NATO Review*, *47*(1), 11–15.

Zhongguangnet. (2008). *Helicopters and assault boats are dispatched urgently into the relief operation*. Retrieved 15th July, 2008, from http://www.sznews.com/news /content/2008-05/17/ content_ 2052018.htm

Zhongguangnet. (2008). *Port of Luzhou*. Retrieved 15th July, 2008, from http://www.cnr.cn/zhuanti1/ gkwlx/zjgk/t20050628_504078120.html

Zolin, R. (2002). *Swift trust in hastily formed networks*. The Hastily Formed Networks Research Group. Retrieved September 4, 2008, from http://www.hfncenter.org/cms/files/swifttrustinHFN10-03-02.pdf

About the Contributors

Gyöngyi Kovács is the Director of the Humanitarian Logistics and Supply Chain Research Institute (HUMLOG Institute) and an acting Professor in Supply Chain Management and Corporate Geography at the Hanken School of Economics in Helsinki, Finland. She serves as a co-editor in chief of the Journal of Humanitarian Logistics and Supply Chain Management and as one of the European editors of the International Journal of Physical Distribution and Logistics Management. Her research interests include humanitarian logistics, sustainable supply chain management, supply chain collaboration, and the abductive research approach in logistics.

Karen M. Spens is Professor of Supply Chain Management and Corporate Geography at Hanken School of Economics in Helsinki, Finland. She has written several book chapters and published in logistics and supply chain journals such as International Journal of Physical Distribution and Logistics Management, International Journal of Logistics Management and Supply Chain Management: An International Journal as well as in other journals such as Disaster Prevention and Management. She has also edited several special issues for different journals, such as Management Research News, and is currently co-editor of the Journal of Humanitarian Logistics and Supply Chain Management. Her research interests include humanitarian logistics, health care related research and methodological issues in logistics and supply chain management. She can be contacted at karen.spens@hanken.fi.

* * *

Ruth Banomyong is currently an Associate Professor in the Department of International Business, Logistics and Transport Management at the Thammasat Business School, Thammasat University in Thailand. He received his PhD in 2001, in the field of International Logistics within the Logistics & Operations Management Section (LOMS) at Cardiff Business School (UK). He was the winner of the James Cooper Cup in 2001 for the best PhD dissertation in logistics from the Chartered Institute of Logistics & Transport (CILT) in the United Kingdom. Since 1995, Ruth has been a consultant for international agencies such as the United Nations Conference on Trade & Development (UNCTAD), the United Nations Economic and Social Commission for Asia and the Pacific (UN-ESCAP), The World Bank, The Asian Development Bank (ADB), The Association of South East Asian Nations (ASEAN), et cetera.

Elizabeth Barber has been involved in logistics for the past decade, specializing in military logistics. She is employed by the University of New South Wales at the Australian Defence Force Academy campus. Her research interests include strategies in supply chains, risk and resilience, performance and performance based logistics contracts as well as the physical flows of military logistics. She has written journal articles and numerous conference papers over the past few years. She lectures both the undergraduate and post graduate logistics students at the Academy.

Anthony Beresford has travelled widely in an advisory capacity within the ports and transport fields in Europe, Africa, Asia and North America. He has been involved in a broad range of transport-related research and consultancy projects including: transport rehabilitation, aid distribution and trade facilitation for UNCTAD and for the Rwandan government, (1995, 1998); cost structures on Multimodal Transport Corridors in Southeast Asia (1998-2006). Most recently he has been working on humanitarian supply chain operations in the context of man-made emergencies and natural disasters. He is a member of the Cardiff-Cranfield Humanitarian Logistics Initiative, a research group developing research interests in the area of humanitarian aid.

Ira Haavisto is a PhD student at Hanken School of Economics, Helsinki Finland. She has a Master's degree in Supply Chain Management and Corporate Geography from the Hanken School of Economics. She has a background of working with development and environmental responsibility for a commercial logistics service provider. She is part of the Humanitarian Logistics and Supply Chain Research Institute (HUMLOG Institute) and focuses in her research on humanitarian logistics. Her research is specifically on humanitarian organizations' performance, and sustainability in humanitarian supply chains.

Odran Hayes is the Team Leader of a Project Management Unit implementing an EU-funded environmental infrastructure programme in Cyprus (2009 to present). He spent over eight years in Kosovo where he was Head of Operations at the European Agency for Reconstruction, responsible for the implementation of EU-funded projects including infrastructure and institutional development. In the early post conflict years he was involved in the monitoring of a substantial EU-funded housing reconstruction programme. He has previously worked for six years in Bosnia for the UK Department for International Development (DFID) and the International Management Group, where he worked on post conflict reconstruction of housing and infrastructure. Mr. Hayes is a mechanical engineer having obtained degrees in engineering and mathematics at Trinity College Dublin before spending 11 years in Namibia as an engineer, mainly in the water sector. More recently he completed an MBA at City College/Sheffield University.

Graham Heaslip is Course Director of the BBS (Hons) degree in Business and Management in NUI Maynooth, and a supervisor on the MA (Leadership, Management and Defence Studies) run in conjunction with the Military College of the Irish Defence Forces'. Graham's research interests are broadly in the intersections between global logistics / supply chain management, humanitarian logistics, and organisational management development. His research based teaching brings together themes of risk, resilience, and complex systems theory with practical management disciplines such as supply chain management, operations management, quality management, and business continuity. His research and consultancy interests span mainstream commercial, defence, and other public service contexts examining supply chain management, operations management, and quality management issues which impact on organisational effectiveness or innovative practices. Prior to entering academia, Graham spent fourteen years working in the Irish Defence Forces both at home and abroad in a variety of logistical appointments, as well as spending time seconded to Humanitarian agencies in a logistical capacity. Graham is concluding his own PhD studies in the area of Civil Military Cooperation / Coordination at the Logistics Institute, University of Hull.

Susanne Hertz is a Professor in Business Administration specialising in Logistics and Supply Chain Management at Jönköping International Business School. Her research field includes supply chains integration, dynamics of alliances in supply chains, change processes in supply chains, logistics providers and logistics services, dynamics of logistics networks, retailing logistics, and humanitarian logistics. Her research appears in: Journal of Supply Chain Management, International Journal of Logistics- Research and application, Industrial Marketing Management, and Journal of Business to Business Marketing, among others. She also takes on the role of reviewer, evaluator, examiner, et cetera for the academic field. She is also member of the advisory board to the Swedish government "Logistics Forum," and one of the founding members of HUMLOG group. Earlier in her career, she spent over a decade in logistics industry managing positions mainly in marketing and strategic planning.

Paul D. Larson is the CN Professor of Supply Chain Management at the University of Manitoba. He is also Head of the SCM Department and Director of the Transport Institute. Dr. Larson earned BSB and MBA degrees at the University of Minnesota, and his Ph.D. at the University of Oklahoma. From 1979 to 1981, he worked with the Ministry of Cooperatives in Fiji, as a United States Peace Corps Volunteer. The Institute for Supply Management (ISM) funded his doctoral dissertation, which won the 1991 Academy of Marketing Science/Alpha Kappa Psi award. He has consulted and conducted executive seminars in Europe, North and South America, Australia, the Caribbean, and China, on logistics and SCM. Paul is a former Associate Editor of the Journal of Business Logistics. His current research interests include supply chain sustainability, supply chain risk management, and humanitarian logistics.

Aristides Matopoulos holds a BSc from Aristotle University of Thessaloniki, an MSc degree from Imperial College, UK, and a PhD from the University of Macedonia, Greece. He currently serves as an Adjunct Lecturer in the University of Macedonia and in City College, The International Faculty of the University of Sheffield in Greece. His research interest focuses on logistics and supply chain management, with particular emphasis on the agri-food industry and the humanitarian sector. He has published more than 40 papers in various outlets, including the Supply Chain Management: An International Journal, Computers and Electronics in Agriculture, and International Journal of Logistics: Research and Applications. Aristides is a member of the Editorial Advisory Board of the recently launched Journal of Humanitarian Logistics and Supply Chain Management and also a member of the International Society of Logistics Engineers, the European Association of Operations Management, and the European Federation of Information Communication Technologies in Agriculture, Food and Environment.

Stephen Pettit graduated with a BSc Honours degree in Maritime Geography from Cardiff University in 1989 and in 1993 was awarded a PhD from the University of Wales. Subsequently he has been involved in a range of transport related research projects and his most recent research work has considered aspects of the logistics of humanitarian aid delivery. He first worked on a project funded by the Chartered Institute of Logistics and Transport through their Seedcorn Grant scheme and co-researched with Dr. Anthony Beresford. Stephen has contributed to many academic papers, conference papers, and reports, and he is a member of the Cardiff-Cranfield Humanitarian Logistics Initiative, a research group developing research interests in the area of humanitarian aid.

Joseph Sarkis is Professor of Management within Clark University's Graduate School of Management. He earned his Ph.D. from the University of Buffalo. He currently researches and teaches in areas of Operations, Supply Chain, and Technology Management as well as the relationship between business and the natural environment. He has over 250 publications in a wide variety of outlets. He is currently editor of Management Research Review and Department Editor for IEEE Transactions on Engineering Management.

Per Skoglund is a Lieutenant Colonel in the Swedish armed forces with a Master in Engineering from Chalmers University of Technology. He currently holds a research position at the Swedish Defence Materiel Administration (FMV), and is affiliated to Jönköping International Business School, as PhD Cand. where he also teaches in military logistics. He has previously held a number of positions within FMV including: Director of Logistics, Head for Competence Centre of Logistics, Head of Directorate for Aeronautical Chief Engineers, and ILS manager in several projects. He is one of the founding members of HUMLOG Group and NODLOREN (Nordic Defence Logistics Research Network).

Apichat Sopadang was born in Chiang Mai, Thailand. He graduated from Chiang Mai University, Thailand in 1987 with a degree in industrial engineering. For several years, he worked as a maintenance planning engineer in Electricity Generator Authority of Thailand (EGAT). He completed his Ph.D. from Clemson University, USA in 2001. Following the completion of his Ph.D., he now works for Chiang Mai University as an Assistant Professor and head of the Supply Chain and Engineering Management research unit. He is a frequent speaker at industry and academic meetings. Dr. Sopadang also served as a consultant of the Asian Development Bank (ADB) and The Japan External Trade Organization (JETRO).

Peter Tatham joined the Royal Navy in 1970 and served in a variety of logistics appointments during his career of some 35 years in which he rose to the rank of Commodore. During his final three years in the Service he also gained an MSc in Defence Logistic Management. Following his retirement from the RN, he joined the staff of Cranfield University, UK, lecturing in Human Systems and Humanitarian and Defence Logistics. Awarded his PhD for research into the issues surrounding the role of shared values within military supply networks in 2009, he has recently joined the faculty of Griffith University, Queensland, Australia as a senior lecturer in business and humanitarian logistics and supply network management. He has published in the field of humanitarian and defence logistics, and his current research interests are in the "softer" aspects of the management of supply networks, including the development and maintenance of interpersonal trust.

David Taylor was, from 1997 to 2010, Senior Research Fellow in the Logistics and Operations Management Section of Cardiff Business School where he worked in the Lean Enterprise Research Centre and was Co-director of the Food Process Innovation Unit. His work at Cardiff University involved the development and testing of innovative approaches to the application of lean concepts and methodologies to supply chain improvement. He worked on a number of major research projects in different industry sectors including automotive, engineering, defense, FMCG, and agri-foods. He is currently a visiting lecturer at in the Centre for Defense Acquisition at Cranfield University and also at Herriot Watt University. In recent years he has been increasingly involved in apply lean thinking and value chain analysis to the humanitarian sector. He is a member of the Chartered Institute of Logistics and Transport 'HELP' committee, the HUMLOG group and is the Director of the Cardiff/Cranfield Humanitarian Logistics Initiative.

Rolando Tomasini is a doctoral student at the Humanitarian Logistics and Supply Chain Research Institute (HUMLOG Institute) in Helsinki, Finland. Previously, he was the Program Manager at the INSEAD Humanitarian Research Group (France). His research focuses on humanitarian logistics and partnerships with the private sector. He has produced several articles and publications through his secondments and collaborations with humanitarian organizations.

Index

A

Access Constraints and Organization 57
accountability 7, 26, 29, 94, 97, 121, 165, 170
ad-hoc channels 35
Air Force freighters 50
Airline Ambassadors International (AAI) 11-12
air response 136, 145
air transport 39, 46, 57, 75, 81, 197-198
American Logistics Aid Network (ALAN) 202
Andaman Sea 36
Area of Operations (AOR) 148
assisted self-help 95, 97, 99
awareness 5-6, 13, 106, 115, 156, 190, 197-198, 202

B

barrier 9, 57, 158, 161, 165, 196, 201-202, 204, 207
Bataan Amphibious Ready Group (ARG) 58
Battle Command Sustainment Support System (BCS3) 142
beneficiary 16, 22, 27, 67, 74-75, 81, 90, 94, 96-98, 102
beneficiary empowerment 102
Blue Scenario 75
border crossings 67, 82
Bosnia 108
BP 176
building awareness 198
business-to-business (B2B) 9

C

C-17 Globemaster aircraft 136
Canadian Foodgrains Bank 202
capability gap 157
Capacities and Vulnerability Analysis (CVA) 96
carbon footprint 68, 75, 79, 82, 84, 89, 202

Centre for Research on the Epidemiology of Disasters 213, 222
Chad 8, 130, 137, 147, 149
Changjiang River Administration of Navigational Affairs 51
Chartered Institute of Logistics and Transport (CILT) 183, 185
China 45, 48-49, 51-54, 59, 61-62, 65, 71, 75, 77, 81, 215-216, 218
Chongqing Communications Committee 51
Civilian aircraft 50
Civil-Military Cooperation (CIMIC) 183, 185
Civil-Military Coordination 129-130, 167, 169, 171
Civil-Military Coordination Section (CMCS) 130-131, 146
Civil-Military Operations Centers (CMOC) 130-131
civil unrest 46, 127
cluster 9, 27-28, 55, 121, 143-144, 154, 171, 176, 182-184, 187-188, 191, 194
Cold War 103, 125-126, 144, 148-149
co-located teams 175, 195
common pipeline 17, 20-30, 132
Communication Environment 179, 185-186, 189
community based approach 102
compatibility 1, 5, 13-15, 142
complementarity 1, 5-6, 14-15
Coordination and Control Department 38-40
Country Logistics Performance 208-210, 212-213, 220
Cyclone Nargis 176, 191, 194

D

Department for Humanitarian Action (DHA) 27, 129, 145
Department of Peacekeeping Operations (DPKO) 130, 146
Department of Political Affairs (DPS) 130